Trends in Complex Analysis, Differential Geometry and Mathematical Physics

Trends in Complex Analysis, Differential Geometry and Mathematical Physics

Proceedings of the 6th International Workshop on Complex Structures and Vector Fields

St. Konstantin, Bulgaria 3 – 6 September 2002

Editors

Stancho Dimiev
Bulgarian Academy of Sciences, Bulgaria

Kouei Sekigawa
Niigata University, Japan

World Scientific
New Jersey • London • Singapore • Hong Kong

Published by

World Scientific Publishing Co. Pte. Ltd.
5 Toh Tuck Link, Singapore 596224
USA office: Suite 202, 1060 Main Street, River Edge, NJ 07661
UK office: 57 Shelton Street, Covent Garden, London WC2H 9HE

British Library Cataloguing-in-Publication Data
A catalogue record for this book is available from the British Library.

TRENDS IN COMPLEX ANALYSIS, DIFFERENTIAL GEOMETRY AND MATHEMATICAL PHYSICS
Proceedings of the Sixth International Workshop on Complex Structures and Vector Fields

ISBN 981-238-452-9

Printed in Singapore by World Scientific Printers (S) Pte Ltd

SIXTH INTERNATIONAL WORKSHOP ON COMPLEX STRUCTURES AND VECTOR FIELDS

September 3–7, 2002, Varna, Bulgaria

CONTRIBUTED COMMUNICATIONS

1. K.-H Spallek
   Spaces and fields. — Reality, truth and fiction —
2. K. Sekigawa
   Notes on Kähler surfaces with distinct constant Ricci eigenvalues
3. G. Ganchev
   Involutive distributions of codimension one in Kaehler manifolds
4. V. Milusheva (G. Ganchev)
   Two-dimensional surfaces in Euclidean space generating semi-symmetric hypersurfaces
5. M. Manev
   On hypercomplex pseudo-Hermitian manifolds
6. S. Maeda
   Circles and Kaehler immersions
7. T. Adachi
   Kaehler magnetic fields on product of complex hyperbolic spaces
8. A. Yamada
   Integrability of almost quaternionic manifolds
9. V. Cristea
   Oka's theorem for superholomorphic functions and superforms
10. E. Corel
    Fuchs-like relation for regular connections on a compact Riemann surface
11. H. Tasaki
    Poincaré formulae of submanifolds in complex space forms
12. H. Kamada
    Time-like killing vector fields on compact scalar-flat Kaehler surfaces

To Professor Alfred Gray,
In Memoriam

PREFACE

This book presents the sixth conference of the international interdisciplinary workshop called in short the sixth international workshop on Complex Structures and Vector Fields. This time many specialists on Complex Analysis, Differential Geometry and Mathematical Physics gathered together in Sunny Beach, Varna, Bulgaria, September 3–7, 2002. Besides some traditional participants from Germany and Russia, there ware different new participants from Japan and France, and especially from Balkan countries like Turkey, Romania and Bulgaria, a natural environment for stimulating the scientific activity in this part of Europe. The published papers reflect the fundamental orientation of the workshop during the last 10 years of its existence, namely the facility of scientific investigations and the communications between mathematicians from different big and small countries.

The book is dedicated to the dear memory of Professor Alfred Gary, unforgettable friendly fellow of many participants of this Workshop.

The Editors would like to express here their gratitude to Professor T. Oguro for his outstanding cooperation and efforts in the arrangement of this volume, and Professors S. Manoff for his help.

Editors

CONTENTS

KÄHLER MAGNETIC FIELDS ON A PRODUCT OF COMPLEX HYPERBOLIC SPACES

TOSHIAKI ADACHI

Department of Mathematics, Nagoya Institute of Technology
Nagoya 466-8555, Japan
E-mail: adachi@math.kyy.nitech.ac.jp

In this paper we make a survey of trajectories for Kähler magnetic fields on a product of complex hyperbolic spaces from the viewpoints of cyclicity, semi-conjugacy of magnetic flows and asymptotic behavior.

1. Introduction

Let $(M, J, \langle\ ,\ \rangle)$ be a Kähler manifold with complex structure J and Riemannian metric $\langle\ ,\ \rangle$. A closed 2-form $\mathbb{B}_\kappa = \kappa\mathbb{B}_J$ which is a constant multiple of the Kähler form $\mathbb{B}_J$ on M is called a *Kähler magnetic field* of force κ. Kähler magnetic fields are natural extension of the notion of a static magnetic field on a Euclidean 3-space, and are canonical object associated to a Kähler manifold. We call a smooth curve γ on M parameterized by its arclength a *trajectory* for $\mathbb{B}_\kappa$ if it satisfies $\nabla_{\dot\gamma}\dot\gamma = \kappa J\dot\gamma$. This curve represents a motion of a unit charged particle with unit speed under the action of $\mathbb{B}_\kappa$.

In the preceding paper [1] we studied trajectories for Kähler magnetic fields on a complex space form, which is a complex projective space, a complex Euclidean space and a complex hyperbolic space. In those three spaces the author is especially interested in a complex hyperbolic space, because the feature of trajectories for Kähler magnetic fields deeply depends on their force and the holomorphic sectional curvature of the base manifold. On the other hand, it was pointed out by Ikawa [10] that on a Hermitian symmetric space N of rank r (≥ 2) every trajectory lies on a r-dimensional totally geodesic flat torus if N is of compact type, and it lies on a totally geodesic r product of real hyperbolic spaces if N is of noncompact type. In this paper we take a product of complex hyperbolic spaces, which is also a typical example of a Hadamard manifold, and study trajectories for Kähler magnetic fields in connection with their force and holomorphic sectional curvatures.

2. Trajectories on a product manifold

Let γ be a smooth curve (which is not necessarily parameterized by its arc length) on a Kähler manifold (M, J) satisfying the equation $\nabla_{\dot\gamma}\dot\gamma = \kappa J\dot\gamma$ for some constant κ. As we have

$$\frac{d}{dt}||\dot\gamma||^2 = 2\langle\nabla_{\dot\gamma}\dot\gamma, \dot\gamma\rangle = 2\kappa\langle J\dot\gamma, \dot\gamma\rangle = 0,$$

we see it has constant speed. When γ is not a point curve, we put $\lambda = ||\dot\gamma||$ $(\neq 0)$, and define a smooth curve $\tilde\gamma$ by $\tilde\gamma(t) = \gamma(t/\lambda)$. We then find

$$\nabla_{\dot{\tilde\gamma}}\dot{\tilde\gamma} = \frac{1}{\lambda^2}\nabla_{\dot\gamma}\dot\gamma = \frac{\kappa}{\lambda^2}J\dot\gamma = \frac{\kappa}{\lambda}J\dot{\tilde\gamma},$$

hence $\tilde\gamma$ is a trajectory for a Kähler magnetic field $\mathbb{B}_{\kappa/\lambda}$. This quite trivial property makes the field of study on Kähler magnetic fields on a product manifold rich.

Let $M = M_1 \times \cdots \times M_q$ be a product of Kähler manifolds (M_i, J), $i = 1, \ldots, q$. If $\gamma = (\gamma_1, \ldots, \gamma_q)$, where γ_i is a smooth curve on M_i, is a trajectory for a Kähler magnetic field $\mathbb{B}_\kappa$ on M, then it satisfies

$$\nabla_{\dot\gamma_i}\dot\gamma_i = \kappa J\dot\gamma_i, \quad i = 1, \ldots, q.$$

Thus γ is of the form $\gamma(t) = \big(\tilde\gamma_1(\lambda_1 t), \ldots, \tilde\gamma_q(\lambda_q t)\big)$, where $\tilde\gamma_i$ is either a trajectory for a Kähler magnetic field $\mathbb{B}_{\kappa/\lambda_i}$ on M_i when $\lambda_i \neq 0$, or a point curve $\tilde\gamma_i(t) = \gamma_i(0)$ when $\lambda_i = 0$, which can be regarded as a trajectory for a Kähler magnetic field of infinite force.

3. Cyclicity of trajectories

We first discuss cyclicity of trajectories on a product of complex hyperbolic spaces. We call a smooth curve γ parameterized by its arclength *closed* if there is a non-zero constant T with $\gamma(t+T) = \gamma(t)$ for all t. For a closed curve we call the minimum positive T with such a property its *length*. A smooth curve is said to be *simple* if it does not have self-intersection points. On a complex hyperbolic space $\mathbb{C}H^n(-\beta)$ of constant holomorphic sectional curvature $-\beta$, every trajectory for $\mathbb{B}_\kappa$ is

1) simple and closed with length $2\pi/\sqrt{\kappa^2 - \beta}$ if $|\kappa| > \sqrt{\beta}$,

2) simple and unbounded in both directions if $|\kappa| \leq \sqrt{\beta}$.

Here we shall call a smooth curve parameterized by its arclength unbounded in both directions if both of the images $\gamma([0, \infty))$ and $\gamma(-\infty, 0])$ are unbounded sets. This leads us to the following:

Proposition 1. *For a product manifold* $M = \mathbb{C}H^{n_1}(-\beta_1) \times \cdots \times \mathbb{C}H^{n_q}(-\beta_q)$ *of complex hyperbolic spaces, we set* $\beta = 1/(\sum_{i=1}^q 1/\beta_i)$.

(1) *If* $|\kappa| > \max_{1\le i\le q}\sqrt{\beta_i}$, *then every trajectory for* $\mathbb{B}_\kappa$ *is simple and bounded.*

(2) *If* $|\kappa| \le \sqrt{\beta}$, *then every trajectory for* $\mathbb{B}_\kappa$ *is simple and unbounded in both directions.*

(3) *When* $\sqrt{\beta} < \kappa \le \max_{1\le i\le q}\sqrt{\beta_i}$, *there exist for* $\mathbb{B}_\kappa$ *both bounded trajectories and trajectories which are unbounded in both directions.*

(4) *For a trajectory* $\gamma = (\gamma_1, \gamma_2, \ldots, \gamma_q)$ *for* $\mathbb{B}_\kappa$ *on* M *we put* $\lambda_i = \|\dot\gamma_i\|$. *Then* γ *is bounded if and only if* $\lambda_i < \kappa/\sqrt{\beta_i}$ $(i = 1, 2, \ldots, q)$. *Under this condition, it is closed if and only if all*

$$\sqrt{(\kappa^2 - \beta_i\lambda_i^2)/(\kappa^2 - \beta_j\lambda_j^2)} \quad (1 \le i < j \le q,\ \textit{with}\ \lambda_i \ne 0, \lambda_j \ne 0)$$

are rational. In this case, its length is the least common multiple of

$$\left\{2\pi/\sqrt{\kappa^2 - \beta_i\lambda_i^2}\ \middle|\ 1 \le i \le q,\ \lambda_i \ne 0\right\}.$$

In the case that M has a complex Euclidean factor, as a trajectory for $\mathbb{B}_\kappa$ on a complex Euclidean space $\mathbb{C}^r$ is a circle of radius $1/|\kappa|$ in the sense of Euclidean geometry, we have the following similar result.

Proposition 2. *Let* M *be a product manifold of the form* $\mathbb{C}^r \times \mathbb{C}H^{n_1}(-\beta_1) \times \cdots \times \mathbb{C}H^{n_q}(-\beta_q)$.

(1) *If* $|\kappa| > \max_{1\le i\le q}\sqrt{\beta_i}$, *then every trajectory for* $\mathbb{B}_\kappa$ *is bounded.*

(2) *For a trajectory* $\gamma = (\gamma_0, \gamma_1, \gamma_2, \ldots, \gamma_q)$ *for* $\mathbb{B}_\kappa$ *on* M *we put* $\lambda_i = \|\dot\gamma_i\|$. *Then* γ *is bounded if and only if* $\lambda_i < \kappa/\sqrt{\beta_i}$ $(i = 1, 2, \ldots, q)$. *Under this condition and the additional condition that* $\lambda_0 \ne 0$, *it is closed if and only if all*

$$\sqrt{(\kappa^2 - \beta_i\lambda_i^2)/(\kappa^2 - \beta_j\lambda_j^2)} \quad \textit{and} \quad |\kappa|/\sqrt{\kappa^2 - \beta_i\lambda_i^2},$$

$1 \le i < j \le q$ *with* $\lambda_i \ne 0$, $\lambda_j \ne 0$ *are rational. In this case, its length is the least common multiple of*

$$\left\{2\pi/\sqrt{\kappa^2 - \beta_i\lambda_i^2}\ \middle|\ 1 \le i \le q,\ \lambda_i \ne 0\right\} \cup \{2\pi/|\kappa|\}.$$

4. Semi-conjugacy of Kähler magnetic flows

For a Kähler magnetic field $\mathbb{B}_\kappa$ we shall call $|\kappa|$ its strength. In this section we study Kähler magnetic fields from their strength by investigating their magnetic flow. Just like the definition of the geodesic flow, we define for a Kähler magnetic field $\mathbb{B}_\kappa$ on M its *Kähler magnetic flow* $\mathbb{B}_\kappa\varphi_t$ on the unit tangent bundle UM by $\mathbb{B}_\kappa\varphi_t(v) = \dot\gamma_v(t)$, where γ_v is a trajectory for $\mathbb{B}_\kappa$ with $\dot\gamma_v(0) = v$.

We call two flows ϕ_t, ψ_t on a smooth manifold X are *smoothly conjugate* in the strong sense if there exist a diffeomorphism f of X and a constant c with $f \circ \phi_t = \psi_{ct} \circ f$. For a complex hyperbolic space $\mathbb{C}H^n(-\beta)$, Kähler magnetic flows are classified into three smoothly conjugacy classes according to their strength([1]):

1) If $|\kappa| < \sqrt{\beta}$, then there exists a diffeomorphism f_κ of $U\mathbb{C}H^n(-\beta)$ which satisfies $f_\kappa \circ \mathbb{B}_\kappa \varphi_t = \varphi_{\sqrt{1-(\kappa^2/\beta)}\,t} \circ f_\kappa$, where φ_t denotes the geodesic flow for $\mathbb{C}H^n(-\beta)$. This guarantees that
$$f_{\nu,\kappa} \circ \mathbb{B}_\kappa \varphi_t = \mathbb{B}_\nu \varphi_{\sqrt{(\beta-\kappa^2)/(\beta-\nu^2)}\,t} \circ f_{\nu,\kappa}$$
with $f_{\nu,\kappa} = f_\nu^{-1} \circ f_\kappa$ for κ, ν with $|\kappa|, |\nu| < \sqrt{\beta}$.

2) If $|\kappa|, |\nu| > \sqrt{\beta}$, then there exists a diffeomorphism $f_{\nu,\kappa}$ of $U\mathbb{C}H^n(-\beta)$ which satisfies
$$f_{\nu,\kappa} \circ \mathbb{B}_\kappa \varphi_t = \mathbb{B}_\nu \varphi_{\mathrm{sgn}(\kappa\nu)\sqrt{(\kappa^2-\beta)/(\nu^2-\beta)}\,t} \circ f_{\nu,\kappa},$$
where $\mathrm{sgn}(c)$ denotes the signature of a real number c.

3) When $|\kappa| = \sqrt{\beta}$, the Kähler magnetic flow $\mathbb{B}_\kappa \varphi_t$ is so called the horocycle flow. They are not conjugate to other Kähler magnetic flows.

We denote by $p : U\mathbb{C}H^n(-\beta) \to \mathbb{C}H^n(-\beta)$ the canonical projection. We should note that the distance $d(p(v), p(f_\kappa(v)))$ between a point $p(v)$ and $p(f_\kappa(v))$ does not depend on v, and only depend on κ and β.

We now concern Kähler magnetic flows for a product of complex hyperbolic spaces. We shall call a smooth flow ϕ_t on X is smoothly *semi-conjugate* to a smooth flow ψ_t on X if there exist a dense open subset Y of X, an injective smooth open map $f : Y \to X$ and a constant c such that

i) Y is invariant under ϕ_t,

ii) $f(Y)$ is invariant under ψ_t,

iii) $f^{-1} : f(Y) \to Y$ is smooth,

iv) $f \circ \phi_t = \psi_{ct} \circ f$.

This means that the restrictions $\phi_t|_Y$ and $\psi_t|_{f(Y)}$ are smoothly conjugate in the strong sense.

For a product $M = \mathbb{C}H^{n_1}(-\beta_1) \times \cdots \times \mathbb{C}H^{n_q}(-\beta_q)$, we consider a dense open subset S of UM which is given by
$$\begin{aligned} S &= S(\mathbb{C}H^{n_1}, \ldots, \mathbb{C}H^{n_q}) \\ &= \{ v = (v_1, \ldots, v_q) \in UM \mid v_i \in T\mathbb{C}H^{n_i},\ v_i \neq 0,\ 1 \le i \le q \}. \end{aligned}$$

This set is invariant under Kähler magnetic flows on UM. We first study Kähler magnetic fields on M of weak strength. By use of a smooth map $\lambda = (\lambda_1, \dots, \lambda_q) : S \to (0,1) \times \cdots \times (0,1)$ defined by

$$\lambda_i(v) = \sqrt{\{1 - (\kappa^2/\beta)\}\|v_i\|^2 + (\kappa^2/\beta_i)}, \quad i = 1, \dots, q,$$

we define a smooth map $G_\kappa : S \to S$ by

$$G_\kappa(v) = \left(\lambda_1(v) g^{(1)}_{\kappa/\lambda_1(v)}\left(\frac{v_1}{\|v_1\|}\right), \dots, \lambda_q(v) g^{(q)}_{\kappa/\lambda_q(v)}\left(\frac{v_q}{\|v_q\|}\right)\right),$$

where $g^{(i)}_\nu = \left(f^{(i)}_\nu\right)^{-1}$ denotes the diffeomorphism of $U\mathbb{C}H^{n_i}$ which gives the conjugacy of Kähler magnetic flow $\mathbb{B}\varphi^{(i)}_t$ and the geodesic flow $\varphi^{(i)}_t$ for $\mathbb{C}H^{n_i}$. We then find that

i) $G_\kappa(S)$ is invariant under the Kähler magnetic flow $\mathbb{B}_\kappa \varphi_t$ for M,

ii) the geodesic flow φ_t for M satisfies $G_\kappa \circ \varphi_{\sqrt{1-(\kappa^2/\beta)}\, t} = \mathbb{B}_\kappa \varphi_t \circ G_\kappa$ on S,

iii) if $|\kappa| \geq |\nu|$, then $G_\kappa(S) \subset G_\nu(S)$.

Therefore we have the following:

Theorem 1. *Let $M = \mathbb{C}H^{n_1}(-\beta_1) \times \cdots \times \mathbb{C}H^{n_q}(-\beta_q)$ be a product of complex hyperbolic spaces. For each κ with $|\kappa| < \sqrt{\beta}$ $\left(= \sqrt{1/(\sum_{i=1}^q 1/\beta_i)}\right)$, the geodesic flow for M is smoothly semi-conjugate to the Kähler magnetic flow $\mathbb{B}_\kappa \varphi_t$.*

Next we study Kähler magnetic fields on M of strong strength. When $|\kappa| \geq |\nu| > \max_{1 \leq i \leq q} \sqrt{\beta_i}$, we define a smooth map $F_{\nu,\kappa} : S \to S$ by

$$F_{\nu,\kappa}(v) = \left(\mu_1(v) f^{(1)}_{\frac{\nu}{\mu_1(v)}, \frac{\kappa}{\|v_1\|}}\left(\frac{v_1}{\|v_1\|}\right), \dots, \mu_q(v) f^{(q)}_{\frac{\nu}{\mu_q(v)}, \frac{\kappa}{\|v_q\|}}\left(\frac{v_q}{\|v_q\|}\right)\right).$$

Here $f^{(i)}_{\kappa_1,\kappa_2}$ is the diffeomorphism of $U\mathbb{C}H^{n_i}$ which gives the conjugacy of Kähler magnetic flows $\mathbb{B}_{\kappa_j}\varphi^{(i)}_t$, $j = 1, 2$, and $\mu = (\mu_1 \dots, \mu_q) : S \to (0,1) \times \cdots \times (0,1)$ is given by

$$\mu_i(v) = \sqrt{\frac{1}{\beta_i}\left\{\nu^2 - \frac{(\nu^2 - \beta)(\kappa^2 - \beta_i\|v_i\|^2)}{\kappa^2 - \beta}\right\}}, \quad i = 1, \dots, q.$$

We then find

i) $F_{\nu,\kappa}(S)$ is invariant under $\mathbb{B}_\nu \varphi_t$,

ii) $F_{\nu,\kappa} \circ \mathbb{B}_\kappa \varphi_t = \mathbb{B}_\nu \varphi_{\mathrm{sgn}(\kappa\nu)\sqrt{(\kappa^2-\beta)/(\nu^2-\beta)}\, t} \circ F_{\nu,\kappa}$ on S,

iii) if $|\kappa_1| \geq |\kappa_2| \geq |\kappa_3| > \max_{1 \leq i \leq q} \sqrt{\beta_i}$, then $F_{\kappa_3,\kappa_1} = F_{\kappa_3,\kappa_2} \circ F_{\kappa_2,\kappa_1}$,

hence obtain the following:

Theorem 2. *Let* $M = \mathbb{C}H^{n_1}(-\beta_1) \times \cdots \times \mathbb{C}H^{n_q}(-\beta_q)$ *be a product of complex hyperbolic spaces. For each pair* κ, ν *with* $|\kappa| \geq |\nu| > \max_{1\leq i\leq q}\sqrt{\beta_i}$, *the Kähler magnetic flow* $\mathbb{B}_\kappa\varphi_t$ *for* M *is smoothly semi-conjugate to the Kähler magnetic flow* $\mathbb{B}_\nu\varphi_t$.

For the sake of comparison we here give a result when a Kähler manifold has a complex Euclidean factor.

Theorem 3. *Let* $M = \mathbb{C}^r \times N$ *be a Kähler manifold with complex Euclidean factor. If* $|\kappa| \geq |\nu| > 0$, *a Kähler magnetic flow* $\mathbb{B}_\kappa\varphi_t$ *for* M *is smoothly semi-conjugate to a Kähler magnetic flow* $\mathbb{B}_\nu\varphi_t$: *We have* $F_{\nu,\kappa}\circ\mathbb{B}_\kappa\varphi_t = \mathbb{B}_\nu\varphi_{\kappa t/\nu}\circ F_{\nu,\kappa}$ *with an injective smooth open map* $F_{\nu,\kappa} : S \to S$ *of the set* $S = \{ v = (v_e, v_n) \in UM \mid v_e \in T\mathbb{C}^r, v_n \in TN,\ v_e \neq 0, v_n \neq 0 \}$.

We should note that a property of semi-conjugacy does not hold in general on Kähler magnetic flows for a manifold with a complex hyperbolic factor. For example, for a product $\mathbb{C}H^n(-\beta) \times \mathbb{C}P^m(\alpha)$ of a complex hyperbolic space and a complex projective space, a property of semi-conjugacy does not hold for any two Kähler magnetic flows. Here we briefly explain the difference between a Euclidean factor and a hyperbolic factor. Kähler magnetic flows except the geodesic flow for a complex Euclidean space $\mathbb{C}^r$ are conjugate each other in the relation $f_{\nu,\kappa}\circ\mathbb{B}_\kappa\varphi_t = \mathbb{B}_\nu\varphi_{\kappa t/\nu}\circ f_{\nu,\kappa}$ with a diffeomorphism $f_{\nu,\kappa}$ of $U\mathbb{C}^r$. Since the time relation is linear with respect to κ, we can define an injective map $F_{\nu,\kappa}$ of the unit tangent bundle by

$$F_{\nu,\kappa}\big((v_e, v_n)\big) = \left(\mu(v_n) f_{\frac{\nu}{\mu(v_n)},\frac{\kappa}{\|v_e\|}}\left(\frac{v_e}{\|v_e\|}\right),\ \frac{\nu}{\kappa}v_n\right)$$

with $\mu(v_n) = \sqrt{1-\nu^2\|v_n\|^2/\kappa^2}$, which just changes the speed of trajectories uniformly on N-component. On the other hand, as the conjugate relation for Kähler magnetic flows for a complex hyperbolic space depends also on the holomorphic sectional curvature of the base manifold, such a simple map does not cause semi-conjugacy.

5. Ideal boundary and trajectories

In this section we study trajectories for a Kähler magnetic field on a product of complex hyperbolic spaces from the viewpoint of the ideal boundary. We quickly review on the ideal boundary of a Hadamard manifold M. We call two geodesic rays $\sigma_1, \sigma_2 : [0,\infty) \to M$ of unit speed *asymptotic* if the distance function $d(\sigma_1(t), \sigma_2(t))$ is bounded for $t \geq 0$. The ideal boundary ∂M of M consists of asymptotic classes of geodesic rays. For given a point $x \in M$ and a boundary point $z \in \partial M$ we have a unique geodesic ray joining them: The exponential map $\exp_x$ at x restricted on the unit tangent space U_xM gives a bijective correspondence between U_xM and ∂M. Therefore we can introduce a topology on $M \cup \partial M$, which is called the cone topology (see [8] and also [11]). The notion of the ideal boundary is quite useful in the study of Hadamard manifolds. If M is negatively curved, two distinct boundary points can be joined by a unique geodesic. By use of Tits metric defined by Gromov, we can get the dimension of the Euclidean factor of a Hadamard manifold from the geometry of its ideal boundary (see [6] and also [7]).

We now come back to trajectories for Kähler magnetic fields. For a curve γ which is unbounded in both directions on a Hadamard manifold M, we define its limit points at infinity by

$$\gamma(\infty) = \lim_{t\to\infty} \gamma(t), \quad \gamma(-\infty) = \lim_{t\to-\infty} \gamma(t) \in \partial M,$$

if they exist in the ideal boundary of M. We shall call γ a *horocycle* if the following conditions hold:

i) it has a single limit point; $\gamma(\infty) = \gamma(-\infty)$,

ii) if γ crosses to a geodesic σ with $\sigma(\infty) = \gamma(\infty)$, then they cross orthogonally at their crossing point.

In view of expressions for trajectories for Kähler magnetic fields on a ball model of a complex hyperbolic space, we find the following properties on unbounded trajectories for $\mathbb{B}_\kappa$ ($|\kappa| \leq \sqrt{\beta}$) on a complex hyperbolic space $\mathbb{C}H^n(-\beta)$ (see [1]).

1) When $\kappa = \pm\sqrt{\beta}$, it is a horocycle.

2) When $|\kappa| < \sqrt{\beta}$, it has distinct limit points.

3) If $|\kappa| \leq \sqrt{\beta}$, for arbitrary points $x \in \mathbb{C}H^n(-\beta)$ and $z \in \partial\mathbb{C}H^n(-\beta)$, we have a unique trajectory γ for $\mathbb{B}_\kappa$ with $\gamma(0) = x$ and $\gamma(\infty) = z$.

4) If $|\kappa| < \sqrt{\beta}$, for arbitrary distinct points $z, w \in \partial\mathbb{C}H^n(-\beta)$, we have a unique trajectory γ for $\mathbb{B}_\kappa$ with $\gamma(-\infty) = z$ and $\gamma(\infty) = w$.

In order to explain more precisely, we denote by σ_v a geodesic with $\dot\sigma_v(0) = v$ and by γ_v a trajectory for $\mathbb{B}_\kappa$ with $\dot\gamma_v(0) = v$. The properties of the diffeomorphism $f_\kappa(|\kappa| < \sqrt{\beta})$ of $U\mathbb{C}H^n(-\beta)$ guarantee the following, which also shows the fourth property:

Proposition 3. *For a trajectory γ_v for $\mathbb{B}_\kappa$ ($|\kappa| < \sqrt{\beta}$) on $\mathbb{C}H^n(-\beta)$, the function $t \mapsto d(\gamma_v(t), \sigma_{f_\kappa(v)}(\sqrt{1-(\kappa^2/\beta)}\,t)$ is constant on $\mathbb{R}$.*

To explain the third property, we define a fiberwise linear isomorphism $\Psi_\kappa : T\mathbb{C}H^n(-\beta) \to T\mathbb{C}H^n(-\beta)$ of the tangent bundle by

$$\Psi_\kappa(v) = \sqrt{1 - \frac{\kappa^2}{\beta}}v + \frac{\kappa}{\sqrt{\beta}}Jv.$$

By an expression of geodesics on a ball model, one can easily get $\sigma_{f_\kappa(v)}(\infty) = \sigma_{\Psi_\kappa(v)}(\infty)$ for κ with $|\kappa| < \sqrt{\beta}$. For $\kappa = \pm\sqrt{\beta}$, it is clear that $\Psi_{\pm\sqrt{\beta}}(v) = \pm Jv$, hence $\Psi_{\pm\sqrt{\beta}} = -\Psi_{\mp\sqrt{\beta}}$. We here note that if we reverse the parameter of a trajectory for a Kähler magnetic field $\mathbb{B}_\kappa$ then it is a trajectory for $\mathbb{B}_{-\kappa}$.

Proposition 4. *Consider a magnetic field* $\mathbb{B}_\kappa$ $(|\kappa| < \sqrt{\beta})$ *on* $\mathbb{C}H^n(-\beta)$. *Then the function* $t \mapsto d\big(\gamma_v(t), \sigma_{\Psi_\kappa(v)}\big(\sqrt{1-(\kappa^2/\beta)}\,t\big)\big)$ *is bounded for* $t \geq 0$. *Hence* $\gamma_v(\infty) = \sigma_{\Psi_\kappa(v)}(\infty)$.

Proposition 5. *For a trajectory* γ_v *for a Kähler magnetic field* $\mathbb{B}_\kappa$ *with* $\kappa = \pm\sqrt{\beta}$ *on* $\mathbb{C}H^n(-\beta)$, *we have the following for* $t > 0$:

1) $d\big(\gamma_v(\pm(2/\sqrt{\beta})\sinh\sqrt{\beta}\,t/2), \gamma_v(0)\big) = t,$

2) $t < d\big(\gamma_v(\pm(2/\sqrt{\beta})\sinh\sqrt{\beta}\,t/2), \sigma_{\Psi_\kappa(v)}(t)\big) < t + 2/\sqrt{\beta}.$

Hence $\gamma_v(\infty) = \gamma_v(-\infty) = \sigma_{\Psi_\kappa(v)}(\infty)$.

We now study trajectories on a product $M = \mathbb{C}H^{n_1}(-\beta_1)\times\cdots\times\mathbb{C}H^{n_q}(-\beta_q)$ of complex hyperbolic spaces. We put $\beta = 1/(\sum_{i=1}^q 1/\beta_i)$. The following is trivial.

Proposition 6.

(1) *If* $|\kappa| < \sqrt{\beta}$, *every trajectory for* $\mathbb{B}_\kappa$ *on* M *has distinct points at infinity.*

(2) *When* $\kappa = \pm\sqrt{\beta}$, *a trajectory* $\gamma = (\gamma_1, \ldots, \gamma_q)$ *for* $\mathbb{B}_{\pm\sqrt{\beta}}$ *has a single point at infinity if and only if* $\|\dot\gamma_i\| = \sqrt{\beta/\beta_i}$, $i = 1, \ldots, q$. *In this case it is a horocycle. Otherwise a trajectory for* $\mathbb{B}_{\pm\sqrt{\beta}}$ *has distinct points at infinity.*

We now focus our attention on horocycle trajectories on M. As a trajectory γ_v for $\mathbb{B}_\kappa$ on M with $v = (v_1, \cdots, v_q)$ is a horocycle if and only if $\|v_{i_j}\|^2 = \kappa^2/\beta_{i_j}$, $j = 1, \ldots, \ell$ and $\|v_i\| < \kappa^2/\beta_i$, $i \neq i_j$ for some positive integer ℓ, we obtain the following:

Proposition 7. *There is a horocycle trajectories for* $\mathbb{B}_\kappa$ *on* M *if and only if* $\beta \leq \kappa^2 \leq \max_{1\leq i\leq q}\sqrt{\beta_i}$.

We now apply this proposition to study the rank of a Hermitian symmetric space of noncompact type. It was pointed out by Ikawa [10] that each trajectory for a Kähler magnetic field on a Hermitian symmetric space of noncompact type of rank r with minimum sectional curvature $-\delta$ lies on some totally geodesic r product $\mathbb{C}H^1(-\delta)\times\cdots\times\mathbb{C}H^1(-\delta)$ of real hyperbolic planes of curvature $-\delta$ (cf. [1] for the case $r=1$). Therefore we can get the following:

Theorem 4. *Let N be a symmetric space of noncompact type.*

(1) *The minimum of the sectional curvature of N coincides with*

$$-\max\left\{\,\kappa^2 \;\middle|\; \mathbb{B}_\kappa \text{ has a horocycle trajectory}\,\right\}.$$

(2) *The number*

$$\frac{\max\left\{\,\kappa^2 \;\middle|\; \mathbb{B}_\kappa \text{ has a horocycle trajectory}\,\right\}}{\min\left\{\,\kappa^2 \;\middle|\; \mathbb{B}_\kappa \text{ has a horocycle trajectory}\,\right\}}$$

is an integer and coincides with the rank of N.

(3) *If each of Kähler magnetic fields $\mathbb{B}_{\kappa_1}$, $\mathbb{B}_{\kappa_2}$ $(|\kappa_1|\le|\kappa_2|)$ has a horocycle trajectory on N, then the rank of N is not smaller than $(\kappa_2/\kappa_1)^2$.*

(4) *Suppose the minimum of the sectional curvature of N is $-\delta$. If there is a Kähler magnetic field $\mathbb{B}_\kappa$ on N which has a horocycle trajectory, then the rank of N is not smaller than δ/κ^2.*

Next we study the property of joining an interior point and a boundary point. We put $S(\infty)$ the set of boundary points $\{\sigma_v(\infty)\;|\;v\in S\}$ associated with $S=S\big(\mathbb{C}H^{n_1}(-\beta_1),\dots,\mathbb{C}H^{n_q}(-\beta_q)\big)$. If we set $S_x=S\cap U_xM$ and $S_x(\infty)=\{\,\sigma_v(\infty)\;|\;v\in S_x\,\}$, it is clear that $S(\infty)=S_x(\infty)$. We first study the case $|\kappa|<\sqrt{\beta}$. By use of the map $\lambda: S\to(0,1)\times\cdots\times(0,1)$ given in section 4 we define a map $\Phi_\kappa: S\to S$ by

$$\Phi_\kappa(v)=\left(\lambda_1(v)\Phi^{(1)}_{\kappa/\lambda_1(v)}\left(\frac{v_1}{\|v_1\|}\right),\cdots,\lambda_q(v)\Phi^{(q)}_{\kappa/\lambda_q(v)}\left(\frac{v_q}{\|v_q\|}\right)\right),$$

where $\Phi^{(i)}_\nu=\big(\Psi^{(i)}_\nu\big)^{-1}$ denotes the fiberwise linear isomorphism of $T\mathbb{C}H^{n_i}$ defined by

$$\Phi^{(i)}_\nu(v_i)=\sqrt{1-\frac{\nu^2}{\beta_i}}\,v_i-\frac{\nu}{\sqrt{\beta_i}}Jv_i.$$

Since we have $|\kappa|/\lambda_i(v)<\sqrt{\beta_i}$, by Proposition 4, we find the function $t\mapsto d\big(\sigma_v(\sqrt{1-(\kappa^2/\beta)}\,t),\gamma_{\Phi_\kappa(v)}(t)\big)$ is bounded for $t\ge 0$.

For the case $\kappa=\pm\sqrt{\beta}$, we define $\Phi_\kappa: S\to S$ by

$$\Phi_\kappa(v)=\left(-\frac{\sqrt{\beta}}{\sqrt{\beta_1}\|v_1\|}Jv_1,\dots,-\frac{\sqrt{\beta}}{\sqrt{\beta_q}\|v_q\|}Jv_q\right).$$

We then find

1) $d\big(p(v), \gamma_{\Phi_\kappa(v)}(\pm(2/\sqrt{\beta})\sinh\sqrt{\beta}\,t/2)\big) = t$ for $t \geq 0$,

2) $\lim_{t\to\infty}\frac{1}{t}d\big(\sigma_v(t), \gamma_{\Phi_\kappa(v)}(\pm(2/\sqrt{\beta})\sinh\sqrt{\beta}\,t/2)\big) = 1$

by Proposition 5. As we have

$$UM = \bigcup_i U\mathbb{C}H^{n_i} \bigcup_{i_1<i_2} S(\mathbb{C}H^{n_{i_1}}, \mathbb{C}H^{n_{i_2}}) \bigcup \cdots \bigcup S(\mathbb{C}H^{n_1}, \cdots, \mathbb{C}H^{n_q}),$$

we obtain the following:

Theorem 5. *We consider a Kähler magnetic field* $\mathbb{B}_\kappa$ *with* $|\kappa| \leq \sqrt{\beta}$ *on a product* $M = \mathbb{C}H^{n_1}(-\beta_1) \times \cdots \times \mathbb{C}H^{n_q}(-\beta_q)$, *where* $\beta = 1/(\sum_{i=1}^q 1/\beta_i)$.

(1) *For arbitrary points* $x \in M$ *and* $z \in \partial M$, *there is a trajectory for* $\mathbb{B}_\kappa$ *on* M *with* $\gamma(0) = x$ *and* $\gamma(\infty) = z$.

(2) *If* $z \in S(\infty)$, *then such a trajectory is unique.*

We note that under the situation in Theorem 5 there exist infinitely many trajectories for $\mathbb{B}_\kappa$ joining x and z when $z \notin S(\infty)$. We consider the case $M = \mathbb{C}H^{n_1}(-\beta_1) \times \mathbb{C}H^{n_2}(-\beta_2)$. For a unit tangent vector $v = (v_1, 0) \in U_xM$ at a point $x = (x_1, x_2)$, we have $\gamma_v(\infty) \in \partial\mathbb{C}H^{n_1} \subset \partial M$. Given a tangent vector $u_2 \in T_{x_2}\mathbb{C}H^{n_2}$ with $0 < \|u_2\| < |\kappa|/\sqrt{\beta_2}$ we put $u = (\sqrt{1-\|u_2\|^2}\,v_1, u_2) \in U_xM$. We find that the distance between $\gamma_v(\sqrt{1-\|u_2\|^2}\,t)$ and $\gamma_u(t)$ is bounded, hence $\gamma_v(\infty) = \gamma_u(\infty)$.

Finally we make mention of the property of joining two distinct points in the ideal boundary. We shall call two points z and w in ∂M are antipodal if there exists a geodesic σ with $\sigma(\infty) = z$ and $\sigma(-\infty) = w$. Theorem 1 and Proposition 3 lead us to the following:

Proposition 8. *We consider a Kähler magnetic field* $\mathbb{B}_\kappa$ *with* $|\kappa| < \sqrt{\beta}$ *on a product* $M = \mathbb{C}H^{n_1}(-\beta_1) \times \cdots \times \mathbb{C}H^{n_q}(-\beta_q)$, *where* $\beta = 1/(\sum_{i=1}^q 1/\beta_i)$.

(1) *Two points* $z, w \in \partial M$ *are antipodal if and only if there is a trajectory* γ *for* $\mathbb{B}_\kappa$ *with* $\gamma(\infty) = z$ *and* $\gamma(-\infty) = w$.

(2) *If two trajectories* $\gamma = (\gamma_1, \ldots, \gamma_q)$ *and* $\rho = (\rho_1, \ldots, \rho_q)$ *satisfy* $\gamma(\infty) = \rho(\infty), \gamma(-\infty) = \rho(-\infty)$ *and these points at infinity are contained in* $S(\infty)$, *then there exists a vector* $(t_1, \ldots, t_q)$ *with* $\rho_i(t) = \gamma_i(t + t_i)$, $i = 1, \ldots, q$.

Acknowledgment

This research is partially supported by Grant-in-Aid for Scientific Research (C) (No. 14540075), The Ministry of Education, Science, Sports and Culture.

References

1. T. Adachi, *Kähler magnetic flows on a manifold of constant holomorphic sectional curvature*, Tokyo J. Math. **18**(1995), 473–483.
2. ______, *A comparison theorem for magnetic Jacobi fields*, Proc. Edinburgh Math. Soc. **40**(1997), 293–308.
3. ______, *Kähler magnetic fields on a complex hyperbolic space*, A Report on KOREA-JAPAN Joint Workshop in Mathematics 2000, Differential Geometry and its Related Topics, H. Naitho ed.
4. ______, *Kähler magnetic flows for a product of complex space forms*, preprint.
5. ______, *Rank of a Hermitian symmetric space of noncompact type and Kähler magnetic fields*, in preperation.
6. T. Adachi and F. Ohtsuka, *The Euclidean factor of a Hadamard manifold*, Proc. A. M. S. **113**(1991), 209–213.
7. W. Ballmann, M. Gromov and V. Schroeder, *Manifolds of nonpositive curvature*, Progress in Math. 61(1985), Birkhäuser, Boston, Basel and Stuttgart.
8. P. Eberlein and B. O'Neill, *Visibility manifolds*, Pacific J. Math. **46**(1973), 45–109.
9. O. Ikawa, *Hamiltonian dynamics of a charged particle*, to appear in Hokkaido J. Math.
10. ______, *Motion of charged particles in homogeneous Kähler and homogeneous Sasakian manifolds*, preprint.
11. K. Shiga, *Hadamard manifolds*, Adv. St. Pure Math. **3**(1984), 239–281.

REAL ANALYTIC ALMOST COMPLEX MANIFOLDS

L. N. APOSTOLOVA
Institute of Mathematics and Informatics
Bulgarian Academy of Sciences
Acad. G. Bonchev Street, Bl. 8
1113 Sofia - Bulgaria
E-mail: liliana@bas.bg

Real analyticity of some classes of almost complex manifolds will be considered. Using the known result for real analyticity of symplectic manifolds the result that every almost Kähler manifold is equivalent in a natural sense to a real analytic one will be proved. The same result for closed $2k$ dimensional almost complex manifolds admitting real cohomology class α such that $\alpha^k \neq 0$ will be proved modulo the Thurston's conjecture that such manifolds are symplectic ones. Some concluding remarks are included.

1. Introduction and preliminaries.

Let M be a C^∞ smooth paracompact $2n$-dimensional manifold and J be an antiinvolutive automorphism of the tangent bundle on M. Then for every point $p \in M$ the restriction J_p of J on the fibre T_pM acts as an antiinvolutive automorphism too, i.e. $J_p : T_pM \to T_pM$ and $J^2 = -Id$. The couple (M, J) is called *an almost complex manifold* and J is called *an almost complex structure* on M. Let g be a riemannian metric on M and let h be defined by the equality $h(X,Y) = 1/2\,(g(X,JY) + g(JX,Y))$ for each two vector fields X, Y defined on an open set in M. Then h is called *an almost hermitian metric* on the almost complex manifold (M, J) and (M, J, h) is called *an almost hermitian manifold.* It is easy to check that the identity $h(JX,Y) = h(Y,JY)$ holds for the almost hermitian metric h. Indeed, $h(JX,Y) := 1/2(g(JX,JY) + g(JJX,Y)) = 1/2(g(-X,-JJY) + g(JX,JY)) = 1/2(g(X,JJY) + g(JX,JY)) =: h(X,JY))$.

The two-form w defined on M as follows: $w(X,Y) = h(X,JY)$ is called *a fundamental two-form* on the almost hermitian manifold M. This is a non-degenerate two-form on M, i.e. $w^n \neq 0$, where $2n$ is the dimension of the

AMS Subject Classification: 53C15, 32H02
Keywords: almost Kähler manifold, symplectic manifold, real analytic manifold

differentiable manifold M. Let us note that every almost complex manifold is even dimensional orientable smooth differentiable manifold. But smooth manifolds with nondegenerate two-form are called *an almost symplectic manifolds*. So every almost hermitian manifold is an almost symplectic manifold.

An almost hermitian manifold is called *an almost Kähler manifold*, if the form w is a closed two-form, i.e. if $dw = 0$. Let us remind also that *a symplectic manifold* is an infinitely smooth manifold, which possesses a closed non-degenerate two-form. So we obtain that every almost Kähler manifold is a symplectic one.

We recall the main result in [KL]. This is the theorem that each symplectic manifold M is C^∞ isomorphic to a real analytic symplectic manifold, unique up to real analytic isomorphism.

In § 2 we will show that every almost Kähler manifold (M, J, h) admits a structure of real analytic manifold $\tilde{M}$, real analytic almost complex structure $\tilde{J}$ and real analytic almost hermitian metric $\tilde{h}$ such that M is diffeomorphic to $\tilde{M}$, J is C^∞ isomorphic to $\tilde{J}$ and the almost hermitian metric $\tilde{h}$ is C^∞ equivalent to the given almost hermitian metric h. Moreover, $(\tilde{M}, \tilde{J}, \tilde{h})$ is the unique triple with such a property. More precisely, if $(\overline{M}, \overline{J}, \overline{h})$ is other real analytic almost hermitian manifold, C^∞ equivalent to (M, J, h), then $(\tilde{M}, \tilde{J}, \tilde{h})$ and $(\overline{M}, \overline{J}, \overline{h})$ are really analytic isomorphic one with other.

In § 3 we prove that each closed $2k$-dimensional almost complex manifold which has a real cohomology class α such that $\alpha^k \neq 0$ is isomorphic to an unique real analytic almost complex manifold, assuming that the following conjecture is true:

Conjecture (Thurston [T]). *Every closed $2k$-manifold which has an almost complex structure τ and a real cohomology class α such that $\alpha^k \neq 0$ has a symplectic structure realizing τ and α.*

In §4 some concluding remarks are included.

2. Real analyticity of almost Kähler manifolds.

Definition 1. We say that an almost complex manifold (M, J) is C^∞ equivalent to an almost complex manifold $(\tilde{M}, \tilde{J})$ if there exists a diffeomorphism $\varphi : \overline{M} \to \tilde{M}$ with mapping φ^* conjugated to the differential $d\varphi$, commuting with the authomorphisms J^* on T^*M and $\tilde{J}^*$ on $T^*\tilde{M}$ respectively, i.e. if the diagram

$$\begin{array}{ccc} T^*\tilde{M} & \xrightarrow{\tilde{J}^*} & T^*\tilde{M} \\ {\scriptstyle \varphi^*}\downarrow & & \downarrow{\scriptstyle \varphi^*} \\ T^*\overline{M} & \xrightarrow[\overline{J}^*]{} & T^*\overline{M} \end{array}$$

is commutative. Here the automorphim J^* is defined on T^*M by the commutative diagram

$$\begin{array}{ccc} T_pM & \xrightarrow{J_p} & T_pM \\ \downarrow & & \downarrow \\ T_p^*M & \xrightarrow[J_p^*]{} & T_p^*M \end{array}$$

where J_p is the restriction of the given almost complex structure on the fiber T_pM and the downarrow is the mapping of the natural duality between the fibers over p of the tangent and cotangent bundles on M.

Definition 2. We say that the almost hermitian manifold (M, J, h) is C^∞ equivalent to the almost hermitian manifold $(\tilde{M}, \tilde{J}, \tilde{h})$ if the almost complex manifolds (M, J) and $(\tilde{M}, \tilde{J})$ are C^∞ equivalent and if φ^* maps the almost hermitian metric $\tilde{h}$ to the almost hermitian metric h, where φ is the diffeomorphism from the definition 1.

Theorem 1. *Let (M, J, h) be C^∞ almost Kähler manifold. Then there exists a real analytic manifold $\tilde{M}$, a real analytic almost complex structure $\tilde{J}$ and a real analytic almost hermitian metric $\tilde{h}$ on $\tilde{M}$ such that $(\tilde{M}, \tilde{J}, \tilde{h})$ is C^∞ equivalent to (M, J, h). Such a manifold $(\tilde{M}, \tilde{J}, \tilde{h})$ is unique up to an real analytic isomorphism.*

Remark. The unicity of the real analytic almost hermitian manifold $(\tilde{M}, \tilde{J}, \tilde{h})$, C^∞ equivalent to the given one (M, J, h), means exactly that if for the almost hermitian manifold (M, J, h) exists another real analytic almost hermitian manifold $(\overline{M}, \overline{J}, \overline{h})$, C^∞ equivalent to (M, J, h), then the real analytic almost hermitian manifolds $(\tilde{M}, \tilde{J}, \tilde{h})$ and $(\overline{M}, \overline{J}, \overline{h})$ are really analytic isomorphic. The last means that there exists a real analytic isomorphism $\psi : \tilde{M} \to \overline{M}$ with mapping ψ^* commuting with the authomorphisms $\tilde{J}^*$ on $T^*\tilde{M}$ and $\overline{J}^*$ on $T^*\overline{M}$, i.e. the diagram

$$\begin{array}{ccc} T^*\widetilde{M} & \xrightarrow{\tilde{J}^*} & T^*\widetilde{M} \\ {\scriptstyle \psi^*}\downarrow & & \downarrow{\scriptstyle \psi^*} \\ T^*\overline{M} & \xrightarrow[\overline{J}^*]{} & T^*\overline{M} \end{array}$$

is commutative. Also the almost hermitian metrics on $\tilde{M}$ and $\overline{M}$ are connected one with other by the diffeomorphism ψ.

Sketch of the proof of Theorem 1. *Existence.* Let us consider the fundamental form w of the given almost Kähler manifold. It is a closed two-form and (M, w) is a symplectic manifold. By the theorem of F. Kutzschebauch and F. Loose [KL] a real analytic almost symplectic manifold $(\tilde{M}, \tilde{w})$ and C^∞ diffeomorphism $\varphi : M \to \tilde{M}$ exist so that $\varphi^* \tilde{w} = w$. As the symplectic form $\tilde{w}$ is a real analytic one, the almost complex structure $\tilde{J}$, determined by $\tilde{w}$ on $\tilde{M}$, will be real analytic too. Also the mapping ${\varphi^*}^{-1}$ of the diffeomorphism φ^{-1} maps the almost complex structure J to the almost complex structure $\tilde{J}$ and ${\varphi^*}^{-1}$ maps the almost hermitian metric h to an almost hermitian metric $\tilde{h}$ on $\tilde{M}$. The just obtained almost Kähler manifold $(\tilde{M}, \tilde{J}, \tilde{h})$ is such that it has a fundamental two-form, coinsiding with the symplectic form $\tilde{w}$.

Uniqueness. Let us assume that $(\tilde{M}, \tilde{J}, \tilde{h})$ and $(\overline{M}, \overline{J}, \overline{h})$ are two real analytic almost Kähler manifolds, C^∞ equivalent to the almost Kähler manifold (M, J, h) and with isomorphisms of the C^∞ equivalence $\tilde{\varphi} : M \to \tilde{M}$ and $\overline{\varphi} : M \to \overline{M}$. Then the real analytic symplectic manifolds $(\tilde{M}, \tilde{w})$ and $(\overline{M}, \overline{w})$ will be equivalent to the symplectic manifold (M, w). Let $\tilde{\omega}$, $\overline{\omega}$ be their fundamental two-forms.

Now we shall use the theorem for the uniqueness in [KL]. According to this theorem there exists a real analytic isomorphism $\psi : \tilde{M} \to \overline{M}$ such that $\psi^* \overline{w} = \tilde{w}$. As in the part of existence we shall consider the almost Kähler manifolds $(\tilde{M}, \tilde{J}, \tilde{h})$ and $(\overline{M}, \overline{J}, \overline{h})$, determined by $\tilde{\varphi}^{-1}$ and $\overline{\varphi}^{-1}$ on $\tilde{M}$ and $\overline{M}$, respectively. Then we shall consider the real analytic almost complex structure J_1 and the real analytic almost hermitian metric h_1 on $\overline{M}$, determined by the isomorphism $\psi : \tilde{M} \to \overline{M}$ above and by $\tilde{J}$ and $\tilde{h}$ on $\tilde{M}$, respectively. They determine a fundamental two-form w_1 on $\overline{M}$, coinsiding with the fundamental two-form $\overline{w}$ of the almost Kähler structure $(\overline{M}, \overline{J}, \overline{h})$, $w_1 = \overline{w}$. Applying the technique used by H. Whitney and F. Bruhat in [WB], it can be proved that $J_1 = \overline{J}$ and $h_1 = \overline{h}$, what completes the proof of the theorem. □

3. Real analyticity of almost complex manifolds, admitting a real cohomology class α such that $\alpha^k \neq 0$.

Theorem 2. *Let (M, J) be a closed $2k$-dimensional C^∞ smooth almost complex manifold, admitting a real cohomology class α such that $\alpha^k \neq 0$. Then there exists a real analytic almost complex manifold $(\tilde{M}, \tilde{J})$ which is C^∞ equivalent to (M, J). Such a manifold $(\tilde{M}, \tilde{J})$ is unique up to real analytic isomorphism.*

Sketch of the poof. Let us determine the symplectic structure w on M, realizing the almost complex structure J and the cohomology class α. This is possible by the Thurston Conjecture [T].

Again with the help of the theorem for existence and uniqueness for the symplectic manifold (M, w) in [KL], the theorem would be proved in a way similar to the given above proof of the Theorem 1. □

4. Concluding remarks.

Here we propose the following

Open Problem. *To prove that every almost complex manifold is C^∞ equivalent to a real analytic almost complex manifold and such a real analytic manifold is unique up to real analytic isomorphism or to find a counter example.*

A possible way to prove the assertion above would be by use the result of Whitney [W] (see also [dR]) from where follows that every smooth manifold M could be imbedded in a Euclidean space as a real analytic submanifold $\tilde{M}$. Then the arising almost complex structure $\tilde{J}$ would be changed, if it is needed, by a real analytic almost complex structure. The technique used by [WB], [M] and [G] would be useful in the proof.

References

[dR] G. de Rham, *Variétés Differentiables*, Paris, Herman & Cie, (1955).

[G] H. Grauert, On Lewy's problem and the imbedding of real-analytic manifolds, *Ann. of Math.*, **68** (1958) 460–472.

[KL] F. Kutzschebauch, F. Loose, Real analytic structures on a symplectic manifold, *P.A.M.S.*, (10) **128** (2000), 3009–3016.

[M] J. Moser, On the volume element of a manifolds, *Trans. Am. Math. Soc.* **130** (1965), 286–294.

[T] W.P. Thurston, Some simple examples of symplectic manifolds, *P.A.M.S.*, (2) **55** (1976), 467–468.

[W] H. Whitney, Differentiable manifolds, *Ann. of Math.* **37** (1936), 645–680.

[WB] H. Whitney and F. Bruhat, Quelques propriétés foundamentales des ensembles analytiques, *Comment. Math. Helv.* **33** (1959), 132–160.

FUCHS-LIKE RELATION FOR REGULAR CONNECTIONS ON A COMPACT RIEMANN SURFACE

EDUARDO COREL

Departament d'Àlgebra i Geometria, Universitat de Barcelona
Gran Via de les Corts Catalanes, 585, 08007 Barcelona
E-mail : ecorel@cerber.mat.ub.es

In this article, we extend the formal algebraic notion of exponents for a regular connection to a vector bundle over a Riemann surface, and obtain the analog of Fuchs' relation for differential systems over $\mathbb{P}^1(\mathbb{C})$.

1. Meromorphic connections on a vector bundle

Let X be a Riemann surface. Denote with $\mathcal{O}_X$ the structural sheaf of X and with Ω_X the sheaf of holomorphic differential 1-forms over X. Let $\mathcal{S} = \{a_1, \ldots, a_t\}$ be a discrete subset of X. Denote further with $\mathcal{O}_X(\mathcal{S})$ the sheaf of meromorphic functions on X having their poles only at the points of $\mathcal{S}$. For any $f \in \mathcal{O}_X(\mathcal{S})$, denote with (f) the divisor of its zeroes et poles (*cf.* [Se], II, n° 3, pp. 18 *seq.*). Let $\mathcal{D} = \mu_1 a_1 + \cdots + \mu_t a_t$ be a divisor over X. Write $\mathcal{O}_\mathcal{D}$ the sheaf of meromorphic functions f such that $(f) \geqslant -\mathcal{D}$. Finally, denote the sheaves of meromorphic differential 1-forms having poles on $\mathcal{S}$ with $\Omega(\mathcal{S}) = \Omega_X \otimes_{\mathcal{O}_X} \mathcal{O}_X(\mathcal{S})$, and having their divisor bounded from below by the divisor $-\mathcal{D}$ with $\Omega_\mathcal{D} = \Omega_X \otimes_{\mathcal{O}_X} \mathcal{O}_\mathcal{D}$, respectively. Let $d : \mathcal{O}_X \longrightarrow \Omega_X$ be the natural exterior derivative and let d also denote its extension to $d : \mathcal{O}_X(\mathcal{S}) \longrightarrow \Omega(\mathcal{S})$.

Let $\mathcal{V}$ be a locally free sheaf of $\mathcal{O}_X(\mathcal{S})$-modules of rank n; it happens to be the sheaf of sections of a meromorphic vector bundle $\mathbb{V}$ of rank n defined over X and having its poles on $\mathcal{S}$. Instead of the bundle $\mathbb{V}$, we will always consider the attached sheaf $\mathcal{V}$. Endow $\mathcal{V}$ with a holomorphic connection ∇, by which we mean a $\mathbb{C}$-linear sheaf homomorphism

$$\nabla : \mathcal{V} \longrightarrow \mathcal{V} \otimes_{\mathcal{O}_X} \Omega_X$$

satisfying Leibniz's rule : for any open subset U of X, any section $s \in \Gamma(U, \mathcal{V})$ and any holomorphic function $f \in \mathcal{O}_X(U)$, one has

$$\nabla(fs) = f\nabla(s) + s \otimes df \in \Gamma(U, \mathcal{V} \otimes_{\mathcal{O}_X} \Omega_X).$$

We will say that $\mathcal{E}$ is a *lattice* of $\mathcal{V}$, if $\mathcal{E}$ is a locally free sheaf of $\mathcal{O}_X$-modules of same rank n which is a subbundle of $\mathcal{V}$. It is in turn the sheaf of holomorphic sections of a *holomorphic* vector subbundle $\mathbb{E}$ of $\mathbb{V}$. The restriction of the connection ∇ to $\mathcal{E}$ *meromorphic* connection over $\mathcal{E}$, that is it satisfies

$$\nabla(\mathcal{E}) \subset \mathcal{E} \otimes_{\mathcal{O}_X} \Omega(\mathcal{S}).$$

Denote with $[\mathcal{S}]$ the divisor $a_1 + \cdots + a_t$. The *Poincaré divisor* of $\mathcal{E}$ is the smallest positive divisor $\mathcal{D}_P = \sum_{i=1}^t p_i a_i$ such that $\nabla(\mathcal{E}) \subset \mathcal{E} \otimes_{\mathcal{O}_X} \Omega_{\mathcal{D}_P + [\mathcal{S}]}$. The integers p_i are called the *Poincaré ranks* of the connection ∇ with respect to the lattice $\mathcal{E}$ at a_i. One denotes them with $p_i = \mathfrak{p}_{\mathcal{E}}^{a_i}(\nabla)$. Define the *height* of the lattice $\mathcal{E}$ as the degree $h(\mathcal{E})$ of its Poincaré divisor, *i.e.* the sum $h(\mathcal{E}) = p_1 + \cdots + p_t = \deg \mathcal{D}_P$ of its Poincaré ranks. If $\mathcal{D}_P = 0$, the lattice $\mathcal{E}$ is said to be *logarithmic*. The connection ∇ is called *regular* if there is a logarithmic lattice, and *irregular* if there isn't.

In case only a holomorphic vector bundle $\mathcal{E}$ over X is given, a *meromorphic connection* on $\mathcal{E}$ will mean a holomorphic connection on $\mathcal{V} = \mathcal{E} \otimes_{\mathcal{O}_X} \mathcal{O}_X(\mathcal{S})$. A convenient reference on all these questions is [Sa].

2. Local connections and exponents

Consider in this section a given point $x \in \mathcal{S}$. Denote with $\mathcal{F}_x$ the stalk of the sheaf $\mathcal{F}$ at x, and fix the following.

The algebra $(\mathcal{O}_X)_x$ of germs of holomorphic functions at x is a local ring, whose maximal ideal will be denoted with $\mathfrak{m}$, while v denotes the corresponding $\mathfrak{m}$-adic valuation. Denote with $\mathcal{O}$ its $\mathfrak{m}$-adic completion, and with K the completion of the algebra $(\mathcal{O}_X(\mathcal{S}))_x$ of germs of meromorphic functions at x, which is the fraction field of $\mathcal{O}$. Denote with $\mathcal{O}(\ell)$ the completion of the module $(\mathcal{O}_{[\ell x]})_x$ of functions bounded from below with $-\ell$ at x. Denote the free $\mathcal{O}$-module of rank 1 of germs of formally holomorphic differential 1-forms with $\Omega = (\Omega_X)_x \otimes_{(\mathcal{O}_X)_x} \mathcal{O}$. Let $\Omega_K = \Omega \otimes_{\mathcal{O}} K$ be the module of germs of formally meromorphic differential 1-forms at x, and denote with $\Omega(\ell) = \Omega \otimes_{\mathcal{O}} \mathcal{O}(\ell)$ the free module of germs of formally meromorphic differential 1-forms having their order bounded from below with $-\ell$ at x.

The formalization $V = \mathcal{V}_x \otimes_{(\mathcal{O}_X(\mathcal{S}))_x} K$ of the stalk at x of the meromorphic bundle is a K-vector space of dimension n, of which the $\mathcal{O}$-module $\Lambda = \mathcal{E}_x \otimes_{(\mathcal{O}_X)_x} \mathcal{O}$ is a lattice (in the sense of [De] or [G-L]). Let us also denote with ∇ the formalized of the connection germ ∇_x.

The triple (V, Λ, ∇) will be called the *formal local model* of $(\mathcal{V}, \mathcal{E}, \nabla)$ at x.

For any lattice Λ of V, the *Poincaré rank of* ∇ *over* Λ denotes the smallest integer $p = \mathfrak{p}_\Lambda(\nabla) \geqslant 0$ such that $\nabla(\Lambda) \subset \Lambda \otimes_{\mathcal{O}} \Omega(p+1)$. Naturally, if (V, Λ, ∇)

is the formal local model of $(\mathcal{V}, \mathcal{E}, \nabla)$ at x, one has

$$\mathfrak{p}_\Lambda(\nabla) = \mathfrak{p}_{\mathcal{E}}^x(\nabla).$$

Denote with Res : $\Omega_K \longrightarrow \mathbb{C}$ the residue map of a differential meromorphic 1-form. If $\dim_K V = 1$, the residue of ∇ is defined as follows.

Let $e \neq 0$ be a vector of V and consider $\omega \in \Omega_K$ such that $\nabla(e) = -e \otimes \omega$. Let $\tilde{e} = \lambda e$ with $\lambda \in K^*$. Then

$$\nabla(\tilde{e}) = -\tilde{e} \otimes \tilde{\omega} \quad \text{holds with} \quad \tilde{\omega} = \omega - \frac{d\lambda}{\lambda}.$$

The residues Res ω and Res $\tilde{\omega}$ are equal if and only if $\lambda \in \mathcal{O}^*$. The vectors e and $\tilde{e}$ then span the same lattice Λ.

Definition 1. With the previous notations, if $\dim_K V = 1$, one defines the *residue of* ∇ *over* Λ as the number $\mathrm{Res}_\Lambda(\nabla) = \mathrm{Res}\,\omega$.

When $\dim_K V > 1$, the previous construction does not work in general.

Definition 2. The lattice Λ is said to be *stable* if $\mathfrak{p}_\Lambda(\nabla) = 0$.

If Λ is stable, $\nabla(\mathfrak{m}\Lambda) \subset (\mathfrak{m}\Lambda) \otimes_{\mathcal{O}} \Omega(1)$ holds. One calls then *residue of the connection* ∇ *over* Λ the map $\mathrm{Res}_\Lambda \nabla$ obtained by composing the quotient map $\bar{\nabla}$ induced by ∇ over $\Lambda/\mathfrak{m}\Lambda$ with the residue map in the following way :

$$\mathrm{Res}_\Lambda \nabla : \Lambda/\mathfrak{m}\Lambda \xrightarrow{\bar{\nabla}} \Lambda \otimes_{\mathcal{O}} \Omega(1) \;/\; \Lambda \otimes_{\mathcal{O}} \Omega(1)$$
$$\simeq (\Lambda/\mathfrak{m}\Lambda) \otimes_{\mathbb{C}} \Omega(1) \xrightarrow{id \otimes \mathrm{Res}} \Lambda/\mathfrak{m}\Lambda.$$

For any lattice only the *residue trace of the connection* ∇ *over the lattice* Λ can be defined as the residue

$$\tau_\Lambda(\nabla) = \mathrm{Res}_{\bigwedge^n \Lambda} \bigwedge^n \nabla,$$

of the connection $\bigwedge^n \nabla$ induced by ∇ over the n-th exterior power $\bigwedge^n V$ of V.

Recall the following result and definitions from [Cor].

Proposition 1. *Let* Λ *be a lattice of* V. *If the connection* ∇ *is regular, then the set of stable sublattices of* Λ *has a unique maximal element.*

Definition 3. One calls *Levelt lattice of* Λ *for the connection* ∇ the largest stable sublattice Λ_L of Λ, and *exponents of the connection* ∇ *on the lattice* Λ, the eigenvalues $(e_i^\Lambda(\nabla))_{i=1,\dots,n}$ of the residue $\mathrm{Res}_{\Lambda_L}(\nabla)$ of the connection ∇ over the Levelt lattice Λ_L.

3. Fuchs' relations

Let $\mathcal{E}$ be a holomorphic vector bundle of rank n over X, endowed with a meromorphic connection ∇ over $\mathcal{E}$ having poles only on $\mathcal{S}$. Embed $\mathcal{E}$ in its canonical meromorphic bundle $\mathcal{V} = \mathcal{E} \otimes_{\mathcal{O}_X} \mathcal{O}_X(\mathcal{S})$.

Theorem 1. *Assume X to be compact. With the preceding notations, the sum of the exponents e_i^x of ∇ on $\mathcal{E}$ at all points $x \in X$ satisfies :*

$$\deg \mathcal{E} - \frac{n(n-1)}{2} h(\mathcal{E}) \leqslant \sum_{x \in X} \sum_{i=1}^{n} e_i^x \leqslant \deg \mathcal{E} - h(\mathcal{E}).$$

Proof. It follows from [Cor] (proof of theorem 4.3), that if (V, Λ, ∇_x) is the formal local model of $(\mathcal{V}, \mathcal{E}, \nabla)$ at x, then we have

$$\tau_\Lambda(\nabla_x) - \frac{n(n-1)}{2} \mathfrak{p}_\Lambda(\nabla_x) \leqslant \sum_{i=1}^{n} e_i^x \leqslant \tau_\Lambda(\nabla_x) - \mathfrak{p}_\Lambda(\nabla_x).$$

Of course, one has $\mathfrak{p}_\Lambda(\nabla_x) = \mathfrak{p}_{\hat{\mathcal{E}}_x}(\nabla_x) = \mathfrak{p}_{\mathcal{E}}^x(\nabla)$ and $\tau_\Lambda(\nabla_x) = \mathrm{Res}_x \bigwedge^n \nabla$. Taking the sum over X yields

$$\sum_{x \in X} \tau_\Lambda(\nabla_x) - \frac{n(n-1)}{2} \sum_{x \in X} \mathfrak{p}_\Lambda(\nabla_x)$$
$$\leqslant \sum_{x \in X} \sum_{i=1}^{n} e_i^x \leqslant \sum_{x \in X} \tau_\Lambda(\nabla_x) - \sum_{x \in X} \mathfrak{p}_\Lambda(\nabla_x).$$

On the one hand, one has $\sum_{x \in X} \mathfrak{p}_\Lambda(\nabla_x) = h(\mathcal{E})$. Since $\bigwedge^n \nabla$ is a meromorphic connection on the bundle $\bigwedge^n \mathcal{E}$ of rank 1, one has the following result which concludes the proof (*cf.* [Bot]). □

Lemma 1. *Let $\mathcal{E}$ be a vector bundle of rank 1 on a compact Riemann surface X. Any meromorphic connection ∇ on $\mathcal{E}$ satisfies*

$$\sum_{x \in X} \mathrm{Res}_x \nabla = \deg \mathcal{E}.$$

The proof that we give for completeness' sake is due to C. Sabbah (*cf.* [Sa], p. 135, Ex. 1.10)

Proof. Assume first that $\mathcal{E}$ is a trivial bundle. Then there exist a global non-zero holomorphic section h of $\mathcal{E}$, and a meromorphic 1-form ω over X such that

$\nabla(h) = -h \otimes \omega$. Since h is an $\mathcal{O}_X$-basis of $\mathcal{E}$, one has $\mathrm{Res}_x \nabla = \mathrm{Res}_x \omega$, hence by the residue theorem

$$\sum_{x \in X} \mathrm{Res}_x \nabla = \sum_{x \in X} \mathrm{Res}_x \omega = 0.$$

The formula is then proved for $\mathcal{E}$ trivial since $\deg \mathcal{E} = 0$.

When $\mathcal{E}$ is not trivial, it contains a global *meromorphic* section s whose divisor D_s satisfies $\deg D_s = \deg \mathcal{E}$. There is an open covering (U_i) of X such that non-zero holomorphic sections $e_i \in \Gamma(U_i, \mathcal{E})$ exist. One can find meromorphic functions s_i on U_i such that $s = s_i e_i$ over U_i. We construct locally a connection over $\mathcal{E}_{|U_i}$ by the formula

$$\tilde{\nabla}(e_i) = -e_i \otimes \frac{ds_i}{s_i}. \tag{1}$$

Accordingly, $\tilde{\nabla}$ is a meromorphic connection over $\mathcal{E}$ such that $\tilde{\nabla}(s) = 0$. Equip then the bundle $\mathit{End}\,\mathcal{E} = \mathcal{E} \otimes_{\mathcal{O}_X} \mathcal{E}^*$ with the connection $\nabla \otimes id_{\mathcal{E}^*} - id_{\mathcal{E}} \otimes \tilde{\nabla}$. The bundle $\mathit{End}\,\mathcal{E}$ is trivial since it admits the global section *id*. According to the first part of the proof,

$$\sum_{x \in X} \mathrm{Res}_x (\nabla \otimes id_{\mathcal{E}^*} - id_{\mathcal{E}} \otimes \tilde{\nabla}) = 0$$

holds. On the one hand, one has $\sum_{x \in X} \mathrm{Res}_x \nabla \otimes id_{\mathcal{E}^*} = \sum_{x \in X} \mathrm{Res}_x \nabla$. On the other hand, according to formula (1) and definition 1, one has the following :

$$\mathrm{Res}_x \tilde{\nabla} = \mathrm{Res}_x \frac{ds_i}{s_i} = v_x s, \quad \text{for} \quad x \in U_i.$$

Hence we find

$$\sum_{x \in X} \mathrm{Res}_x \nabla = \sum_{x \in X} \mathrm{Res}_x \tilde{\nabla} = \sum_{x \in X} v_x s = \deg D_s = \deg \mathcal{E}. \qquad \square$$

Theorem 2. *Let $\mathcal{E}$ be a holomorphic vector bundle equipped with a regular connection ∇. There exists a unique logarithmic maximal sublattice $\mathcal{E}_L$ of $\mathcal{E}$.*

This result relies in an essential way on the following lemma (*cf.* [Ma]), which expresses the well-known fact that at a regular singularity, formal solutions are in fact convergent.

Lemma 2. *Let Λ be a lattice of $V = \mathbb{C}(\{z\})^n$, equipped with a regular connection ∇. The Levelt lattice $\hat{\Lambda}_L$ of $\hat{\Lambda} = \Lambda \otimes_{\mathbb{C}\{z\}} \mathbb{C}[[z]]$ satisfies*

$$\hat{\Lambda}_L = \left(\hat{\Lambda}_L \cap V\right) \otimes_{\mathbb{C}\{z\}} \mathbb{C}[[z]].$$

In particular, the lattice $\Lambda_L = \hat{\Lambda}_L \cap V$ of V is the largest stable convergent sublattice of Λ. One calls it the (convergent) Levelt lattice *of Λ.*

Proof of the theorem. Embed $\mathcal{E}$ in its meromorphic bundle $\mathcal{V} = \mathcal{E} \otimes_{\mathcal{O}_X} \mathcal{O}_X(\mathcal{S})$. The connection ∇ extends to $\mathcal{V}$ as a holomorphic connection. Cover X with the open set $U = X \backslash \mathcal{S}$ and for any $x \in \mathcal{S}$, an connected and simply connected open set $U_x \ni x$ such that if $x \neq x'$ are two points of $\mathcal{S}$, then $U_x \cap U_{x'} = \emptyset$. For any $x \in \mathcal{S}$, the bundle $\mathcal{E}_{|U_x}$ is trivial (see [Fo], ch. 30). At any $x \in \mathcal{S}$, form the local convergent model $(\mathcal{V}_x, \mathcal{E}_x, \nabla)$, and construct the Levelt convergent lattice $(\mathcal{E}_x)_L$ of $\mathcal{E}_x$. There exist n local sections of $(\mathcal{E}_x)_L$ that span it as an $(\mathcal{O}_X)_x$-module. Reducing if necessary the open set U_x, the lattice $(\mathcal{E}_x)_L$ can be pulled back over U_x as a convergent lattice $\mathcal{E}_L^x$. Since the stalks $\mathcal{E}$ and $\mathcal{E}_L^x$ coincide over $U \cap U_x$, the local bundles $\mathcal{E}_L^x$ can be glued together along $\mathcal{E}_{|U}$ into a global holomorphic bundle $\mathcal{E}_L$ over X, which is a lattice of $\mathcal{V}$, a sublattice of $\mathcal{E}$, which is logarithmic with poles over $\mathcal{S}$ and maximal in $\mathcal{E}$. □

Definition 4. The lattice $\mathcal{E}_L$ is called the *(convergent) Levelt lattice* of $\mathcal{E}$.

Corollary 1. *The exponents of ∇ over $\mathcal{E}$ at x (in the sense of definition 3) are the eigenvalues of the residue at x of ∇ over the convergent Levelt lattice $\mathcal{E}_L$ of $\mathcal{E}$.*

Corollary 2. *Let $\mathcal{E}$ be a holomorphic vector bundle equipped with a regular connection ∇. Let $\mathcal{E}_L$ be its Levelt lattice. The sum over X of the exponents e_i^x of ∇ over $\mathcal{E}$ at all points $x \in X$ satisfies the following relation*

$$\sum_{x \in X} \sum_{i=1}^{n} e_i^x = \deg \mathcal{E}_L.$$

Proof. According to definition 3, one has $e_i^x = e_i^{(\hat{\mathcal{E}}_x)_L}(\nabla)$. Theorem 1 then yields

$$\deg \mathcal{E}_L - \frac{n(n-1)}{2} h(\mathcal{E}_L) \leqslant \sum_{x \in X} \sum_{i=1}^{n} e_i^x \leqslant \deg \mathcal{E}_L - h(\mathcal{E}_L).$$

Since the height of a logarithmic lattice is zero, the proof is done. □

References

B-V. D.G. Babbitt, V.S. Varadarajan, Local moduli for meromorphic differential equations, *Astérisque*, vol. **169–170**, S.M.F., 1989.

B. D. Bertrand, Travaux récents sur les points singuliers des équations différentielles linéaires, *Sém. Bourbaki*, n°538, Juin 1979.

Bo. A.A. Bolibrukh, The 21st Hilbert problem for Linear Fuchsian Systems, *Proc. Steklov Inst. of Math.*, Vol 206(5), Amer. Math. Soc., 1995.

Bot. R. Bott, A residue formula for vector bundles, *J. Diff. Geom.*, 1967, pp. 311–330.

Cor. E. Corel, Relations de Fuchs pour les systèmes différentiels réguliers, *Bull. S.M.F.*, **129**, 2001, pp. 189–210.

Cor2. ———, On Fuchs' relation for linear differential systems, to appear in *Compositio Math.*

De. P. Deligne, *Équations différentielles à points singuliers réguliers*, Lect. Notes in Math., vol. 163, Springer-Verlag, 1970.

FK. H.M. Farkas and I. Kra, *Riemann Surfaces*, Springer-Verlag, New York, 1991.

Fo. O. Forster, *Lectures on Riemann surfaces*, GTM 81, Springer-Verlag, 1981.

GH. P. Griffiths and J. Harris, *Principles of Algebraic Geometry*, Wiley-Interscience, New York, 1994.

G-L. R. Gérard and A.H.M. Levelt, Invariants mesurant l'irrégularité en un point singulier d'un système d'équations différentielles linéaires, *Ann. Inst. Fourier*, Grenoble, **23, 1** (1973), pp. 157–195.

Le1. A.H.M. Levelt, Hypergeometric functions, II, *Nederl. Akad. Wetensch.*, Proc. Ser. A, **64** (1961), pp. 373–385.

Ma. B. Malgrange, Connexions méromorphes, II : le réseau canonique, *Invent. Math.* **124** (1996), pp. 367–387.

M. Y. Manin, Moduli fuchsiani, *Ann. Sc. Norm. Sup. Pisa*, **19** (1) Serie III, 1965, pp. 113–126.

Po. E.G.C. Poole, *Introduction to the theory of linear differential equations*, Dover, New-York, 1960.

Sa. C. Sabbah, *Déformations isomonodromiques et variétés de Frobenius. Une introduction*, Publ. École Polytechnique, 2000.

Se. J.P. Serre, *Groupes algébriques et corps de classes*, Hermann, Paris, 1959.

Va. V.S. Varadarajan, Meromorphic differential equations, *Expo. Math.* 9, 1991, pp. 97–188.

OKA'S THEOREM FOR SUPERHOLOMORPHIC FUNCTIONS AND SUPERFORMS

VALENTIN GABRIEL CRISTEA
"Valahia" University Targoviste
Email: valentin_cristea@yahoo.com

The Oka's theorem for superholomorphic functions and superforms is given.

Let V [4], [6] be a supervector space and V^* [4] be the dual supervector space. We consider $\mathbf{C}_L$ the Grassmann algebra over the complex numbers with generators $1^L, \nu_1^L, \ldots, \nu_L^L$ with L positive integer and the relations

$$1^L \cdot \nu_i^L = \nu_i^L \cdot 1^L = \nu_i^L \quad i = 1, \ldots, L,$$
$$\nu_i^L \cdot \nu_j^L = -\nu_j^L \cdot \nu_i^L \quad i, j = 1, \ldots, L$$

and $\mathbf{C}_L^{m,n} = (\mathbf{C}_L)_0{}^m \oplus (\mathbf{C}_L)_1{}^n$ with $L > n$ where $\mathbf{C}_L = (\mathbf{C}_L)_0 \oplus (\mathbf{C}_L)_1$ [15].

Definition 1 ([13]). A function $f : \mathbf{C}_L{}^{m,n} \to \mathbf{C}_L$ is called a superholomorphic or superdifferentiable function if there exist $f_\mu \in \mathcal{H}(\mathbf{C}^m, \mathbf{C})$ holomorphic functions such that:

$$f(z, \zeta) = \sum_{\mu \in M_n} f_\mu(z) \zeta^\mu$$

where $M_n = \{ (\mu_1, \ldots, \mu_n) \mid 1 \leq \mu_1 < \ldots < \mu_n \leq n \}$ [8].

As it follows in [8] for each μ in M_L,

$$\nu_\mu^L = \nu_{\mu_1}^L \cdots \nu_{\mu_k}^L, \quad k = 1, \ldots, L$$

and

$$\nu_\phi^L = 1^L$$

Keywords: superholomorphic functions, superforms, supervector space, complex superholomorphic supermanifolds
Mathematics Subject Classification (1991): 58A50

a typical element b of $\mathbf{C}_L$ may be expressed as

$$b = \sum_{\mu \in M_L} b^\mu \cdot \nu^L_\mu,$$

where the coefficients b^μ are complex numbers.

With the norm on $\mathbf{C}_L$ defined by

$$||b|| := \sum_{\mu \in M_L} |b^\mu|,$$

$\mathbf{C}_L$ is a Banach algebra [12].

Let M be a Hausdorff topological space. [13] (a) An (m,n) chart on M over $\mathbf{C}_L$ is a pair (U,ψ) with U an open set of M and ψ a homeomorphism of U onto an open subset of ${\mathbf{C}_L}^{m,n}$. (b) An (m,n) superholomorphic structure on M over $\mathbf{C}_L$ is a collection $\{(U_\alpha, \psi_\alpha) \mid \alpha \in \Lambda\}$ of (m,n) charts on M such that (*I*) $M = \cup_{\alpha\in\Lambda} U_\alpha$ and (*II*) for each pair α, β in Λ the mapping $\psi_\beta \circ {\psi_\alpha}^{-1}$ is a superholomorphic function of $\psi_\alpha(U_\alpha \cap U_\beta)$ onto $\psi_\beta(U_\alpha \cap U_\beta)$, (*III*) the collection $\{(U_\alpha, \psi_\alpha) \mid \alpha \in \Lambda\}$ is a maximal collection of open charts for which (*I*) and (*II*) hold.

Definition 2 ([13]). An (m,n)-dimensional complex superholomorphic supermanifold over $\mathbf{C}_L$ is a Hausdorff topological space M with an (m,n) holomorphic structure over $\mathbf{C}_L$.

Example 1. ${\mathbf{C}_L}^{m,n}$ is an (m,n)-dimensional complex superholomorphic supermanifold.

For a given (m,n)-dimensional supermanifold M, there is a natural projection onto an underlying m-dimensional conventional manifold M_0. A supermanifold is said to be simply connected if that is the case for its underlying manifold.

Definition 3 ([4]). A subset M' of a complex superholomorphic supermanifold M of dimension (m,n) is called a complex superholomorphic sub-supermanifold of dimension (m',n'), $m \geq m'$, $n \geq n'$, if M' is contained in the union of a set $\{(U,\psi)\}$ of charts each of which has the property that, for all $(z,\zeta) \in U \cap M'$, $\psi(z,\zeta) = (z^1,\ldots,z^{m'},a^{m'+1},\ldots,a^m,\zeta^1,\ldots,\zeta^{n'},\eta^{n'+1},\ldots,\eta^n)$ where $(a^{m'+1},\ldots,a^m,\eta^{n'+1},\ldots,\eta^n)$ is a fixed element of ${\mathbf{C}_L}^{m-m',n-n'}$ depending on the chart in question.

The pairs $\{(U',\psi')\}$ where $U' = U \cap M'$ and $\psi'(z,\zeta) = (z^1,\ldots,z^{m'},$ $\zeta^1,\ldots,\zeta^{n'})$ constitute an atlas for M'.

Example 2. Let us consider ${\mathbf{C}_L}^{m,n}$ the (m,n)-dimensional supereuclidean space. ${\mathbf{C}_L}^{m,n}$ is an (m,n)-dimensional superdifferentiable supermanifold over

$\mathbf{C}_L$ from the Definition 2. We consider the following subset of ${\mathbf{C}_L}^{m,n}$, ${S_L}^{m-1,n-2} = \{ (z,\zeta) \in {\mathbf{C}_L}^{m,n} \mid (z^1)^2 + \cdots + (z^m)^2 + 2\cdot\zeta^1)\cdot\zeta^{r+1}) + \cdots + 2\cdot\zeta^r)\cdot\zeta^n) = 1 + 2\cdot\nu_1)\cdot\nu_{r+1}) + \cdots + \nu_r)\cdot\nu_n) \}$ and we conclude that ${S_L}^{m-1,n-2}$ is an $(m-1, n-2)$-dimensional complex superholomorphic sub-supermanifold of ${\mathbf{C}_L}^{m,n}$, $n = 2\cdot r$.

Let M be an (m,n)-dimensional complex superholomorphic supermanifold, U an open subset of M and let $\psi(z,\zeta) = (z^1,\ldots,z^m,\zeta^1,\ldots,\zeta^n)$ be the supercomplex coordinates on U. Let us consider 1-superforms as in [4] and let $A^{p,q,p_1,q_1}(M)$ be the space of all superforms of type (p,q,p_1,q_1) on M, and let $A(M) = \sum_{p,q,p_1,q_1\geq 0} A^{p,q,p_1,q_1}(M)$ be the space of all superforms with supercomplex values on M. Then, any element of $A^{p,q,p_1,q_1}(M)$ is written in this form

$$\omega(a) = \sum_{\substack{j_1<\cdots<j_p,\ k_1<\cdots<k_q,\\ {j_1}^1<\cdots<{j_{p_1}}^1,\ {k_1}^1<\cdots<{k_{q_1}}^1}} \omega_{JKJ'K'}\cdot dz^J \wedge d\bar{z}^K \wedge d\zeta^{J'} \wedge d\bar{\zeta}^{K'}$$

with $J = (j_1 < \cdots < j_p)$, $K = (k_1 < \cdots < k_q)$, $J^1 = ({j_1}^1 < \cdots < {j_{p_1}}^1)$, $K^1 = ({k_1}^1 < \cdots < {k_{q_1}}^1)$ for all $a \in M$ and

$$d\omega(a) = \sum_{\substack{j_1<\cdots<j_p,\ k_1<\cdots<k_q,\\ {j_1}^1<\cdots<{j_{p_1}}^1,\ {k_1}^1<\cdots<{k_{q_1}}^1}} \omega_{JKJ'K'} \wedge dz^J \wedge d\bar{z}^K \wedge d\zeta^{J'} \wedge d\bar{\zeta}^{K'},$$

$$d\omega = \partial\omega + \bar{\partial}\omega + \partial_1\omega + \bar{\partial}_1\omega$$

with $\partial\omega \in A^{p+1,q,p_1,q_1}(M)$, $\bar{\partial}\omega \in A^{p,q+1,p_1,q_1}(M)$, $\partial_1\omega \in A^{p,q,p_1+1,q_1}(M)$, and $\bar{\partial}_1\omega \in A^{p,q,p_1,q_1+1}(M)$, where $\partial^2 = 0$, $\bar{\partial}^2 = 0$, $\partial\bar{\partial} + \bar{\partial}\partial = 0$, ${\partial_1}^2 = 0$, $\bar{\partial}_1^2 = 0$ and $\partial_1\bar{\partial}_1 + \bar{\partial}_1\partial_1 = 0$.

Definition 4. Let Ω be an open set of ${\mathbf{C}_L}^{m,n}$. Ω is called $(\bar{\partial},\bar{\partial}_1)$-acyclic if for all superform ω of type (p,q,p_1,q_1) on Ω with $p \geq 0$, $q \geq 0$, $p_1 \geq 1$, $q_1 \geq 1$, such that $\bar{\partial}\omega + \bar{\partial}_1\omega = 0$, there are two 1-superforms ω' and ω'' of type $(p, q-1, p_1, q_1)$ and (p,q,p_1,q_1-1) respectively, such that $\bar{\partial}\omega' + \bar{\partial}_1\omega'' = \omega$.

Theorem 1. *Let* $D = \{(z,\zeta) \in \mathbf{C}_L \mid \|(z,\zeta)\| < 1\}$ *the unity disk in* $\mathbf{C}_L$. *Let* Ω *be an open set of* ${\mathbf{C}_L}^{m,n}$, *such that* $\Omega \times D$ *is* $(\bar{\partial},\bar{\partial}_1)$*-acyclic. If* f *is superholomorphic on* Ω, *then the set* $\Omega_f = \{(z^1,\ldots,z^m,\zeta^1,\ldots,\zeta^n) \in \Omega \mid \|f((z^1,\ldots,z^m,\zeta^1,\ldots,\zeta^n))\| < 1\}$ *is* $(\bar{\partial},\bar{\partial}_1)$*-acyclic. For every* 1*-superform* ω *of type* (p,q,p_1,q_1) *on* Ω_f, $p \geq 0$, $q \geq 0$, $p_1 \geq 0$, $q_1 \geq 0$, *such that* $\bar{\partial}\omega + \bar{\partial}_1\omega = 0$, *there are two* 1*-superforms* ω' *and* ω'' *of type* (p,q,p_1,q_1) *on* $\Omega \times D$, *with* $\bar{\partial}\omega' + \bar{\partial}_1\omega'' = 0$, *such that* $u^*(\omega' + \omega'') = \omega$, *where* $u : \Omega_f \to \Omega \times D$ *is the application* $(z^1,\ldots,z^m,\zeta^1,\ldots,\zeta^n) \to (z^1,\ldots,z^m,\zeta^1,\ldots,\zeta^n, f(z^1,\ldots,z^m,\zeta^1,\ldots,\zeta^n))$.

Proof. We will prove the last statement of our conclusion. Let ω be the 1-superform of type (p, q, p_1, q_1) on Ω_f, $p \geq 0$, $q \geq 0$, $p_1 \geq 0$, $q_1 \geq 0$, such that $\bar{\partial}\omega + \bar{\partial}_1\omega = 0$. We note that u is superholomorphic, injective and proper and thus u_* is injective in every point. Then $V = u(\Omega_f)$ is a closed superholomorphic sub-supermanifold of $\Omega \times D$. Let $\pi : \Omega \times D \to \Omega$ be the projection $(z^1, \ldots, z^m, \zeta^1, \ldots, \zeta^n, z, \zeta) \to (z^1, \ldots, z^m, \zeta^1, \ldots, \zeta^n)$ and $U = \pi^{-1}(\Omega_f) = \Omega_f \times D$. We get that U is a neighborhood of V in $\Omega \times D$. Let U' be a neighborhood of V in $\Omega \times D$ with $\bar{U}' \subset U$. Let α be a function on $\Omega \times D$ such that: $\alpha((z^1, \ldots, z^m, \zeta^1, \ldots, \zeta^n, z, \zeta)) = 0$ if $(z^1, \ldots, z^m, \zeta^1, \ldots, \zeta^n, z, \zeta) \notin U'$ and $\alpha((z^1, \ldots, z^m, \zeta^1, \ldots, \zeta^n, z, \zeta)) = 1$ if $(z^1, \ldots, z^m, \zeta^1, \ldots, \zeta^n, z, \zeta) \in W$ where W is a neighborhood of V. The superform $\omega_0 = (\pi)^*(\omega)$ is of type (p, q, p_1, q_1) on Ω. We have $u^*(\omega_0) = (\pi\ circu)^*(\omega) = \omega$ because $\pi\ circu = id$ on Ω_f. We define a superform ω_1 on $\Omega \times D$ by $\omega_1 = \alpha \cdot \omega_0$ on U and $\omega_1 = 0$ for all $(z^1, \ldots, z^m, \zeta^1, \ldots, \zeta^n, z, \zeta) \notin U$. Then ω_1 is a superform of type (p, q, p_1, q_1) on $\Omega \times D$ and $u^*(\omega_1) = \omega$. Let us define $\omega_2(((z^1, \ldots, z^m, \zeta^1, \ldots, \zeta^n, z, \zeta) = 0$ if $(z^1, \ldots, z^m, \zeta^1, \ldots, \zeta^n, z, \zeta) \in V$ and

$$\begin{aligned}&\omega_2((z^1, \ldots, z^m, \zeta^1, \ldots, \zeta^n, z, \zeta))\\&= ((z^1, \ldots, z^m, \zeta^1, \ldots, \zeta^n, z, \zeta)\\&\qquad - -f((z^1, \ldots, z^m, \zeta^1, \ldots, \zeta^n, z, \zeta)))^{-1}(\bar{\partial}\omega + \bar{\partial}_1\omega)\end{aligned}$$

if $(z^1, \ldots, z^m, \zeta^1, \ldots, \zeta^n, z, \zeta) \in \Omega \times D - V$. Then ω_2 is a superform of type $(p, q+1, p_1, q_1+1)$ on $\Omega \times D$ and $(\bar{\partial}\omega_2 + \bar{\partial}_1\omega_2) = 0$. From $\Omega \times D$ is $(\bar{\partial}, \bar{\partial}_1)$-acyclic we get that there are two superforms $\omega_3{}'$ and $\omega_3{}''$ of type (p, q, p_1, q_1) on $\Omega \times D$ with $(\bar{\partial}\omega_3{}' + \bar{\partial}_1\omega_3{}'') = \omega_2$. Let ω' be

$$\begin{aligned}\omega' = (1/2) \cdot \omega_1 + ((z^1, \ldots, z^m, \zeta^1, \ldots, \zeta^n, z, \zeta)\\ - f((z^1, \ldots, z^m, \zeta^1, \ldots, \zeta^n, z, \zeta))) \cdot \omega_3{}',\end{aligned}$$

let ω'' be

$$\begin{aligned}\omega'' = (1/2) \cdot \omega_1 + ((z^1, \ldots, z^m, \zeta^1, \ldots, \zeta^n, z, \zeta)\\ - f((z^1, \ldots, z^m, \zeta^1, \ldots, \zeta^n, z, \zeta))) \cdot \omega_3{}'';\end{aligned}$$

ω' and ω'' are of type (p, q, p_1, q_1) on $\Omega \times D$. Moreover,

$$\begin{aligned}&(\bar{\partial}\omega' + \bar{\partial}_1\omega'')\\&= (1/2) \cdot \bar{\partial}\omega_1 + ((z^1, \ldots, z^m, \zeta^1, \ldots, \zeta^n, z, \zeta)\\&\qquad - f((z^1, \ldots, z^m, \zeta^1, \ldots, \zeta^n, z, \zeta))) \cdot \bar{\partial}\omega_3{}'\\&\quad + (1/2) \cdot \bar{\partial}_1\omega_1 + ((z^1, \ldots, z^m, \zeta^1, \ldots, \zeta^n, z, \zeta)\\&\qquad - f((z^1, \ldots, z^m, \zeta^1, \ldots, \zeta^n, z, \zeta))) \cdot \bar{\partial}_1\omega_3{}''\end{aligned}$$

$$
\begin{aligned}
&= (1/2)\cdot(\bar{\partial}\omega_1 + \bar{\partial}_1\omega_1) \\
&\quad + ((z^1,\ldots,z^m,\zeta^1,\ldots,\zeta^n,z,\zeta) \\
&\qquad - f((z^1,\ldots,z^m,\zeta^1,\ldots,\zeta^n,z,\zeta)))\cdot(\bar{\partial}\omega_3{}' + \bar{\partial}_1\omega_3{}'') \\
&= (1/2)\cdot(\bar{\partial}\omega_1 + \bar{\partial}_1\omega_1) \\
&\quad + ((z^1,\ldots,z^m,\zeta^1,\ldots,\zeta^n,z,\zeta) \\
&\qquad - f((z^1,\ldots,z^m,\zeta^1,\ldots,\zeta^n,z,\zeta)))\cdot\omega_2 = 0
\end{aligned}
$$

and $u^*(\omega' + \omega'') = u^*(\omega_1)$ because

$$((z^1,\ldots,z^m,\zeta^1,\ldots,\zeta^n,z,\zeta) - f((z^1,\ldots,z^m,\zeta^1,\ldots,\zeta^n,z,\zeta))) = 0$$

if $(z^1,\ldots,z^m,\zeta^1,\ldots,\zeta^n,z,\zeta) \in V$. Thus $u^*(\omega' + \omega'') = u^*(\omega)$. From $\Omega \times D$ is $(\bar{\partial},\bar{\partial}_1)$-acyclic and the above result we get that Ω_f is $(\bar{\partial},\bar{\partial}_1)$-acyclic. This ends the proof. □

Corollary 1. *With the notations from the Theorem 1, if $\Omega \times D^{m,n}$ is $(\bar{\partial},\bar{\partial}_1)$-acyclic for all $m, n \geq 0$ then $\Omega_f \times D^{m,n}$ is $(\bar{\partial},\bar{\partial}_1)$-acyclic for all $m, n \geq 0$.*

Proof. Let us consider $g((z,\ldots,z,(,\ldots,(,z,()) = f((z,\ldots,z,(,\ldots,())$ and $\Omega' = \Omega \times D^{m,n}$. Then from the Theorem 1 and from the above consideration we get that $\Omega_f \times D^{m,n} = \Omega'_g$. This ends the proof. □

References

1. Bartocci, C., Bruzzo, U., Hernandez-Ruiperez, D., *The Geometry of Supermanifolds*, Mathematics and Its Applications, Volume 71, 1991.
2. Berezin, F.A., *The method of second quantization*, Academic Press, New York, 1966.
3. Bruzzo, U., Cianci, R., *Class. Quantum Grav.* 1, 213 (1984).
4. Dewitt, B., *Supermanifolds*, Cambridge, Univ. Press, Cambridge, London, 1984.
5. Inoue, A., Maeda, Y., Foundations of Calculus on Supereuclidean Space based on a Frechet-Grassmann Algebra, *Kodai Math. J.* 14 (1991).
6. Jadczyk, A., Pilch, K., *Commun. Math. Phys.* 78, 373 (1981).
7. Kobayashi, S., Nomizu, K., *Foundations of Differential Geometry* (I), Interscience, New York-London 1963.
8. Kostant, B., *Graded manifolds, Graded Lie Theory and Prequantization*, Lect. Notes in Math. no. 570, Springer-Verlag 1977.
9. Leites, D.A., Introduction to the theory of supermanifolds, *Russ. Math. Surv.*, 35 (1980) 1–64.
10. Manin, Yu.I., *Gauge Field Theory and Complex Geometry*, A Series of Comprehensive Studies in Mathematics 289, Springer-Verlag 1988.
11. Narasimhan, R., *Analysis on Real and Complex Manifolds*, Elsevier Science Publishers S.V., Theta Foundation Publishing House, Bucharest, 2001.

12. Rogers, A., A global theory of supermanifolds, *J. Math. Phys.* 21(6), 1352–1365 June 1980.
13. ______, Graded Manifolds, Supermanifolds and Infinite-Dimensional Grassmann Algebras, *Commun. Math. Phys.* 105, 375–384 (1986).
14. Rudin, W., *Real and complex analysis*, Third Edition 1987, 1974, 1966 by McGraw-Hill, Inc.
15. Scheunert, M., *The theory of Lie superalgebras*, Lect. Notes in Math. no. 716, Springer-Verlag 1979.

INVOLUTIVE DISTRIBUTIONS OF CODIMENSION ONE IN KAEHLER MANIFOLDS *

G. GANCHEV

Bulgarian Academy of Sciences, Institute of Mathematics and Informatics, Acad. G. Bonchev Str. bl. 8, 1113, Sofia, Bulgaria
E-mail: ganchev@math.bas.bg

In the present paper we study involutive distributions of codimension one in a Kähler manifold and their integral submanifolds which carry a natural almost contact Riemannian structure. Seven basic geometric classes of involutive distributions are obtained with respect to the structural group $U(n-1) \times I(2)$. Some basic classes of involutive distributions are characterized in terms of the almost contact Riemannian structure of their integral submanifolds.

1. Preliminaries

Let (M, g, η) be an n-dimensional Riemannian manifold with unit 1-form η. The Lie algebra of all C^∞-vector fields on M will be denoted by $\mathcal{X}M$ and T_pM will stand for the tangent space to M at any point $p \in M$. The unit vector field ξ corresponding to the 1-form η is determined by $g(\xi, X) = \eta(X)$, $X \in \mathcal{X}M$. The distribution Δ associated with η is given by

$$\Delta(p) = \{x \in T_pM \mid \eta(x) = 0\}, \quad p \in M.$$

The fundamental structures (g, η) on M generate the following ortogonal decomposition of the tangent spaces

$$T_pM = \Delta(p) \oplus \text{span}\,\{\xi\}, \quad p \in M.$$

This implies the structural group of the manifold (M, g, η) is $O(n-1) \times 1$. As a rule, X, Y will stand for vectors in T_pM, $p \in M$ (vector fields in $\mathcal{X}M$) and x, y will denote vectors in $\Delta(p)$, $p \in M$ (vector fields in $\mathcal{X}\Delta$). For any vector X we have the unique decomposition

$$X = x + \eta(X)\xi, \quad X \in \mathcal{X}M, \quad x \in \mathcal{X}\Delta.$$

*This research is financially supported by Contract No 136/2002, "L. Karavelov" Civil Engineering Higher School
Keywords: involutive distribution of codimension one, real hypersurface of a Kähler manifold
MS classification: 53C55, 53C12

In this paper we consider involutive distributions Δ, i.e.

$$x, y \in \mathcal{X}\Delta \Rightarrow [x, y] \in \mathcal{X}\Delta.$$

Let ∇ be the Levi-Civita connection of the metric g. We define the following 1-form associated with the distribution Δ (or with the 1-form η):

$$\theta(X) = g(\nabla_\xi \xi, X), \quad X \in \mathcal{X}M.$$

Since $\theta(\xi) = g(\nabla_\xi \xi, \xi) = 0$, it follows that $\theta(X) = \theta(x)$, $X \in \mathcal{X}M$.

The involutive distributions are characterized [2] by the following

Lemma 1.1. *The distribution Δ of η is involutive iff $d\eta = \eta \wedge \theta$.*

Thus any involutive distribution Δ generates:

1) a symmetric tensor field of type (0,2) on Δ:

$$(\nabla_x \eta)y = (\nabla_y \eta)x, \quad x, y \in \Delta \tag{1}$$

2) a 1-form on Δ:

$$\theta(x) = g(\nabla_\xi \xi, x), \quad x \in \Delta. \tag{2}$$

Now, let (M, g, J, η) be a 2n-dimensional Kähler manifold with metric g, complex structure J and unit 1-form η. The unit vector field corresponding to η and the distribution of η are denoded by ξ and Δ, respectively. The complex structure J gives rise to the unit vector field $J\xi$ and the corresponding unit 1-form

$$\tilde{\eta}(X) = g(J\xi, X) = -\eta(JX), \quad X \in \mathcal{X}M.$$

The distribution Δ_0 of codimension 2 associated with η and $\tilde{\eta}$ is given by:

$$\Delta_0(p) = \{x_0 \in T_pM \mid \eta(x_0) = \tilde{\eta}(x_0) = 0\}, \quad p \in M.$$

The structures (g, J, η) generate the following orthogonal decomposition of the tangent spaces

$$T_pM = \Delta_0(p) \oplus \text{span}\{\xi\} \oplus \text{span}\{J\xi\}, \quad p \in M;$$
$$\Delta(p) = \Delta_0(p) \oplus \text{span}\{J\xi\}, \quad p \in M.$$

This implies the structural group of the manifold (M, g, J, η) is $U(n-1) \times I(2)$, where $I(2)$ is the identity matrix of order 2.

Further we associate the following functions and 1-forms with the distribution Δ:

$$p = g(\nabla_\xi \xi, J\xi), \quad p^* = g(\nabla_{J\xi} J\xi, \xi); \tag{3}$$

$$\begin{aligned} \theta(X) &= g(\nabla_\xi \xi, X) - p\tilde{\eta}(X), \\ \theta^*(X) &= g(\nabla_{J\xi} J\xi, X) - p^*\eta(X), \ X \in \mathcal{X}M. \end{aligned} \tag{4}$$

We denote by X, Y vectors in T_pM, $p \in M$; by x, y vectors in Δ and by x_0, y_0 vectors in Δ_0. Any tangent vector X is decomposed in a unique way as follows:

$$X = x_0 + \tilde{\eta}(X)J\xi + \eta(X)\xi.$$

It is clear that

$$\theta(X) = \theta(x_0), \quad \theta^*(X) = \theta^*(x_0), \quad X \in T_pM, \, p \in M.$$

Lemma 1.1 implies in a straightforward way

Lemma 1.2. *The distribution Δ is involutive iff*

$$(\nabla_{x_0}\eta)y_0 = (\nabla_{y_0}\eta)x_0, \quad (\nabla_{x_0}\eta)J\xi = \theta^*(Jx_0), \quad x_0, y_0 \in \Delta_0.$$

Thus any involutive distribution Δ generates:
1) a symmetric tensor field of type (0,2) on Δ_0:

$$(\nabla_{x_0}\eta)y_0 = (\nabla_{y_0}\eta)x_0, \quad x_0, y_0 \in \Delta_0; \tag{5}$$

2) two 1-forms on Δ_0:

$$\theta(x_0) = g(\nabla_\xi \xi, x_0), \quad \theta^*(x_0) = g(\nabla_{J\xi}J\xi, x_0), \quad x_0 \in \Delta_0; \tag{6}$$

3) two functions on M:

$$p = g(\nabla_\xi \xi, J\xi), \quad p^* = g(\nabla_{J\xi}J\xi, \xi). \tag{7}$$

Then the condition characterising an involutive distribution Δ becomes

$$d\eta = \eta \wedge \theta + p\eta \wedge \tilde{\eta}.$$

In the next section we give the background of our considerations.

2. Involutive distributions in a Riemannian manifold

Let (M, g, η) be a Riemannian manifold with unit 1-form η. If the distribution Δ is involutive, then the covariant derivative $\nabla\eta$ has the properties:

$$(\nabla_X\eta)\xi = 0, \quad X \in \mathcal{X}M; \quad (\nabla_x\eta)y = (\nabla_y\eta)x, \quad x, y \in \mathcal{X}\Delta. \tag{8}$$

Let now (V, g, η) be an n-dimensional Euclidean space with inner product g and unit 1-form η. The unit vector field corresponding to η and the nullity space of η are denoted by ξ and Δ, respectively. We consider the linear space $\mathcal{L}$ of all tensors L of type (0,2) over V which have the properties implied by (8):

$$L(X, \xi) = 0, \quad X \in V; \quad L(x, y) = L(y, x), \quad x, y \in \Delta. \tag{9}$$

We associate the following 1-form with the tensor L:

$$\theta(X) = L(\xi, X), \quad X \in V.$$

It follows from (9) that θ is a 1-form on Δ:

$$\theta(x) = L(\xi, x), \quad x \in \Delta; \quad \theta(\xi) = 0.$$

The subgroup of $O(n)$ preserving the structure (g, η) is the group $O(n-1)\times 1$. This group acts on the linear space $\mathcal{L}$ in the standard way [4,6]:

$$a(L)(X,Y) = L(a^{-1}X, a^{-1}Y), \quad L \in \mathcal{L}, \quad a \in O(n-1) \times 1.$$

The scalar product g in V induces a scalar product $<,>$ in $\mathcal{L}$ in the usual way:

$$\langle L_1, L_2 \rangle = g^{ik} g^{jl} L_1(e_i, e_j) L_2(e_k, e_l),$$

where $\{e_1, \ldots, e_n\}$ is an orthonormal basis of V.

Taking into account that

$$X = x + \eta(X)\xi, \quad Y = y + \eta(Y)\xi,$$

we consider the following tensors:

$$L_1(X,Y) = \frac{\operatorname{tr} L}{n-1} g(x,y);$$
$$L_2(X,Y) = L(x,y) - \frac{\operatorname{tr} L}{n-1} g(x,y);$$
$$L_3(X,Y) = \eta(X)\theta(Y).$$

These tensors determine the following subspaces of $\mathcal{L}$:

$$\mathcal{L}_1 : \{L \in \mathcal{L} \mid L = L_1\};$$
$$\mathcal{L}_2 : \{L \in \mathcal{L} \mid L = L_2\};$$
$$\mathcal{L}_3 : \{L \in \mathcal{L} \mid L = L_3\}.$$

For any $L \in \mathcal{L}$ we have $L = L_1 + L_2 + L_3$. It is easy to check that the subspaces $\mathcal{L}_1$, $\mathcal{L}_2$, $\mathcal{L}_3$ are mutually orthogonal and invariant under the action of the group $O(n-1) \times 1$. The action of $O(n-1) \times 1$ on the space $\mathcal{L}_1 \oplus \mathcal{L}_2$ coinsides with the action of $O(n-1)$. Taking into account that the decomposition $\mathcal{L}_1 \oplus \mathcal{L}_2$ is irreducible under the action of $O(n-1)$ we obtain:

Proposition 2.1 (Decomposition into basic classes).

$$\mathcal{L} = \mathcal{L}_1 \oplus \mathcal{L}_2 \oplus \mathcal{L}_3$$

where $\mathcal{L}_1, \mathcal{L}_2$ *and* $\mathcal{L}_3$ *are mutually orthogonal, invariant under the action of* $O(n-1) \times 1$ *and irreducible factors.*

The basic subspaces $\mathcal{L}_1$, $\mathcal{L}_2$ and $\mathcal{L}_3$ generate further $2^3 = 8$ invariant subspaces of $\mathcal{L}$:

$$\mathcal{L}_0, \quad \mathcal{L}_1, \quad \mathcal{L}_2, \quad \mathcal{L}_3, \quad \mathcal{L}_2 \oplus \mathcal{L}_3, \quad \mathcal{L}_1 \oplus \mathcal{L}_3, \quad \mathcal{L}_1 \oplus \mathcal{L}_2, \quad \mathcal{L},$$

the space $\mathcal{L}_0$ being the zero subspace of $\mathcal{L}$, i.e. $L = 0$.

Now let (M, g, η) be a Riemannian manifold with unit 1-form η and set $V(p) = T_pM$, $p \in M$ and consider the corresponding linear space $\mathcal{L}(p)$ over $V(p)$, $p \in M$. According to (8) and (9) we can set $L(p) = \nabla\eta$. Applying the above considerations to the tensor $L(p) = \nabla\eta$ we obtain the following basic classes of involutive distributions Δ on M:

1) $\mathcal{F}_0 : \quad (\nabla_x \eta)y = 0, \quad \theta = 0;$

2) $\mathcal{F}_1 : \quad (\nabla_x \eta)y = \dfrac{\operatorname{div}\xi}{n-1} g(x,y), \quad \theta = 0;$

3) $\mathcal{F}_2 : \quad \operatorname{div}\xi = 0, \quad \theta = 0;$

4) $\mathcal{F}_3 : \quad (\nabla_x \eta)y = 0;$

5) $\mathcal{F}_2 \oplus \mathcal{F}_3 : \operatorname{div}\xi = 0;$

6) $\mathcal{F}_1 \oplus \mathcal{F}_3 : (\nabla_x \eta)y = \dfrac{\operatorname{div}\xi}{n-1} g(x,y);$

7) $\mathcal{F}_1 \oplus \mathcal{F}_2 : \quad \theta = 0;$

8) $\mathcal{F}$: no conditions for $(\nabla_x \eta)y$ and θ.

Further we consider involutive distributions in Euclidean space and characterise the eight classes of such distributions in terms of their integral submanifolds.

Let $(\mathbf{E}^n, g, \eta)$ be the standard Euclidean space with unit 1-form η and suppose the distribution Δ of η is involutive. Then $\mathbf{E}^n$ is foliated locally by the integral submanifolds Q^{n-1} of Δ, which are hypersurfaces in $\mathbf{E}^n$. Taking into account that ξ is the unit normal field and $(\nabla_x \eta)y$ is the second fundamental tensor of any Q^{n-1} we obtain the following characterization for the invariant subclasses of $\mathcal{F}$ in terms of the integral submanifolds of Δ and the integral curves of ξ.

Proposition 2.2. *Let Q^{n-1} be any integral submanifold of the involutive distribution Δ of codimension one in $\mathbf{E}^n$. Then the invariant subclasses of $\mathcal{F}$ are characterised locally as follows:*

1) $\mathcal{F}_0$: All Q^{n-1} are parallel hyperplanes;

2) $\mathcal{F}_1$: All Q^{n-1} are parallel (concentric) hyperspheres;

3) $\mathcal{F}_2$: All Q^{n-1} are parallel minimal hyperspheres;

4) $\mathcal{F}_3$: All Q^{n-1} are hyperplanes;

5) $\mathcal{F}_2 \oplus \mathcal{F}_3$: *All* Q^{n-1} *are minimal hypersurfaces;*

6) $\mathcal{F}_1 \oplus \mathcal{F}_3$: *All* Q^{n-1} *are hyperspheres;*

7) $\mathcal{F}_1 \oplus \mathcal{F}_2$: *All* Q^{n-1} *are parallel hypersurfaces;*

8) $\mathcal{F}$: Q^{n-1} *are arbitrary hypersurfaces.*

Here we use the notion "parallel hypersurfaces" in the usual manner for hypersurfaces in Euclidean space.

Now let (M, g, η) be a Riemannian manifold with unit 1-form η whose distribution Δ is involutive. A C^∞-function on M is said to be *a propre function of* Δ [2] if $du \neq 0$ and $du = ||du||\eta$, where $||du||$ is the usual tensor norm of du.

Finally we shall characterise the geometric classes of involutive distributions in M in terms of PDE for their propre functions.

Let u be a propre function for Δ. We denote $T = grad\, u$ and find for any $X \in \mathcal{X}M$:

$$x = X - \eta(X)\xi = X - \frac{du(X)}{||du||^2}T.$$

Further we calculate

$$\theta = d(\ln ||du||) - \frac{T(\ln ||du||)}{||du||^2} du;$$

$$\nabla_T T = ||du||^2 \nabla_\xi \xi + T(\ln ||du||)T, \quad \nabla_T u_j = ||du||^2 \partial_j(\ln ||du||);$$

$$\nabla_X T = ||du|| \nabla_x \xi - \frac{T(\ln ||du||)}{||du||^2} du(X)T + X(\ln ||du||)T + \frac{du(X)}{||du||^2}\nabla_T T;$$

$$\operatorname{div} T = ||du|| \operatorname{div} \xi + T(\ln ||du||).$$

Taking into account Proposition 2.1 we obtain

Proposition 2.3. *Let u be any propre function of the involutive distribution Δ of codimension one in $\mathbf{E}^n$. Then the invariant subclasses of $\mathcal{F}$ are characterised locally by the following conditions for ∇du:*

1) $\mathcal{F}_0$:

$$\nabla_i u_j = \frac{T(\ln ||du||)}{||du||^2} u_i u_j.$$

2) $\mathcal{F}_1$:

$$\nabla_i u_j = \frac{\operatorname{div} T - T(\ln ||du||)}{n-1} g_{ij} - \frac{\operatorname{div} T - nT(\ln ||du||)}{(n-1)||du||^2} u_i u_j.$$

3) $\mathcal{F}_2$:

$$\nabla_T u_j = \operatorname{div} T u_j.$$

4) $\mathcal{F}_3$:

$$\nabla_i u_j = -\frac{T(\ln ||du||)}{||du||^2} u_i u_j + \partial_i(\ln ||du||) u_j + \partial_j(\ln ||du||) u_i.$$

5) $\mathcal{F}_2 \oplus \mathcal{F}_3$:

$$\operatorname{div} T = T(\ln ||du||).$$

6) $\mathcal{F}_1 \oplus \mathcal{F}_3$:

$$\nabla_i u_j = \frac{\operatorname{div} T - T(\ln ||du||)}{n-1} g_{ij} - \frac{\operatorname{div} T + (n-2)T(\ln ||du||)}{(n-1)||du||^2} u_i u_j$$
$$+ \partial_i(\ln ||du||) u_j + \partial_j(\ln ||du||) u_i.$$

7) $\mathcal{F}_1 \oplus \mathcal{F}_2$:

$$\nabla_T u_j = T(\ln ||du||) u_j.$$

8) $\mathcal{F}$: *no condition for* $\nabla_i u_j$.

3. Involutive distributions of codimension one in Kähler manifolds

Let (M, g, J, η) be a $2n$-dimensional Kähler manifold with unit 1-form η. If the distribution Δ of η is involutive, then the covariant derivative $\nabla\eta$ satisfies the equalities:

$$(10) \qquad \begin{aligned} &(\nabla_X \eta)\xi = 0, \quad X \in \mathcal{X}M; \\ &(\nabla_{x_0}\eta) y_0 = (\nabla_{y_0}\eta) x_0, \quad (\nabla_{x_0}\eta) J\xi = \theta^*(Jx_0); \quad x_0, y_0 \in \mathcal{X}\Delta_0. \end{aligned}$$

Let now (V, g, J, η) be a Hermitian vector space with unit 1-form η. The unit vector corresponding to η and the nullity space of η are denoted by ξ and Δ, respectively. Further $\tilde{\eta}$ denotes the 1-form corresponding to the vector $J\xi$ and Δ_0 denotes the vector space of vectors perpendicular to ξ and $J\xi$.

We denote by $\mathcal{L}$ the linear space of all tensors L over V of type (0,2) satisfying the properties (10) of $\nabla\xi$:

$$L(X, \xi) = 0, \quad X \in V;$$
$$L(x_0, y_0) = L(y_0, x_0), \quad L(x_0, J\xi) = L(J\xi, x_0), \quad x_0, y_0 \in \Delta_0.$$

Let $\{e_i\}$, $i = 1, \ldots, 2(n-1)$ be an orthonormal basis for the space Δ_0 and g_0 be the restriction of the scalar product onto Δ_0. Then we consider the trace of the tensor L on Δ_0:

$$\operatorname{tr}_0 L = g_0^{ij} L(e_i, e_j).$$

Here (g_0^{ij}) is the inverse matrix of $(g_0(e_i, e_j))$.

Further we associate the following scalars and 1-forms with the tensor L:

$$\begin{aligned} &p = L(\xi, J\xi), \quad p^* = -L(J\xi, J\xi); \\ &\theta(X) = L(\xi, X) - p\tilde{\eta}(X), \quad \theta^*(X) = L(J\xi, JX) + p^*\eta(X). \end{aligned} \tag{11}$$

It follows from (11) that

$$\theta(x_0) = L(\xi, x_0), \quad \theta(J\xi) = \theta(\xi) = 0;$$
$$\theta^*(x_0) = L(J\xi, Jx_0), \quad \theta^*(J\xi) = \theta^*(\xi) = 0.$$

The subgroup of the unitary group $U(n)$ preserving the structure (g, J, η) is the group $U(n) \times I(2)$, where $I(2)$ denotes the unit matrix of order 2.

Taking into account the representations

$$X = x_0 + \tilde{\eta}(X)J\xi + \eta(X)\xi, \quad Y = y_0 + \tilde{\eta}(Y)J\xi + \eta(Y)\xi$$

we introduce the following tensors (projection operators):

$$L_1(X,Y) = \frac{L(x_0, y_0) - L(Jx_0, Jy_0)}{2};$$
$$L_2(X,Y) = \frac{\operatorname{tr}_0 L}{2(n-1)} g(x_0, y_0);$$
$$L_3(X,Y) = \frac{L(x_0, y_0) + L(Jx_0, Jy_0)}{2} - \frac{\operatorname{tr}_0 L}{2(n-1)} g(x_0, y_0);$$
$$L_4(X,Y) = \theta^*(JX)\tilde{\eta}(Y) + \theta^*(JY)\tilde{\eta}(X);$$
$$L_5(X,Y) = \eta(X)\theta(Y);$$
$$L_6(X,Y) = -p^*\tilde{\eta}(X)\tilde{\eta}(Y);$$
$$L_7(X,Y) = p\eta(X)\tilde{\eta}(Y).$$

These tensors determine the following subspaces of $\mathcal{L}$:

$$\mathcal{L}_i = \{L \in \mathcal{L} \mid L = L_i\}, \quad i = 1, \dots, 7.$$

For any tensor $L \in \mathcal{L}$ we have $L = L_1 + \cdots + L_7$.

We note that the tensors $g(x_0, y_0)$, $\tilde{\eta}(X)\tilde{\eta}(Y)$ and $\eta(X)\tilde{\eta}(Y)$ are the invariant under the action of $U(n-1) \times I(2)$ tensors in the space $\mathcal{L}$. The action of $U(n-1) \times I(2)$ on the space $\mathcal{L}_1 \oplus \mathcal{L}_2 \oplus \mathcal{L}_3$ reduces to the action of $U(n-1)$. Taking into account that the decomposition $\mathcal{L}_1 \oplus \mathcal{L}_2 \oplus \mathcal{L}_3$ is irreducible under the action of $U(n-1)$ we obtain

Proposition 3.1 (Decomposition into basic classes).

$$\mathcal{L} = \mathcal{L}_1 \oplus \cdots \oplus \mathcal{L}_7,$$

where the factors $\mathcal{L}_i$ $(i = 1, \dots, 7)$ are mutually orthogonal, invariant under the action of $U(n-1) \times I(2)$ and irreducible.

The basic subspaces $\mathcal{L}_i$ $(i = 1, \dots, 7)$ further generate $2^7 = 256$ invariant subspaces of $\mathcal{L}$. The zero subspace of $\mathcal{L}$ generated by $L = 0$ is denoted by $\mathcal{L}_0$.

Now let (M, g, J, η) be a Kähler manifold with unit 1-form η. We set $V(p) = T_pM$, $p \in M$ and consider the corresponding linear space $\mathcal{L}(p)$ of the tensors over $V(p)$ having the symmetries (11). According to (11) we can set $L(p) = \nabla\eta$. Applying the above considerations to the tensor $L(p) = \nabla\eta$ we obtain the following basic classes of involutive distributions Δ on M:

1) $\mathcal{F}_1 : (\nabla_{x_0}\eta)y_0 + (\nabla_{Jx_0}\eta)Jy_0 = 0, \theta = \theta^* = 0, p = p^* = 0;$

2) $\mathcal{F}_2 : (\nabla_{x_0}\eta)y_0 = \dfrac{\operatorname{div}_0\xi}{2(n-1)}g(x_0, y_0),, \theta = \theta^* = 0, p = p^* = 0;$

3) $\mathcal{F}_3 : (\nabla_{x_0}\eta)y_0 = (\nabla_{Jx_0}\eta)Jy_0, \operatorname{div}_0\xi = 0, \theta = \theta^* = 0, p = p^* = 0;$

4) $\mathcal{F}_4 : (\nabla_{x_0}\eta)y_0 = 0, \theta = 0, \theta^* \neq 0, p = p^* = 0;$

5) $\mathcal{F}_5 : (\nabla_{x_0}\eta)y_0 = 0, \theta \neq 0, \theta^* = 0, p = p^* = 0;$

6) $\mathcal{F}_6 : (\nabla_{x_0}\eta)y_0 = 0, \theta = \theta^* = 0, p = 0, p^* \neq 0;$

7) $\mathcal{F}_7 : (\nabla_{x_0}\eta)y_0 = 0, \theta = \theta^* = 0, p \neq 0, p^* = 0.$

We denote by $\mathcal{F}_0$ the zero class:

$$\mathcal{F}_0 : \nabla\eta = 0.$$

Further we consider the integral submanifolds of the involutive distribution Δ. All considerations are local.

Let Q^{2n-1} be an integral submanifold of Δ, i.e. Q^{2n-1} is a real hypersurface in M whose tangent spaces are $\Delta(p)$, $p \in Q^{2n-1}$. We denote by $N = -\xi$ the unit normal vector field at the points of Q^{2n-1}. Then the second fundamental tensor h of Q^{2n-1} is given by

$$h(x, y) = (\nabla_x\eta)y, \quad x, y \in \Delta(p),\ p \in Q^{2n-1}.$$

The real hypersurface Q^{2n-1} carries a natural almost contact Riemannian structure $(\varphi, \tilde{\xi}, \tilde{\eta}, g)$ [1,5,7]:

$$\tilde{\xi} = -JN = J\xi, \quad \tilde{\eta}(x) = g(J\xi, x), \quad x \in \Delta;$$
$$\varphi(x) = Jx + \tilde{\eta}(x)\xi, \quad x \in \Delta.$$

Taking into account (5), (6) and (7) we find the components of h:

$$h(x_0, y_0) = (\nabla_{x_0}\eta)y_0, \quad x_0, y_0 \in \Delta_0;$$
$$h(x_0, J\xi) = \theta^*(Jx_0), \quad x_0 \in \Delta_0;$$
$$h(J\xi, J\xi) = -p^*.$$

From Theorem 2 in [3] and Proposition 3.1 we obtain the following statements:

Proposition 3.2. *Let Q^{2n-1} be any integral submanifold of the involutive distribution Δ in a Kähler manifold (M^{2n}, g, J, η). Then the following conditions are equivalent:*

(i) All Q^{2n-1} are W_1-manifolds;

(ii) $(\nabla_{x_0}\eta)y_0 = 0, \quad \theta^* \neq 0, \quad x_0, y_0 \in \Delta_0;$

(iii) Δ is in the class $\mathcal{F}_4 \oplus \mathcal{F}_5 \oplus \mathcal{F}_6 \oplus \mathcal{F}_7$.

Proposition 3.3. *Let Q^{2n-1} be any integral submanifold of the involutive distribution Δ in a Kähler manifold (M^{2n}, g, J, η). Then the following conditions are equivalent:*

(i) All Q^{2n-1} are W_2-manifolds;

(ii) $(\nabla_{x_0}\eta)y_0 = \dfrac{\operatorname{div}_0\xi}{2(n-1)} g(x_0, y_0), \quad \theta^* = 0, \quad x_0, y_0 \in \Delta_0;$

(iii) Δ is in the class $\mathcal{F}_2 \oplus \mathcal{F}_4 \oplus \mathcal{F}_6 \oplus \mathcal{F}_7$.

Proposition 3.4. *Let Q^{2n-1} be any integral submanifold of the involutive distribution Δ in a Kähler manifold (M^{2n}, g, J, η). Then the following conditions are equivalent:*

(i) All Q^{2n-1} are W_4-manifolds;

(ii) $(\nabla_{x_0}\eta)y_0 = (\nabla_{Jx_0}\eta)Jy_0, \quad \operatorname{div}_0\xi = 0, \quad \theta^* = 0, \quad x_0, y_0 \in \Delta_0;$

(iii) Δ is in the class $\mathcal{F}_3 \oplus \mathcal{F}_4 \oplus \mathcal{F}_6 \oplus \mathcal{F}_7$.

Proposition 3.5. *Let Q^{2n-1} be any integral submanifold of the involutive distribution Δ in a Kähler Manifold (M^{2n}, g, J, η). Then the following conditions are equivalent:*

(i) All Q^{2n-1} are W_6-manifolds;

(ii) $(\nabla_{x_0}\eta)y_0 + (\nabla_{Jx_0}\eta)Jy_0 = 0, \quad \theta^* = 0, \quad x_0, y_0 \in \Delta_0;$

(iii) Δ is in the class $\mathcal{F}_1 \oplus \mathcal{F}_4 \oplus \mathcal{F}_6 \oplus \mathcal{F}_7$.

References

1. Blair D., *Contact manifolds in Riemannian geometry.* Lecture Notes in Math. **509**, (Springer Verlag, Berlin, 1976).
2. Ganchev G. and V. Mihova, *Riemannian manifolds of quasi-constant sectional curvatures.* J. Reine und Angew. Math. **522**(2000), 119–141.
3. Ganchev G. and M. Hristov, *Real hypersurfaces of a Káhler manifold (The sixteen classes).* Perspectives of Complex Analysis, Differential Geometry and Mathematical Physics. (World Scientific, Singapore, 2001), 147–158.
4. Gray A. and L. Hervella, *The sixteen classes of almost Hermitian manifolds and their linear invariants.* Ann di mat. pura ed appl. (IV), CXXIII(1980), 35–58.

5. Tashiro Y., *On contact structure of hypersurfaces in complex manifolds, I.* Tohoku math. J. **15**(1963), 62–78.
6. Tricerri F. and L. Vanhecke, *Curvature tensors on almost Hermitian manifolds.* Trans. Am. Math. Soc., **267**(1981), 365–398.
7. Yano K. and M. Kon, *CR-submanifolds of Kaehlerian and Sasakian manifolds.* Progress in Mathematics, **30** (Birkhäuser, Boston, 1983).

ON THE THEORY OF TWO-DIMENSIONAL SURFACES IN EUCLIDEAN SPACE*

G. GANCHEV

Bulgarian Academy of Sciences, Institute of Mathematics and Informatics,
Acad. G. Bonchev Str. bl. 8, 1113, Sofia, Bulgaria
E-mail: ganchev@math.bas.bg

V. MILOUSHEVA

"L. Karavelov" Civil Engineering Higher School, Department of Mathematics,
32 Suhodolska Str., 1373, Sofia, Bulgaria
E-mail: milousheva@vsu.bg

Using the conformal representation of a two-dimensional surface in n-dimensional Euclidean space, we obtain conditions under which the normal connection of the surface is flat. We introduce the notion of geometric normal vector field and prove that the normal connection of a two-dimensional surface is flat iff the surface admits geometric normal frame fields. We find the relation between two geometric normal frame fields.

1. Preliminaries

Let M^2 be a two-dimensional surface in Euclidean space E^n defined in a domain $D \subset R^2$ by the vector-valued function

$$z = z(u, v), \quad (u, v) \in D.$$

We assume that M^2 is regular in D, i.e. the vectors $z_u(u, v)$ and $z_v(u, v)$ are linearly independent at any point of M^2. If g is the standard metric in E^n, the coefficients of the first fundamental form of M^2 are denoted by

$$E(u, v) = g(z_u, z_u); \quad F(u, v) = g(z_u, z_v); \quad G(u, v) = g(z_v, z_v).$$

The parameters (u, v) of M^2 are called *conformal* (*isothermal*), if

$$g(z_u, z_u) = g(z_v, z_v); \quad g(z_u, z_v) = 0,$$

*The research is financially supported by Contract No 136/2002, "L. Karavelov" Civil Engineering Higher School, Bulgaria
Keywords: conformal parameters; torse-forming vector field; geometric normals
MS classification: 53A07; 53A10; 53B20

i.e. $E = G$, $F = 0$.

The first fundamental form of M^2 with respect to conformal parameters (u, v) is given by

$$ds^2 = \rho(u,v)\,(du^2 + dv^2), \quad \rho(u,v) > 0.$$

The possibility of introducing conformal parameters on an arbitrary regular surface is expressed in the theorem of Korn - Lichtenstein [2, 3], in which the weakest conditions on the surface under which conformal parameters exist are stated.

To formulate this theorem we must recall that a real function $f(u,v)$ in a domain $D \subset R^2$ is said to satisfy the Hölder condition of order α $(0 < \alpha \leq 1)$, if the inequality

$$|f(u,v) - f(u',v')| < Cr^\alpha$$

holds good for arbitrary points (u,v) and (u',v') of D, where C is a constant and r is the Euclidean distance between the points. The theorem of Korn-Lichtenstein is formulated in [1] as follows:

Suppose in a domain $D \subset R^2$ the functions E, F, G satisfy the Hölder condition of order α $(0 < \alpha < 1)$. Then every point of D has a neighbourhood whose local coordinates are isothermal parameters.

This means that for each point of a regular surface of class $C^{1,\alpha}$ $(0 < \alpha < 1)$ there is a neighbourhood, in which the surface has a conformal representation of class $C^{1,\alpha}$.

In what follows we consider surfaces of class C^∞ in E^n. Since our considerations are local, we can assume that they are represented locally by conformal parameters of class C^∞.

Using the conformal representation we obtain derivative formulas for a surface M^2 in E^n and find equivalent conditions under which the normal connection of M^2 is flat (Proposition 2.1). For a surface with flat normal connection we construct geometric normal frame fields. More over, we prove that the normal connection of a surface M^2 is flat iff M^2 admits geometric normal frame fields (Theorem 2.2). Finally, we prove that any two geometric normal frame fields are connected by a constant orthogonal matrix (Proposition 2.2).

2. Two-dimensional surfaces in E^n

Let $M^2 : \; z = z(u,v), \; (u,v) \in D$ be a regular C^∞-surface in Euclidean space E^n, defined in a domain $D \subset R^2$, represented by conformal parameters (u,v), i.e.

$$g(z_u, z_u) = g(z_v, z_v) = E(u,v), \quad E(u,v) > 0;$$
$$g(z_u, z_v) = 0.$$

The tangent space T_pM^2 at a point $p = z(u,v)$ of M^2 is $T_pM^2 = \mathrm{Span}\{z_u, z_v\}$. We denote by Δ_p^{n-2} the normal space of M^2 at the point p. The normal distribution of M^2 is denoted by $\Delta^{n-2} : p \to \Delta_p^{n-2}$. Let $\{b_1, b_2, \dots, b_{n-2}\}$ ($b_\alpha = b_\alpha(u,v)$, $\alpha = 1, \dots, n-2$) be a local orthonormal frame field of Δ^{n-2}.

Then we have the following representation formulas of the second derivatives of $z(u,v)$:

$$\begin{aligned} z_{uu} &= \frac{E_u}{2E} z_u - \frac{E_v}{2E} z_v + \sum_{\alpha=1}^{n-2} c_{11}^\alpha \, b_\alpha; \\ z_{uv} &= \frac{E_v}{2E} z_u + \frac{E_u}{2E} z_v + \sum_{\alpha=1}^{n-2} c_{12}^\alpha \, b_\alpha; \\ z_{vv} &= -\frac{E_u}{2E} z_u + \frac{E_v}{2E} z_v + \sum_{\alpha=1}^{n-2} c_{22}^\alpha \, b_\alpha, \end{aligned} \tag{1}$$

where $c_{ij}^\alpha(u,v)$ ($i,j = 1,2$, $\alpha = 1, \dots, n-2$) are functions of (u,v), determined as follows:

$$\begin{aligned} c_{11}^\alpha &= g(z_{uu}, b_\alpha); \\ c_{12}^\alpha &= g(z_{uv}, b_\alpha); \qquad \alpha = 1, \dots, n-2. \\ c_{22}^\alpha &= g(z_{vv}, b_\alpha); \end{aligned}$$

For the derivatives of the normal vectors b_α we obtain

$$\begin{aligned} (b_\alpha)_u &= \frac{\partial b_\alpha}{\partial u} = -\frac{c_{11}^\alpha}{E} z_u - \frac{c_{12}^\alpha}{E} z_v + \sum_{\beta=1}^{n-2} p_\alpha^\beta \, b_\beta; \\ (b_\alpha)_v &= \frac{\partial b_\alpha}{\partial v} = -\frac{c_{12}^\alpha}{E} z_u - \frac{c_{22}^\alpha}{E} z_v + \sum_{\beta=1}^{n-2} q_\alpha^\beta \, b_\beta; \end{aligned} \qquad \alpha = 1, \dots, n-2, \tag{2}$$

where the functions $p_\alpha^\beta(u,v)$ and $q_\alpha^\beta(u,v)$ are determined as follows

$$\begin{aligned} p_\alpha^\beta &= g((b_\alpha)_u, b_\beta); \\ q_\alpha^\beta &= g((b_\alpha)_v, b_\beta); \end{aligned} \qquad \alpha, \beta = 1, \dots, n-2.$$

Let ∇' be the Levi-Civita connection of the standard metric g (the canonical connection) in E^n. Denoting by $W = \dfrac{z_u}{\sqrt{E}}$ and $\xi = \dfrac{z_v}{\sqrt{E}}$, from equalities (1) we obtain the following derivative formulas for M^2 with respect to conformal parameters (u, v):

$$(3)\qquad \begin{aligned} \nabla'_W W &= & -\frac{E_v}{2E\sqrt{E}}\xi + \frac{c_{11}^{\alpha}}{E}\, b_\alpha; \\ \nabla'_W \xi &= \frac{E_v}{2E\sqrt{E}} W & + \frac{c_{12}^{\alpha}}{E}\, b_\alpha; \\ \nabla'_\xi W &= & \frac{E_u}{2E\sqrt{E}}\xi + \frac{c_{12}^{\alpha}}{E}\, b_\alpha; \\ \nabla'_\xi \xi &= -\frac{E_u}{2E\sqrt{E}} W & + \frac{c_{22}^{\alpha}}{E}\, b_\alpha, \end{aligned}$$

where summation convention is assumed.

Let ∇ be the induced connection on $(M^2,\, g)$ and σ – the second fundamental tensor of M^2. We denote by $\mathcal{X}M^2$ the Lie algebra of all C^∞ vector fields on M^2. The Riemannian curvature tensor R of ∇ is given by the equalities

$$(4)\qquad \begin{aligned} R(X,Y)Z &= \nabla_X\nabla_Y Z - \nabla_Y\nabla_X Z - \nabla_{[X,Y]}Z, \quad X,Y,Z \in \mathcal{X}M^2; \\ R(X,Y,Z,U) &= g(R(X,Y)Z,U), \quad X,Y,Z,U \in \mathcal{X}M^2. \end{aligned}$$

Taking into account the Gauss formula

$$\nabla'_X Y = \nabla_X Y + \sigma(X,Y), \quad X,Y \in \mathcal{X}M^2,$$

from (3) and (4) we calculate the Riemannian curvature K of M^2:

$$K = R(W,\, \xi,\, \xi,\, W) = \frac{E_u^2 + E_v^2}{2E^3} - \frac{E_{uu} + E_{vv}}{2E^2},$$

i.e.

$$K = -\frac{1}{2E}\Delta \ln E.$$

Now, let's consider the normal curvatures R_{b_α}, $\alpha = 1, \dots, n-2$ of M^2, corresponding to the normal vector fields b_α. According to the Weingarten formula

$$\nabla'_X b_\alpha = -A_{b_\alpha}(X) + D_X b_\alpha, \quad X \in \mathcal{X}M^2$$

we have

$$(5)\qquad \begin{aligned} \nabla'_W b_\alpha &= -A_{b_\alpha}(W) + D_W b_\alpha; \\ \nabla'_\xi b_\alpha &= -A_{b_\alpha}(\xi) + D_\xi b_\alpha. \end{aligned}$$

On the other hand, the equalities (2) imply

$$(6)\qquad \begin{aligned} \nabla'_W b_\alpha &= -\frac{c_{11}^{\alpha}}{E} W - \frac{c_{12}^{\alpha}}{E}\xi + \frac{p_\alpha^\beta}{\sqrt{E}}\, b_\beta; \\ \nabla'_\xi b_\alpha &= -\frac{c_{12}^{\alpha}}{E} W - \frac{c_{22}^{\alpha}}{E}\xi + \frac{q_\alpha^\beta}{\sqrt{E}}\, b_\beta; \end{aligned} \qquad \alpha = 1, \dots, n-2.$$

Since the normal curvatures R_{b_α} are given by

$$R_{b_\alpha}(W,\xi) = D_W D_\xi b_\alpha - D_\xi D_W b_\alpha - D_{[W,\xi]} b_\alpha,$$

then the formulas (5) and (6) give the following result:

$$(7) \qquad R_{b_\alpha}(W,\xi) = -\frac{1}{E}\left(\frac{\partial}{\partial v}(p_\alpha^\gamma) - \frac{\partial}{\partial u}(q_\alpha^\gamma) + p_\alpha^\beta q_\beta^\gamma - q_\alpha^\beta p_\beta^\gamma\right) b_\gamma,$$
$$\alpha = 1,\dots,n-2.$$

Let ω and η be the one-forms corresponding to W and ξ, respectively:

$$\omega(X) = g(W,X); \quad \eta(X) = g(\xi,X); \quad X \in \mathcal{X}M^2.$$

For an arbitrary $X \in \mathcal{X}M^2$ we have the unique representation

$$(8) \qquad X = \omega(X)W + \eta(X)\xi.$$

So, using equalities (6) together with (8) we obtain

$$\nabla'_X b_\alpha = -\frac{1}{E}\left(c_{11}^\alpha \omega(X) + c_{12}^\alpha \eta(X)\right) W - \frac{1}{E}\left(c_{12}^\alpha \omega(X) + c_{22}^\alpha \eta(X)\right)\xi$$
$$+ \frac{1}{\sqrt{E}}\left(p_\alpha^\beta \omega(X) + q_\alpha^\beta \eta(X)\right) b_\beta,$$

which implies

$$D_X b_\alpha = \frac{1}{\sqrt{E}}\left(p_\alpha^\beta \omega(X) + q_\alpha^\beta \eta(X)\right) b_\beta, \quad \alpha = 1,\dots,n-2.$$

We denote by θ_α^β the one-forms, determined by the equality

$$\theta_\alpha^\beta(X) = \frac{1}{\sqrt{E}}\left(p_\alpha^\beta \omega(X) + q_\alpha^\beta \eta(X)\right), \quad \alpha,\beta = 1,\dots,n-2.$$

Hence, we obtain

$$(9) \qquad D_X b_\alpha = \theta_\alpha^\beta(X)\, b_\beta, \quad \alpha = 1,\dots,n-2.$$

Taking into account (9) we obtain the following formulas for the normal curvatures R_{b_α}, $\alpha = 1,\dots,n-2$:

$$(10) \quad R_{b_\alpha}(X,Y) = \left(d\theta_\alpha^\gamma(X,Y) - \theta_\alpha^\beta \wedge \theta_\beta^\gamma(X,Y)\right) b_\gamma, \quad \alpha = 1,\dots,n-2.$$

The normal connection D is said to be flat, if $R_{b_\alpha} = 0$, $\alpha = 1,\dots,n-2$. This condition does not depend on the normal frame field $\{b_1,\dots,b_{n-2}\}$. So, taking into account the equalities (7) and (10) we obtain the following

Proposition 2.1. *Let M^2 be a regular C^∞-surface in Euclidean space E^n. Then the following conditions are equivalent:*

(i) *the normal connection of* M^2 *is flat;*

(ii) $d\theta_\alpha^\gamma = \theta_\alpha^\beta \wedge \theta_\beta^\gamma$;

(iii) $\dfrac{\partial}{\partial v}(p_\alpha^\gamma) - \dfrac{\partial}{\partial u}(q_\alpha^\gamma) + p_\alpha^\beta q_\beta^\gamma - q_\alpha^\beta p_\beta^\gamma = 0.$

From now on we consider regular C^∞-surfaces in E^n with flat normal connection. For such surfaces we shall prove the following

Theorem 2.1. *Let* $M^2 : z = z(u,v)$, $(u,v) \in D$ *be a regular* C^∞*-surface in Euclidean space* E^n. *Then the following conditions are equivalent:*

(i) *the normal connection of* M^2 *is flat;*

(ii) *for any point* $(u_0, v_0) \in D$ *and a unit normal vector* N_0 *at* $p = z(u_0, v_0)$ *there exists a unique normal vector field* $N(u,v)$, *defined in a subset* $D_0 \subset D$, *such that*

$$N_u,\ N_v \in T_pM^2, \quad N(u_0,v_0) = N_0, \quad g(N,N) = 1.$$

Proof. Suppose that the surface $M^2 : z = z(u,v)$, $(u,v) \in D$ is represented by conformal parameters (u,v) and the equalities (1), (2) hold good. If $N(u,v)$ is an arbitrary normal vector field of M^2, then

$$N(u,v) = \sum_{\alpha=1}^{n-2} f^\alpha(u,v)\, b_\alpha(u,v), \tag{11}$$

where $f^\alpha(u,v)$, $\alpha = 1, \ldots, n-2$ are functions on M^2.

Differentiating (11) and using (2) we obtain

$$\begin{aligned} N_u &= \nabla'_{z_u} N = -\frac{1}{E}\left(c_{11}^\alpha f^\alpha z_u + c_{12}^\alpha f^\alpha z_v\right) + \left(\frac{\partial f^\beta}{\partial u} + f^\alpha p_\alpha^\beta\right) b_\beta; \\ N_v &= \nabla'_{z_v} N = -\frac{1}{E}\left(c_{12}^\alpha f^\alpha z_u + c_{22}^\alpha f^\alpha z_v\right) + \left(\frac{\partial f^\beta}{\partial v} + f^\alpha q_\alpha^\beta\right) b_\beta. \end{aligned} \tag{12}$$

From equalities (12) it follows that N_u, $N_v \in \mathrm{Span}\{z_u, z_v\}$ iff the functions $f^1(u,v), \ldots, f^{n-2}(u,v)$ form a solution of the following system of partial differential equations:

$$\begin{aligned} &\frac{\partial f^\beta}{\partial u} + p_\alpha^\beta f^\alpha = 0; \\ &\frac{\partial f^\beta}{\partial v} + q_\alpha^\beta f^\alpha = 0; \end{aligned} \qquad \beta = 1, \ldots, n-2. \tag{13}$$

According to the theory of partial differential equations the integrability conditions of the system (13) are exactly the conditions (iii) in Proposition 2.1.

I. Let M^2 be a surface of flat normal connection. Then, according to Proposition 2.1, the integrability conditions of (13) are fulfilled. Hence, if $N_0 = \sum_{\alpha=1}^{n-2} f_0^\alpha \, b_\alpha(u_0, v_0)$ is a unit normal vector at a point $p = z(u_0, v_0)$ of M^2, then there exists a unique solution $f^1(u,v), \ldots, f^{n-2}(u,v)$ of (13), defined in a subset $D_0 \subset D$ ($(u_0, v_0) \in D_0$), such that $f^\alpha(u_0, v_0) = f_0^\alpha$, $\alpha = 1, \ldots, n-2$. Then, according to equalities (12), the vector $N(u,v)$, defined by (11), satisfies the conditions

$$N_u, \, N_v \in T_pM^2, \quad N(u_0, v_0) = N_0.$$

Since $g(N_u, N) = 0$, $g(N_v, N) = 0$, then

$$\frac{\partial}{\partial u} g(N,N) = 0; \quad \frac{\partial}{\partial v} g(N,N) = 0,$$

which imply that $g(N,N) = const = g(N_0, N_0) = 1$.

II. Conversely, if there exists a normal vector field $N(u,v)$ on M^2, such that $N_u, N_v \in T_pM^2$, then the integrability conditions of (13) hold good. According to Proposition 2.1 it follows that the normal connection of M^2 is flat. □

Remark. If N is a normal vector field, satisfying $\nabla'_{z_u} N, \nabla'_{z_v} N \in T_zM^2$, then using the representation (8) for an arbitrary vector field $X \in \mathfrak{X}M^2$ we have $\nabla'_X N \in \mathfrak{X}M^2$. We call such normal vector fields *geometric normals* of M^2. The geometric normals are analogous to the torse-forming vector fields along a curve c, introduced by K. Yano [4]. We recall that a unit vector field N, defined along a curve c is said to be a *torse-forming vector field* along c, if $\nabla'_{\dot{c}} N = k\dot{c}$ for some function k. For each regular curve in E^n there exist torse-forming vector fields. While a regular two-dimensional surface M^2 in E^n admits geometric normal vector fields if and only if M^2 is of flat normal connection.

Now we can prove the main theorem in our paper.

Theorem 2.2. *Let M^2 be a regular C^∞-surface in Euclidean space E^n. Then the following conditions are equivalent:*

(i) *the normal connection of M^2 is flat;*

(ii) *M^2 admits locally geometric normal frame fields.*

Proof. Let M^2 be of flat normal connection and $\{b_1^0, \ldots, b_{n-2}^0\}$ be an orthonormal frame field at a point $p \in M^2$. Denote by $b_\alpha(u,v)$, $\alpha = 1, \ldots, n-2$ the local geometric normals from Theorem 2.1 (defined in a neighbourhood U of p)

and satisfying the conditions $b_\alpha(u_0, v_0) = b^0_\alpha$. Since for each $\alpha = 1, \ldots, n-2$, $(b_\alpha)_u, (b_\alpha)_v \in Span\{z_u, z_v\}$, then we have

$$\frac{\partial}{\partial u} g(b_\alpha, b_\beta) = 0; \quad \frac{\partial}{\partial v} g(b_\alpha, b_\beta) = 0; \quad \alpha, \beta = 1, \ldots, n-2,$$

which imply $g(b_\alpha, b_\beta) = const = g(b^0_\alpha, b^0_\beta)$. Hence, the frame filed $\{b_1(u,v), \ldots, b_{n-2}(u,v)\}$ is orthogonal at every point of U.

The converse follows from Theorem 2.1. □

Now we shall clear up the relation between two geometric normal frame fileds.

Proposition 2.2. *Let M^2 be a regular C^∞-surface in E^n with flat normal connection. Then any two geometric normal frame fields are connected by a constant orthogonal matrix.*

Proof. Let $\{b_1, \ldots, b_{n-2}\}$ and $\{\bar{b}_1, \ldots, \bar{b}_{n-2}\}$ be two geometric normal frame fields of M^2, connected by an orthogonal matrix $(\tau_k^\alpha)_{k,\alpha=1}^{n-2}$ as follows:

$$\bar{b}_k = \tau_k^\alpha \, b_\alpha, \quad k = 1, \ldots, n-2. \tag{14}$$

Differentiating (14) we obtain

$$\begin{aligned} \left(\bar{b}_k\right)_u &= \frac{\partial}{\partial u}(\tau_k^\alpha)\, b_\alpha + \tau_k^\alpha \,(b_\alpha)_u; \\ \left(\bar{b}_k\right)_v &= \frac{\partial}{\partial v}(\tau_k^\alpha)\, b_\alpha + \tau_k^\alpha \,(b_\alpha)_v; \end{aligned} \qquad k = 1, \ldots, n-2.$$

Since for geometric normal vector fields we have $(b_\alpha)_u, (b_\alpha)_v, \left(\bar{b}_k\right)_u, \left(\bar{b}_k\right)_v \in$ Span$\{z_u, z_v\}$, then we get

$$\frac{\partial}{\partial u}(\tau_k^\alpha) = 0; \quad \frac{\partial}{\partial v}(\tau_k^\alpha) = 0; \quad k, \alpha = 1, \ldots, n-2.$$

Hence, $\tau_k^\alpha = const$, $k, \alpha = 1, \ldots, n-2$. □

Let M^2 be a regular C^∞-surface in E^n with flat normal connection and $\{b_1, \ldots, b_{n-2}\}$ be a geometric normal frame field of M^2. For the normal vectors b_α, $\alpha = 1, \ldots, n-2$ we have the following formulas:

$$\begin{aligned} (b_\alpha)_u &= -a^\alpha_{11} z_u - a^\alpha_{12} z_v; \\ (b_\alpha)_v &= -a^\alpha_{12} z_u - a^\alpha_{22} z_v; \end{aligned} \qquad \alpha = 1, \ldots, n-2.$$

Then the derivative formulas of M^2 take the form:

$$
(15) \qquad
\begin{aligned}
\nabla'_W W &= \phantom{\frac{E_v}{2E\sqrt{E}}W} -\frac{E_v}{2E\sqrt{E}}\xi + a_{11}^{\alpha}\, b_{\alpha};\\
\nabla'_W \xi &= \frac{E_v}{2E\sqrt{E}}W \phantom{-\frac{E_v}{2E\sqrt{E}}\xi} + a_{12}^{\alpha}\, b_{\alpha};\\
\nabla'_\xi W &= \phantom{\frac{E_v}{2E\sqrt{E}}W} \frac{E_u}{2E\sqrt{E}}\xi + a_{12}^{\alpha}\, b_{\alpha};\\
\nabla'_\xi \xi &= -\frac{E_u}{2E\sqrt{E}}W \phantom{-\frac{E_v}{2E\sqrt{E}}\xi} + a_{22}^{\alpha}\, b_{\alpha}.
\end{aligned}
$$

Since the normal mean curvature vector H of M^2 is given by the equality

$$H = \frac{1}{2}\operatorname{trace}\sigma,$$

then, using (15) and the Gauss formula, we obtain

$$(16) \qquad H = \frac{a_{11}^{\alpha} + a_{22}^{\alpha}}{2}\, b_{\alpha}.$$

So, the mean curvature ν of M^2 is

$$\nu = \sqrt{H^2} = \frac{1}{2}\sqrt{\textstyle\sum_{\alpha=1}^{n-2}(a_{11}^{\alpha} + a_{22}^{\alpha})^2}.$$

Let H_0 be the unit normal mean curvature vector, i.e. $H_0 = \dfrac{H}{\nu}$. Then we get

$$H_0 = \frac{a_{11}^{\alpha} + a_{22}^{\alpha}}{\sqrt{\sum_{\beta=1}^{n-2}(a_{11}^{\beta} + a_{22}^{\beta})^2}}\, b_{\alpha}.$$

The equality (16) implies that M^2 is a minimal surface in E^n $(H = 0)$ iff

$$a_{11}^{\alpha} + a_{22}^{\alpha} = 0, \quad \alpha = 1, \dots, n-2.$$

Now we shall find the conditions under which the normal mean curvature vector H is a geometric normal of M^2. Since $(b_\alpha)_u$, $(b_\alpha)_v \in \operatorname{Span}\{z_u, z_v\}$, then from (16) it follows that H_u, $H_v \in \operatorname{Span}\{z_u, z_v\}$ iff

$$\frac{\partial}{\partial u}(a_{11}^{\alpha} + a_{22}^{\alpha}) = 0; \quad \frac{\partial}{\partial v}(a_{11}^{\alpha} + a_{22}^{\alpha}) = 0; \quad \alpha = 1, \dots, n-2.$$

Hence, H is a geometric normal of M^2 iff

$$a_{11}^{\alpha} + a_{22}^{\alpha} = \text{const}, \quad \alpha = 1, \dots, n-2.$$

References

1. Chern, S-S., *An elementary proof of the existence of isothermal parameters on a surface*, Proc. Am. Math. Soc. **6** (1955) 771–782.
2. Korn, A., *Zwei Anwendungen der Methode der sukzessiven Annäherungen*, Schwarz-Festschrift. Springer (1914) 215–229.
3. Lichtenstein, L., *Zur Theorie der konformen Abbildung. Konforme Abbildung nichtanalytischer singularitätenfreier Flächenstücke auf ebene Gebiete*, Bull. Acad. Sci. Cracovie, Cl. Sci. Math. Nat. A (1916) 192–217.
4. Yano, K., *On the torse-forming directions in Riemannian spaces*, Proc. Imp. Acad., Tokyo **20** (1944) 340–345.

ON HYPERCOMPLEX PSEUDO-HERMITIAN MANIFOLDS

KOSTADIN GRIBACHEV

Faculty of Mathematics and Informatics, University of Plovdiv,
236 Bulgaria Blvd., Plovdiv 4004, Bulgaria
E-mail: costas@pu.acad.bg

MANCHO MANEV

Faculty of Mathematics and Informatics, University of Plovdiv,
236 Bulgaria Blvd., Plovdiv 4004, Bulgaria
E-mail: mmanev@pu.acad.bg

STANCHO DIMIEV

Institute of Mathematics and Informatics, Bulgarian Academy of Sciences,
Acad. G. Bonchev Str., Bl. 8, Sofia 1113, Bulgaria,
E-mail: sdimiev@math.bas.bg

The class of the hypercomplex pseudo-Hermitian manifolds is considered. The flatness of the considered manifolds with the 3 parallel complex structures is proved. Conformal transformations of the metrics are introduced. The conformal invariance and the conformal equivalence of the basic types manifolds are studied. A known example is characterized in relation to the obtained results.

Introduction

This paper is a continuation of the same authors's paper [4] which is inspired by the seminal work [1] of D. V. Alekseevsky and S. Marchiafava. We follow a parallel direction including skew-Hermitian metrics with respect to the almost hypercomplex structure.

In the first section we give some necessary facts concerning the almost hypercomplex pseudo-Hermitian manifolds introduced in [4].

In the second one we consider the special class of (integrable) hypercomplex pseudo-Hermitian manifolds, namely pseudo-hyper-Kähler manifolds. Here we expose the proof of the mentioned in [4] statement that each pseudo-hyper-Kähler manifold is flat.

The third section is fundamental for this work. A study of the group of conformal transformations of the metric is initiated here. The conformal invariant classes and the conformal equivalent class to the class of the pseudo-hyper-Kähler manifolds are found.

Finally, we characterize a known example in terms of the conformal transformations.

1. Preliminaries

1.1. *Hypercomplex pseudo-Hermitian structures in a real vector space*

Let V be a real $4n$-dimensional vector space. By $\left\{\frac{\partial}{\partial x^i}, \frac{\partial}{\partial y^i}, \frac{\partial}{\partial u^i}, \frac{\partial}{\partial v^i}\right\}$, $i = 1, 2, \ldots, n$, is denoted a (local) basis on V. Each vector x of V is represented in the mentioned basis as follows

$$x = x^i\frac{\partial}{\partial x^i} + y^i\frac{\partial}{\partial y^i} + u^i\frac{\partial}{\partial u^i} + v^i\frac{\partial}{\partial v^i}. \qquad (1)$$

A standard complex structure on V is defined as in [5]:

$$\begin{aligned}
&J_1\frac{\partial}{\partial x^i} = \frac{\partial}{\partial y^i}, \quad J_1\frac{\partial}{\partial y^i} = -\frac{\partial}{\partial x^i}, \quad J_1\frac{\partial}{\partial u^i} = -\frac{\partial}{\partial v^i}, \quad J_1\frac{\partial}{\partial v^i} = \frac{\partial}{\partial u^i};\\
&J_2\frac{\partial}{\partial x^i} = \frac{\partial}{\partial u^i}, \quad J_2\frac{\partial}{\partial y^i} = \frac{\partial}{\partial v^i}, \quad J_2\frac{\partial}{\partial u^i} = -\frac{\partial}{\partial x^i}, \quad J_2\frac{\partial}{\partial v^i} = -\frac{\partial}{\partial y^i};\\
&J_3\frac{\partial}{\partial x^i} = -\frac{\partial}{\partial v^i}, \quad J_3\frac{\partial}{\partial y^i} = \frac{\partial}{\partial u^i}, \quad J_3\frac{\partial}{\partial u^i} = -\frac{\partial}{\partial y^i}, \quad J_3\frac{\partial}{\partial v^i} = \frac{\partial}{\partial x^i}.
\end{aligned} \qquad (2)$$

The following properties about J_α are direct consequences of (2)

$$\begin{aligned}
&J_1^2 = J_2^2 = J_3^2 = -Id,\\
&J_1J_2 = -J_2J_1 = J_3, \quad J_2J_3 = -J_3J_2 = J_1, \quad J_3J_1 = -J_1J_3 = J_2.
\end{aligned} \qquad (3)$$

If $x \in V$, i.e. $x(x^i, y^i, u^i, v^i)$ then according to (2) and (3) we have

$$J_1x(-y^i, x^i, v^i, -u^i), \quad J_2x(-u^i, -v^i, x^i, y^i), \quad J_3x(v^i, -u^i, y^i, -x^i). \qquad (4)$$

Definition 1.1 ([1]). A triple $H = (J_1, J_2, J_3)$ of anticommuting complex structures on V with $J_3 = J_1J_2$ is called *a hypercomplex structure* on V.

A bilinear form f on V is defined as ordinary, $f : V \times V \to \mathbb{R}$. We denote by $\mathcal{B}(V)$ the set of all bilinear forms on V. Each f is a tensor of type $(0, 2)$, and $\mathcal{B}(V)$ is a vector space of dimension $16n^2$.

Let J be a given complex structure on V. A bilinear form f on V is called *Hermitian* (respectively, *skew-Hermitian*) with respect to J if the identity $f(Jx, Jy) = f(x, y)$ (respectively, $f(Jx, Jy) = -f(x, y)$) holds true.

Definition 1.2 ([1]). A bilinear form f on V is called *an Hermitian bilinear form* with respect to $H = (J_\alpha)$ if it is Hermitian with respect to any complex structure $J_\alpha, \alpha = 1, 2, 3$, i.e.

$$f(J_\alpha x, J_\alpha y) = f(x, y) \qquad \forall\, x, y \in V. \tag{5}$$

We denote by $L_0 = \mathcal{B}_H(V)$ the set of all Hermitian bilinear forms on V. The notion of pseudo-Hermitian bilinear forms is introduced by the following

Definition 1.3 ([4]). A bilinear form f on V is called *a pseudo-Hermitian bilinear form with respect to* $H = (J_1, J_2, J_3)$, if it is Hermitian with respect to J_α and skew-Hermitian with respect to J_β and J_γ, i.e.

$$f(J_\alpha x, J_\alpha y) = -f(J_\beta x, J_\beta y) = -f(J_\gamma x, J_\gamma y) = f(x, y) \quad \forall\, x, y \in V, \tag{6}$$

where (α, β, γ) is a circular permutation of $(1, 2, 3)$.

We denote $f \in L_\alpha \subset \mathcal{B}(V)$ $(\alpha = 0, 1, 2, 3)$ when f satisfies the conditions (5) and (6), respectively.

In [1] is introduced a pseudo-Euclidian metric g with signature $(2n, 2n)$ as follows

$$g(x, y) := \sum_{i=1}^{n} \left(-x^i a^i - y^i b^i + u^i c^i + v^i d^i\right), \tag{7}$$

where $x(x^i, y^i, u^i, v^i)$, $y(a^i, b^i, c^i, d^i) \in V$, $i = 1, 2, \ldots, n$. This metric satisfies the following properties

$$g(J_1 x, J_1 y) = -g(J_2 x, J_2 y) = -g(J_3 x, J_3 y) = g(x, y). \tag{8}$$

This means that the pseudo-Euclidean metric g belongs to L_1.

The form $g_1 : g_1(x, y) = g(J_1 x, y)$ coincides with the Kähler form Φ which is Hermitian with respect to J_α, i.e.

$$\Phi(J_\alpha x, J_\alpha y) = \Phi(x, y), \quad \alpha = 1, 2, 3, \qquad \Phi \in L_0.$$

The attached to g associated bilinear forms $g_2 : g_2(x, y) = g(J_2 x, y)$ and $g_3 : g_3(x, y) = g(J_3 x, y)$ are symmetric forms with the properties

$$\begin{aligned} -g_2(J_1 x, J_1 y) &= -g_2(J_2 x, J_2 y) = g_2(J_3 x, J_3 y) = g_2(x, y), \\ -g_3(J_1 x, J_1 y) &= g_3(J_2 x, J_2 y) = -g_3(J_3 x, J_3 y) = g_3(x, y), \end{aligned} \tag{9}$$

i.e. $g_2 \in L_3$, $g_3 \in L_2$.

It follows that the Kähler form Φ is Hermitian regarding H and the metrics g, g_2, g_3 are pseudo-Hermitian of different types with signature $(2n, 2n)$.

Now we recall the following notion:

Definition 1.4 ([4]). The structure $(H,G) := (J_1, J_2, J_3, g, \Phi, g_2, g_3)$ is called *a hypercomplex pseudo-Hermitian structure* on V.

1.2. *Structural tensors on an almost (H, G)-manifold*

Let (M, H) be an almost hypercomplex manifold [1]. We suppose that g is a symmetric tensor field of type $(0,2)$. If it induces a pseudo-Hermitian inner product in T_pM, $p \in M$, then g is called *a pseudo-Hermitian metric on* M. The structure $(H,G) := (J_1, J_2, J_3, g, \Phi, g_2, g_3)$ is called *an almost hypercomplex pseudo-Hermitian structure on* M or in short *an almost* (H,G)*-structure on* M. The manifold M equipped with H and G, i.e. (M,H,G), is called *an almost hypercomplex pseudo-Hermitian manifold*, or in short *an almost* (H,G)*-manifold.* [4]

The 3 tensors of type $(0,3)$ $F_\alpha : F_\alpha(x,y,z) = g\big((\nabla_x J_\alpha)\,y, z\big)$, $\alpha = 1, 2, 3$, where ∇ is the Levi-Civita connection generated by g, is called *structural tensors of the almost* (H,G)*-manifold.* [4]

The structural tensors satisfy the following properties:

$$\begin{aligned} F_1(x,y,z) &= F_2(x,J_3y,z) + F_3(x,y,J_2z),\\ F_2(x,y,z) &= F_3(x,J_1y,z) + F_1(x,y,J_3z),\\ F_3(x,y,z) &= F_1(x,J_2y,z) - F_2(x,y,J_1z); \end{aligned} \tag{10}$$

$$\begin{aligned} F_1(x,y,z) &= -F_1(x,z,y) = -F_1(x,J_1y,J_1z),\\ F_2(x,y,z) &= F_2(x,z,y) = F_2(x,J_2y,J_2z),\\ F_3(x,y,z) &= F_3(x,z,y) = F_3(x,J_3y,J_3z). \end{aligned} \tag{11}$$

Let us recall the Nijenhuis tensors $N_\alpha(X,Y) = \frac{1}{2}\,[[J_\alpha, J_\alpha]]\,(X,Y)$ for almost complex structures J_α and $X, Y \in \mathfrak{X}(M)$, where

$$[[J_\alpha, J_\alpha]]\,(X,Y) = 2\big\{[J_\alpha X, J_\alpha Y] - J_\alpha\,[J_\alpha X, Y] - J_\alpha\,[X, J_\alpha Y] - [X,Y]\big\}.$$

It is well known that the almost hypercomplex structure $H = (J_\alpha)$ is a hypercomplex structure if $[[J_\alpha, J_\alpha]]$ vanishes for each $\alpha = 1, 2, 3$. Moreover it is known that one almost hypercomplex structure H is hypercomplex if and only if two of the structures J_α ($\alpha = 1, 2, 3$) are integrable. This means that two of the tensors N_α vanish. [1]

We recall also the following definitions. Since g is Hermitian metric with respect to J_1, according to [3] the class $\mathcal{W}_4$ is a subclass of the class of Hermitian manifolds. If (H,G)-manifold belongs to $\mathcal{W}_4$, with respect to J_1, then the almost complex structure J_1 is integrable and

$$\begin{aligned} F_1(x,y,z) = \frac{1}{2(2n-1)}\,&[g(x,y)\theta_1(z) - g(x,z)\theta_1(y)\\ &-g(x,J_1y)\theta_1(J_1z) + g(x,J_1z)\theta_1(J_1y)]\,, \end{aligned} \tag{12}$$

where $\theta_1(\cdot) = g^{ij}F_1(e_i, e_j, \cdot) = \delta\Phi(\cdot)$ for the basis $\{e_i\}_{i=1}^{4n}$, and δ — the coderivative.

On other side the metric g is a skew-Hermitian with respect to J_2 and J_3, i.e. $g(J_2x, J_2y) = g(J_3x, J_3y) = -g(x, y)$. A classification of all almost complex manifolds with skew-Hermitian metric (Norden metric or B-metric) is given in [2]. One of the basic classes of integrable almost complex manifolds with skew-Hermitian metric is $\mathcal{W}_1$. It is known that if an almost (H, G)-manifold belongs to $\mathcal{W}_1(J_\alpha)$, $\alpha = 2, 3$, then J_α is integrable and the following equality holds

$$F_\alpha(x, y, z) = \frac{1}{4n}\left[g(x, y)\theta_\alpha(z) + g(x, z)\theta_\alpha(y) + g(x, J_\alpha y)\theta_\alpha(J_\alpha z) + g(x, J_\alpha z)\theta_\alpha(J_\alpha y)\right], \tag{13}$$

where $\theta_\alpha(z) = g^{ij}F_\alpha(e_i, e_j, z)$, $\alpha = 2, 3$, for an arbitrary basis $\{e_i\}_{i=1}^{4n}$.

When (12) is satisfied for (M, H, G), we say that $(M, H, G) \in \mathcal{W}(J_1)$. In the case, (M, H, G) satisfies (13) for $\alpha = 2$ or $\alpha = 3$, we say $(M, H, G) \in \mathcal{W}(J_2)$ or $(M, H, G) \in \mathcal{W}(J_3)$. Let us denote the class $\mathcal{W} := \bigcap_{\alpha=1}^{3} \mathcal{W}(J_\alpha)$.

The next theorem gives a sufficient condition an almost (H, G)-manifold to be integrable.

Theorem 1.1 ([4]). *Let (M, H, G) belongs to the class $\mathcal{W}(J_\alpha) \bigcap \mathcal{W}(J_\beta)$. Then (M, H, G) is of class $\mathcal{W}(J_\gamma)$ for all cyclic permutations (α, β, γ) of $(1, 2, 3)$.*

Let us remark that necessary and sufficient conditions (M, H, G) to be in $\mathcal{W}$ are

$$\theta_\alpha \circ J_\alpha = -\frac{2n}{2n-1}\theta_1 \circ J_1, \qquad \alpha = 2, 3. \tag{14}$$

2. Pseudo-hyper-Kähler manifolds

Definition 2.1 ([4]). A pseudo-Hermitian manifold is called a *pseudo-hyper-Kähler manifold*, if $\nabla J_\alpha = 0$ ($\alpha = 1, 2, 3$) with respect to the Levi-Civita connection generated by g.

It is clear, then $F_\alpha = 0$, ($\alpha = 1, 2, 3$) holds or the manifold is Kählerian with respect to J_α, i.e. $(M, H, G) \in \mathcal{K}(J_\alpha)$.

Immediately we obtain that if (M, H, G) belongs to $\mathcal{K}(J_\alpha) \bigcap \mathcal{W}(J_\beta)$ then $(M, H, G) \in \mathcal{K}(J_\gamma)$ for all cyclic permutations (α, β, γ) of $(1, 2, 3)$.

Then the following sufficient condition for a $\mathcal{K}$-manifold is valid.

Theorem 2.1 ([4]). *If $(M, H, G) \in \mathcal{K}(J_\alpha) \bigcap \mathcal{W}(J_\beta)$ then M is a pseudo-hyper-Kähler manifold ($\alpha \neq \beta \in \{1, 2, 3\}$).*

Let (M^{4n}, H, G) be a pseudo-hyper-Kähler manifold and ∇ be the Levi-Civita connection generated by g. The curvature tensor seems as follows

$$R(X,Y)Z = \nabla_X \nabla_Y Z - \nabla_Y \nabla_X Z - \nabla_{[X,Y]} Z, \tag{15}$$

and the corresponding tensor of type $(0,4)$ is

$$R(X,Y,Z,W) = g\left(R(X,Y)Z, W\right), \quad \forall\, X,Y,Z,W \in \mathfrak{X}(M). \tag{16}$$

Lemma 2.1. *The curvature tensor of a pseudo-hyper-Kähler manifold has the following properties:*

$$\begin{aligned} R(X,Y,Z,W) &= R(X,Y,J_1Z,J_1W) = R(J_1X,J_1Y,Z,W) \\ &= -R(X,Y,J_2Z,J_2W) = -R(J_2X,J_2Y,Z,W) \\ &= -R(X,Y,J_3Z,J_3W) = -R(J_3X,J_3Y,Z,W), \end{aligned} \tag{17}$$

$$\begin{aligned} R(X,Y,Z,W) &= R(X,J_1Y,J_1Z,W) \\ &= -R(X,J_2Y,J_2Z,W) = -R(X,J_3Y,J_3Z,W). \end{aligned} \tag{18}$$

Proof. The equality (17) is valid, because of (15), (16), the condition $\nabla J_\alpha = 0$ $(\alpha = 1,2,3)$, the equality (8) and the properties of the curvature $(0,4)$-tensor.

To prove (18), we will show at first that the property $R(X,J_2Y,J_2Z,W) = -R(X,Y,Z,W)$ holds. Indeed, from (17) we get

$$R(J_2X,Y,Z,W) = R(X,J_2Y,Z,W), \quad R(X,Y,J_2Z,W) = R(X,Y,Z,J_2W)$$

and $\mathfrak{S}_{X,Y,Z} R(X,Y,J_2Z,J_2W) = 0$, where $\mathfrak{S}_{X,Y,Z}$ denotes the cyclic sum regarding X, Y, Z. In the last equality we replace Y by J_2Y and W by J_2W. We get

$$-R(X,J_2Y,J_2Z,W) - R(J_2Y,Z,J_2X,W) + R(Z,X,Y,W) = 0. \tag{19}$$

Replacing Y by Z, and inversely, we get

$$-R(X,J_2Z,J_2Y,W) - R(J_2Z,Y,J_2X,W) + R(Y,X,Z,W) = 0. \tag{20}$$

As we have

$$-R(J_2Z,Y,J_2X,W) = -R(Z,J_2Y,J_2X,W) = R(J_2Y,Z,X,W),$$

with the help of (19) and (20) we obtain

$$\begin{aligned} &- R(X,J_2Y,J_2Z,W) - R(X,J_2Z,J_2Y,W) \\ &\qquad + R(Z,X,Y,W) + R(Y,X,Z,W) = 0. \end{aligned} \tag{21}$$

According to the first Bianchi identity and (17), we obtain

$$\begin{aligned}-R(X,J_2Z,J_2Y,W) &= R(J_2Z,J_2Y,X,W)+R(J_2Y,X,J_2Z,W)\\ &= -R(Z,Y,X,W)-R(X,J_2Y,J_2Z,W).\end{aligned}$$

Then the equality (21) seem as follows

$$-2R(X,J_2Y,J_2Z,W)+R(Z,X,Y,W)-R(X,Y,Z,W)+R(Y,Z,X,W)=0$$

By the first Bianchi identity the equality is transformed in the following

$$-2R(X,J_2Y,J_2Z,W)-2R(X,Y,Z,W)=0,$$

which is equivalent to

$$R(X,J_2Y,J_2Z,W)=-R(X,Y,Z,W). \tag{22}$$

As the tensor R has the same properties with respect to J_3, and to J_2, it follows that the next equality holds, too.

$$R(X,J_3Y,J_3Z,W)=-R(X,Y,Z,W). \tag{23}$$

Using (22) and (23) for $J_1=J_2J_3$ we get successively that

$$\begin{aligned}R(X,Y,Z,W) &= R(X,J_1Y,J_1Z,W)\\ &= R(X,J_2(J_3Y),J_2(J_3Z),W)=-R(X,J_3Y,J_3Z,W),\end{aligned}$$

which completes the proof of (18). □

Now we will prove a theorem which gives us a geometric characteristic of the pseudo-hyper-Kähler manifolds.

Theorem 2.2. *Each pseudo-hyper-Kähler manifold is a flat pseudo-Riemannian manifold with signature* $(2n,2n)$.

Proof. Lemma 2.1 implies the properties

$$\begin{aligned}-R(X,Y,Z,W) &= R(X,J_1Y,Z,J_1W)\\ &= R(X,J_2Y,Z,J_2W)=R(X,J_3Y,Z,J_3W).\end{aligned} \tag{24}$$

As $J_1=J_2J_3$, we also have the following

$$\begin{aligned}R(X,J_1Y,Z,J_1W) &= R(X,J_2(J_3Y),Z,J_2(J_3W))\\ &= -R(X,J_3Y,Z,J_3W)=R(X,Y,Z,W).\end{aligned}$$

Comparing (24) with the last equality we receive

$$-R(X,Y,Z,W)=R(X,J_1Y,Z,J_1W)=R(X,Y,Z,W),$$

or $R\equiv 0$. □

3. Conformal transformations of the pseudo-Hermitian metric

The usual conformal transformation $c : \bar{g} = e^{2u}g$, where u is a differential function on M^{4n}, is known. Since $g_\alpha(\cdot,\cdot) = g(J_\alpha\cdot,\cdot)$, the conformal transformation of g causes the same changes of the pseudo-Hermitian metrics g_2, g_3 and the Kähler form $\Phi \equiv g_1$. Then we say that it is given a conformal transformation c of G to $\bar{G}$ determined by $u \in \mathcal{F}(M)$. These conformal transformations form a group denoted by C. The hypercomplex pseudo-Hermitian manifolds (M, H, G) and $(M, H, \bar{G})$ we call C-equivalent manifolds or conformal-equivalent manifolds.

Let ∇ and $\bar{\nabla}$ be the Levi-Civita connections determined by the metrics g and $\bar{g}$, respectively. The known condition for a Levi-Civita connection implies the following relation

$$\bar{\nabla}_X Y = \nabla_X Y + \mathrm{d}u(X)Y + \mathrm{d}u(Y)X - g(X,Y)\mathrm{grad}(u). \tag{25}$$

Using (25) and the definitions of structural tensors for ∇ and $\bar{\nabla}$ we obtain

$$\begin{aligned}\bar{F}_1(X,Y,Z) = e^{2u}\,[F_1(X,Y,Z) - g(X,Y)\mathrm{d}u(J_1Z) + g(X,Z)\mathrm{d}u(J_1Y)\\ +g(J_1X,Y)\mathrm{d}u(Z) - g(J_1X,Z)\mathrm{d}u(Y)],\end{aligned} \tag{26}$$

$$\begin{aligned}\bar{F}_\alpha(X,Y,Z) = e^{2u}\,[F_\alpha(X,Y,Z) + g(X,Y)\mathrm{d}u(J_\alpha Z) + g(X,Z)\mathrm{d}u(J_\alpha Y)\\ -g(J_\alpha X,Y)\mathrm{d}u(Z) - g(J_\alpha X,Z)\mathrm{d}u(Y)]\end{aligned} \tag{27}$$

for $\alpha = 2, 3$. The last two equalities imply the following relations for the corresponding structural 1-forms

$$\bar{\theta}_1 = \theta_1 - 2(2n-1)\mathrm{d}u \circ J_1, \qquad \bar{\theta}_\alpha = \theta_\alpha + 4n\mathrm{d}u \circ J_\alpha, \qquad \alpha = 2, 3. \tag{28}$$

Let us denote the following $(0,3)$-tensors.

$$\begin{aligned}P_1(x,y,z) = F_1(x,y,z) - \frac{1}{2(2n-1)}\,[g(x,y)\theta_1(z) - g(x,z)\theta_1(y)\\ -g(x,J_1y)\theta_1(J_1z) + g(x,J_1z)\theta_1(J_1y)]\,,\end{aligned} \tag{29}$$

$$\begin{aligned}P_\alpha(x,y,z) = F_\alpha(x,y,z) - \frac{1}{4n}\,[g(x,y)\theta_\alpha(z) + g(x,z)\theta_\alpha(y)\\ +g(x,J_\alpha y)\theta_\alpha(J_\alpha z) + g(x,J_\alpha z)\theta_\alpha(J_\alpha y)]\,, \quad \alpha = 2, 3.\end{aligned} \tag{30}$$

According to (12) and (13) it is clear that

$$(M, H, G) \in \mathcal{W}(J_\alpha) \iff P_\alpha = 0 \quad (\alpha = 1, 2, 3).$$

The equalities (26)–(28) imply the following two interconnections

$$\bar{P}_\alpha = e^{2u} P_\alpha, \quad \alpha = 1,2,3; \tag{31}$$

$$\bar{\theta}_\alpha \circ J_\alpha + \frac{2n}{2n-1}\bar{\theta}_1 \circ J_1 = \theta_\alpha \circ J_\alpha + \frac{2n}{2n-1}\theta_1 \circ J_1, \quad \alpha = 2,3. \tag{32}$$

From (31) we receive that each of $\mathcal{W}(J_\alpha)$ $(\alpha = 1,2,3)$ is invariant with respect to the conformal transformations of C, i.e. they are C-invariant classes. Having in mind also (32), we state the validity of the following

Theorem 3.1. *The class $\mathcal{W}$ of hypercomplex pseudo-Hermitian manifolds is C-invariant.*

Now we will determine the class of the (locally) C-equivalent $\mathcal{K}$-manifolds. Let us denote the following subclass $\mathcal{W}^0 := \{\mathcal{W} \mid d(\theta_1 \circ J_1) = 0\}$.

Theorem 3.2. *A hypercomplex pseudo-Hermitian manifold belongs to $\mathcal{W}^0$ if and only if it is C-equivalent to a pseudo-hyper-Kähler manifold.*

Proof. Let (M,H,G) bc a pscudo-hyper-Kähler manifold, i.e. $(M,H,G) \in \mathcal{K}$. Then $F_\alpha = \theta_\alpha = 0$ (α = 1, 2, 3). Hence (28) has the form

$$\bar{\theta}_1 = -2(2n-1)\mathrm{d}u \circ J_1, \qquad \bar{\theta}_\alpha = 4n\mathrm{d}u \circ J_\alpha, \quad \alpha = 2,3. \tag{33}$$

From (26), (27) and (33) and having in mind (12) and (13) we obtain that $(M,H,\bar{G})$ is a $\mathcal{W}$-manifold. According to (33) the 1-forms $\bar{\theta}_\alpha \circ J_\alpha$ (α = 1, 2, 3) are closed. Because of (14) the condition $\mathrm{d}(\bar{\theta}_1 \circ J_1) = 0$ is sufficient.

Conversely, let $(M,H,\bar{G})$ be a $\mathcal{W}$-manifold with closed $\bar{\theta}_1 \circ J_1$. Because of (14) the 1-forms $\bar{\theta}_\alpha \circ J_\alpha$ (α = 2, 3) are closed, too. We determine the function u as a solution of the differential equation $\mathrm{d}u = -\frac{1}{2(2n-1)}\bar{\theta}_1 \circ J_1$. Then by an immediate verification we state that the transformation $c^{-1} : g = e^{-2u}\bar{g}$ converts $(M,H,\bar{G})$ into $(M,H,G) \in \mathcal{K}$. This completes the proof. □

Let us remark the following inclusions

$$\mathcal{K} \subset \mathcal{W}^0 \subset \mathcal{W} \subset \mathcal{W}(J_\alpha), \quad \alpha = 1,2,3.$$

Let R, ρ, τ and $\bar{R}$, $\bar{\rho}$, $\bar{\tau}$ be the curvature tensors, the Ricci tensors, the scalar curvatures corresponding to ∇ and $\bar{\nabla}$, respectively. The following tensor is curvature-like, i.e. it has the same properties as R.

$$\begin{aligned}\psi_1(S)(X,Y,Z,U) = g(Y,Z)S(X,U) - g(X,Z)S(Y,U)\\ + g(X,U)S(Y,Z) - g(Y,U)S(X,Z),\end{aligned}$$

where S is a symmetric tensor.

Having in mind (25) and (15), we obtain

Proposition 3.1. *The following relations hold for the C-equivalent (H,G)-manifolds*

$$\bar{R} = e^{2u}\{R - \psi_1(S)\},$$
$$\bar{\rho} = \rho - \mathrm{tr}Sg - 2(2n-1)S, \qquad \bar{\tau} = e^{-2u}\{\tau - 2(4n-1)\mathrm{tr}S\}, \tag{34}$$

where

$$S(Y,Z) = S(Z,Y) = (\nabla_Y \mathrm{d}u)\,Z + \mathrm{d}u(Y)\mathrm{d}u(Z) - \frac{1}{2}\mathrm{d}u(\mathrm{grad}(u))g(Y,Z). \tag{35}$$

If (M,H,G) is a C-equivalent $\mathcal{W}$-manifold to a $\mathcal{K}$-manifold, i.e. $(M,H,G) \in \mathcal{W}^0$, then Proposition 3.1 implies

Corollary 3.1. *A $\mathcal{W}^0$-manifold has the following curvature characteristic*

$$R = \frac{1}{2(2n-1)}\left\{\psi_1(\rho) - \frac{\tau}{4n-1}\pi_1\right\},$$

where $\pi_1(X,Y,Z,U) = \frac{1}{2}\psi_1(g) = g(Y,Z)g(X,U) - g(X,Z)g(Y,U)$.

It is well known that the C-invariant tensor of each pseudo-Riemannian manifold is the so-called Weil tensor W. From (34) we receive immediately

$$\bar{W} = e^{2u}W, \qquad W = R - \frac{1}{2(2n-1)}\left\{\psi_1(\rho) - \frac{\tau}{4n-1}\pi_1\right\}. \tag{36}$$

Let us remark that the vanishing of the Weil tensor is a necessary and sufficient condition a pseudo-Riemannian manifold to be conformal equivalent to a flat manifold with dimension greater than 3.

This is confirmed by the combining of Theorem 2.2, Theorem 3.2 and Corollary 3.1, i.e. $(M,H,G) \in \mathcal{W}^0$ iff $W = 0$ on (M,H,G).

Since each conformal transformation determines uniquely a symmetric tensor S by (35) then it takes an interest in the consideration S as a bilinear form on T_pM belonging to each of the components L_α, ($\alpha = 0$, 1, 2, 3).

Let $S \in L_0$. In view of (5) $\mathrm{tr}S = 0$ holds and according to (34) we receive $\bar{\tau} = e^{-2u}\tau$ and an invariant tensor $W_0 = R - \frac{1}{2(2n-1)}\psi_1(\rho)$. When W_0 vanishes on (M,H,G) then the curvature tensor has the form $R = \frac{1}{2(2n-1)}\psi_1(\rho)$.

In the cases when $S \in L_\alpha$ ($\alpha = 1$, 2, 3) we consider (M,H,G) as an $\mathcal{W}^0$-manifold. Then according to Theorem 2.2 and Theorem 3.2 we have $\bar{R} = 0$ on the C-equivalent $\mathcal{K}$-manifold of (M,H,G).

Now let $S \in L_1$. By reason of $g \in L_1$ we have a cause for the consideration of the possibility $S = \lambda g$. Hence $\lambda = \frac{\mathrm{tr}S}{4n} = \frac{\tau}{8n(4n-1)}$. Then having in mind (34)

$R = \frac{\tau}{4n(4n-1)}\pi_1$ holds true. From here it is clear that if $S \in L_1$ then (M, H, G) is an Einstein manifold.

Let us consider the case when $S \in L_2$. Then according to (6) tr$S$ vanishes, and from (34) τ vanishes, too. Because of $g_3 \in L_2$ we consider $S = \lambda g_3$, whence $\lambda = -\frac{\mathrm{tr}(S \circ J_3)}{4n}$. Then (34) implies $R = \frac{\mathrm{tr}(S \circ J_3)}{4n}\pi_3^{J_3}$, where $\pi_3^{J_3}$ is the following tensor π_3 with respect to the complex structure $J = J_3$

$$\pi_3(X, Y, Z, U) = -\pi_1(X, Y, JZ, U) - \pi_1(X, Y, Z, JU).$$

It is known [2] that π_3 is a Kähler curvature-like tensor, i.e. it satisfies the property $\pi_3(X, Y, JZ, JU) = -\pi_3(X, Y, Z, U)$. Therefore in this case R is Kählerian with respect to J_3 and the tensor $R^{*J_3} : R^{*J_3}(X, Y, Z, U) = R(X, Y, Z, J_3U)$ is curvature-like. Then we obtain immediately

$$R = \frac{\tau(R^{*J_3})}{8n(2n-1)}\pi_3^{J_3}, \qquad \rho = -\frac{\tau(R^{*J_3})}{4n}g_3.$$

Hence if $S \in L_2$ then (M, H, G) is a $*$-Einstein manifold with respect to J_3.

By an analogous way, in the case when $S \in L_3$ we receive that (M, H, G) is a $*$-Einstein manifold with respect to J_2.

4. A 4-dimensional pseudo-Riemannian spherical manifold with (H, G)-structure

In [4] is considered a hypersurface S_2^4 in $\mathbb{R}_2^5$ by the equation

$$-\left(z^1\right)^2 - \left(z^2\right)^2 + \left(z^3\right)^2 + \left(z^4\right)^2 + \left(z^5\right)^2 = 1, \tag{37}$$

where $Z\left(z^1, z^2, z^3, z^4, z^5\right)$ is the positional vector of $p \in S_2^4$.

Let $\left(u^1, u^2, u^3, u^4\right)$ be local coordinates of p on S_2^4. The hypersurface S_2^4 is defined by the scalar parametric equations:

$$\begin{gathered} z^1 = \sinh u^1 \cos u^2, \quad z^2 = \sinh u^1 \sin u^2, \quad z^3 = \cosh u^1 \cos u^3 \cos u^4, \\ z^4 = \cosh u^1 \cos u^3 \sin u^4, \quad z^5 = \cosh u^1 \sin u^3. \end{gathered} \tag{38}$$

Further we consider the manifold on $\tilde{S}_2^4 = S_2^4 \backslash \{(0,0,0,0,\pm 1)\}$, i.e. we omit two points for which $\{u^1 \neq 0\} \cap \{u^3 \neq (2k+1)\pi/2, k \in \mathbb{Z}\}$. The tangent space $T_p\tilde{S}_2^4$ of $\tilde{S}_2^4$ in the point $p \in \tilde{S}_2^4$ is determined by the vectors $z_i = \frac{\partial Z}{\partial u^i}$ $(i = 1, 2, 3, 4)$. The vectors z_i are linearly independent on $\tilde{S}_2^4$, defined by (38), and $T_p\tilde{S}_2^4$ has a basis (z_1, z_2, z_3, z_4) in every point $p \in \tilde{S}_2^4$.

The restriction of $\langle \cdot, \cdot \rangle$ from $\mathbb{R}_2^5$ to $\tilde{S}_2^4$ is a pseudo-Riemannian metric g on $\tilde{S}_2^4$ with signature $(2, 2)$. The non-zero components $g_{ij} = \langle z_i, z_j \rangle$ are

$$g_{11} = -1, \; g_{22} = -\sinh^2 u^1, \; g_{33} = \cosh^2 u^1, \; g_{44} = \cosh^2 u^1 \cos^2 u^3. \tag{39}$$

The hypersurface $\tilde{S}_2^4$ is equipped with an almost hypercomplex structure $H = (J_\alpha), (\alpha = 1, 2, 3)$, where the non-zero components of the matrix of J_α with respect to the local basis $\left\{\frac{\partial}{\partial u^i}\right\}_{i=1}^{4}$ are

$$\begin{aligned}
(J_1)_2^1 &= -\frac{1}{(J_1)_1^2} = -\sinh u^1, & (J_1)_4^3 &= -\frac{1}{(J_1)_3^4} = \cos u^3,\\
(J_2)_3^1 &= -\frac{1}{(J_2)_1^3} = -\cosh u^1, & (J_2)_4^2 &= -\frac{1}{(J_2)_2^4} = -\coth u^1 \cos u^3,\\
(J_3)_4^1 &= -\frac{1}{(J_3)_1^4} = \cosh u^1 \cos u^3, & (J_3)_2^3 &= -\frac{1}{(J_3)_3^2} = \tanh u^1.
\end{aligned} \tag{40}$$

Theorem 4.1 ([4]). *The spherical pseudo-Riemannian 4-dimensional manifold, defined by* (38)*, admits a hypercomplex pseudo-Hermitian structure on* $\tilde{S}_2^4$*, determined by* (40) *and* (39)*, with respect to which it is of the class* $\mathcal{W}(J_1)$ *but it does not belong to* $\mathcal{W}$ *and it has a constant sectional curvature* $k = 1$.

Let us consider a conformal transformation determined by the function u which is a solution of the equation $\mathrm{d}u = -\frac{1}{2(2n-1)}(\theta_1 \circ J_1)$, where the nonzero component of θ_1 with respect to the local basis $\left\{\frac{\partial}{\partial u^i}\right\}$ $(i = 1, 2, 3, 4)$ is $\theta_1\left(\frac{\partial}{\partial u^2}\right) = \frac{2\sinh^2 u^1}{\cosh u^1}$.

Since $\tilde{S}_2^4$ has a constant sectional curvature then the Weil tensor vanishes, i.e. $\tilde{S}_2^4$ is C-equivalent to a flat $\mathcal{K}(J_1)$-manifold. If we admit that it is in $\mathcal{K}$, then according to Theorem 3.2 we obtain that the manifold $(\tilde{S}_2^4, H, G) \in \mathcal{W}$ which is a contradiction. Therefore the considered manifold is C-equivalent to a flat $\mathcal{K}(J_1)$-manifold, but it is not a pseudo-hyper-Kähler manifold. By direct verification we state that the tensor S of this conformal transformation belongs to L_1. Therefore $(\tilde{S}_2^4, H, G)$ is an Einstein manifold.

References

1. D.V. Alekseevsky and S. Marchiafava, *Quaternionic structures on a manifold and subordinated structures*, Ann. Mat. Pura Appl. (IV), CLXXI (1996), 205–273.
2. G. Ganchev and A. Borisov, *Note on the almost complex manifolds with a Norden metric*, Compt. rend. Acad. bulg. Sci., **39** (1986), no. 5, 31–34.
3. A. Gray and L. M. Hervella, *The sixteen classes of almost Hermitian manifolds and their linear invariants*, Ann. Mat. Pura Appl. (IV), CXXIII (1980), 35–58.
4. K. Gribachev, M. Manev and S. Dimiev, *Almost hypercomplex pseudo-Hermitian manifolds*, Boll. Unione Mat. Ital. Sez. A, (to appear)
5. A. Sommese, *Quaternionic manifolds*, Math. Ann. **212** (1975), 191–214.
6. J. Wolf, *Spaces of constant curvature*, University of California, Berkley, California, 1972.

HERMITIAN-LIKE CURVATURES OF THE FOUR BASIC CLASSES OF REAL HYPERSURFACES OF A COMPLEX SPACE FORM*

MILEN HRISTOV

University of Veliko Tarnovo, Faculty of Pedagogics,
Department of Algebra and Geometry,
1 Teodosiy Tirnovski Str, 5000, Veliko Tarnovo, Bulgaria
E-mail: milenjh@yahoo.com

Any real hypersurface of a Kaehler manifold carries a natural almost contact metric structure. There are four basic classes of real hypersurfaces of a Kaehler manifold with respect to the induced almost contact metric structure. In this paper we study the basic classes of real hypersurfaces of a complex space form in terms of their Hermitian-like curvatures.

1. Introduction

It is well known the Chern (often called Hermitian) connection in the differential geometry of the almost Hermitian manifolds. The preserving of the complex structure by the symmetric connection, associated to the Chern connection and having the same geodesics, is necessery and sufficient condition to be the almost complex structure integrable, or the almost Hermitian manifold to be an Hermitian manifold. The odd-dimensional analogue of the Chern connection over an almost contact metric manifold $M^{2n+1}(\varphi, \xi, \eta, g)$ in [10] is introduced. This connection, called *Hermitian-like natural connection* and the objects, related to it, in terms of coordinate complex frame fields in [10] are described. The Hermitian-like natural connection plays the same rule as the Chern connection – to be an indicator of the class of the integrable almost contact metric manifolds (sometimes called almost contact metric manifolds of Hermitian type). This class, in the sence of the classification scheme of Alexiev-Ganchev [1], is exactly the direct summ $W_3 \oplus W_5 \oplus W_9 \oplus W_{10}$ [11] and it is characterised by the conditions the contact form η to be closed and the structure tensor field φ to be torsion-free:

*The research is financially supported by Contract No 136/2002, "L. Karavelov" Civil Engineering Higher School
Keywords: Kaehler manifold, Chern connection, contact manifold, real hypersurface
MS classification: 53D15 53C25

$[\varphi, \varphi] \equiv 0$. It is well known fact, that any real hypersurface M^{2n+1} of an almost Hermitian manifold $\overline{M}^{2n+2}(J, G)$ carries a natural almost contact metric structure (φ, ξ, η, g). When the ambiant manifold $\overline{M}^{2n+2}(J, G)$ is a Kaehler manifold, $n \geq 2$, then there are four basic classes of real hypersurfaces $M^{2n+1}(\varphi, \xi, \eta, g)$ – the classes W_1, W_2, W_4, W_6 (in the sence of [1]) and the sixteen possible classes of such hypersurfaces with respect to the second fundamental form in [7] are characterized. In the case $n = 1$, the class W_4 is empty [5] and there are 8 classes of real hypersurfaces of a four dimensional Kaehler manifold. The class $W_2 \oplus W_4$ consits of the quasi Sasakian manifolds [6], the class $W_1 \oplus W_6$ belongs to the class of partially integrable manifolds [11] and examples of such real hypersurfaces in [7] are given. The class $W_1 \oplus W_2$ belongs to the class of conformal almost contact metric manifolds [2] and examples of such hypersurfaces are easely obtained by using apropriate contact conformal transformations of the structure. Many autors have been studied the real hypersurfaces of a complex space form $\overline{M}^{2n+2}(c)$ – Kaehler manifold with pointwise costant holomorphic sectional curwature c. From point of view of [7] the identities for the Riemannian curvature tensor over the four basic classes of such hypersurfaces in [8] are described. The purpose of this paper is to introduse the identities for the Hermitian-like curvature tensor (the curvature, generated from the Hermitian-like natural connection) of the four basic classes of real hypersurfaces of a complex space form.

2. Preliminaries

Let $M^{2n+1}(\varphi, \xi, \eta, g)$ be an almost contact metric manifold of real dimension $2n+1$ and almost contact metric structure (φ, ξ, η, g) , consisting of: φ – smooth tensor field of type (1,1), smooth vector field ξ, differential form η and Riemannian definit metric g, satisfying the well known conditions

$$\varphi^2 = -id + \eta \otimes \xi, \quad \eta(\xi) = 1, \quad g(\varphi X, \varphi Y) = g(X, Y) - \eta(X)\eta(Y),$$

and as consequences $\eta \circ \varphi = 0$, $\varphi(\xi) = 0$, $\operatorname{rank} \varphi = 2n$, $\eta(X) = g(X, \xi)$, $X, Y \in \mathcal{X}M$.

Let $\Phi : \Phi(x, y) = g(x, \varphi y)$, x, $y \in D = \ker \eta$ be the fundamental 2-form for the structure, ∇ be the Levi-Civita connection of g, $F = -\nabla\Phi$: $F(X, Y, Z) = g((\nabla_X \varphi)Y, Z)$ be the fundamental tensor field of type (0,3) and $N = [\varphi, \varphi] + 2d\eta \otimes \xi$ be the generalized Nijenhuis tensor field.

A linear connection D on $M^{2n+1}(\varphi, \xi, \eta, g)$ is said to be *natural* if $Dg = 0$, $D\varphi = 0$. As a consequence $D\xi = 0$, $D\eta = 0$ are valid too [3, 10]. The torsion tensor field T of a natural connection D satisfies [3, 10]

$$2d\eta(X,Y) = \eta(T(X,Y)) = T(X,Y,\xi),$$
$$3d\Phi(X,Y,Z) = T(X,Y,\varphi Z) + T(Y,Z,\varphi X) + T(Z,X,\varphi Y),$$
$$N(X,Y,Z) = T(X,Y,Z) - T(\varphi X,\varphi Y,Z) - T(\varphi X,Y,\varphi Z) - T(X,\varphi Y,\varphi Z),$$

where $T(X,Y,Z) = g(T(X,Y),Z)$, for all $X, Y, Z \in \mathcal{X}M$.

The linear connection $\overset{0}{D}$, defined by

$$\overset{0}{D}_X Y = \nabla_X Y - \frac{1}{2}\varphi(\nabla_X\varphi)Y + [(\nabla_X\eta)Y]\xi - \frac{1}{2}\eta(Y)\nabla_X\xi,$$

is a natural connection an it is called *basic natural connection* [10].

The odd-dimensional analogue of the Chern connection is given by the definition

Definition 1 ([10]). A linear connection D over an almost contact metric manifold $M^{2n+1}(\varphi,\xi,\eta,g)$ is said to be Hermitian-like natural connection, if it is a natural connection, whose torsion tensor field T is of "*pure*" type, i.e. satisfies

$$T(hX,hY,hZ) + T(\varphi X,\varphi Y,hZ) = 0,$$
$$2[T(\xi,hY,hZ) + T(\xi,\varphi Y,\varphi Z)] = T(hY,hZ,\xi) + T(\varphi Y,\varphi Z,\xi).$$

(hX, hY, hZ are the restrictions of X, Y, $Z \in \mathcal{X}M$ to the contact distribution $D = \ker\eta$.)

Theorem 1 ([10]). *For any almost contact metric manifold $M^{2n+1}(\varphi,\xi,\eta,g)$ there exists unique Hermitian-like natural connection D. The defining conditions for D and for the corresponding torsion tensor field T are as follows*

$$g(D_XY,Z) = g(\overset{0}{D}_XY,Z)\frac{1}{4}[3d\Phi(\varphi X,\varphi Y,\varphi Z) + 3d\Phi(\varphi X,hY,hZ)] -$$
$$- \frac{1}{4}\eta(X)[3d\Phi(\xi,\varphi Y,hZ) - 3d\Phi(\xi,hY,\varphi Z)],$$
$$T(X,Y,Z) = \frac{1}{2}F(X,Y,\varphi Z) - \frac{1}{2}F(Y,X,\varphi Z)$$
$$+ \frac{1}{2}[\eta\wedge(\nabla_*\eta)Z](X,Y) + 2d\eta(X,Y)\eta(Z)$$
$$+ \frac{1}{4}[6d\Phi(\varphi X,\varphi Y,\varphi Z) + 3d\Phi(\varphi X,hY,hZ) + 3d\Phi(hX,\varphi Y,hZ)]$$
$$- \frac{1}{4}\{\eta\wedge[3d\Phi(\xi,\varphi *,hZ) - 3d\Phi(\xi,h*,\varphi Z)]\}(X,Y).$$

for all X, Y, $Z \in \mathcal{X}M$.

It is also true

Theorem 2 ([10]). *Let $M^{2n+1}(\varphi,\xi,\eta,g)$ be an almost contact metric manifold and let $\overset{*}{D} = D - \frac{1}{2}T$ be the symmetric connection, associated to the Hermitian-like natural connection D with torsion T. Then $\overset{*}{D}\varphi \equiv 0$ iff $M^{2n+1}(\varphi,\xi,\eta,g)$ is integrable.*

Following [11], we consider the complexification of the tangential distribution:

$$T^C M = D^C \oplus \{\xi\},$$

where D^C is the complexification of the almost Hermitian distribution $(D = \ker \eta; \varphi)$ and $\{\xi\} = (\operatorname{Im}\eta).\xi$ is the vertical distribution. We have $D^C = D^{1,0} \oplus D^{0,1}$, where $D^{1,0}(D^{0,1})$ is $+i(-i)$ – eigenspace of φ and

$$D^{1,0} = \operatorname{span}\{Z_\alpha = e_\alpha - i\varphi e_\alpha\}_{\alpha\in I}, \quad D^{0,1} = \operatorname{span}\{Z_{\bar\alpha} = e_\alpha + i\varphi e_\alpha\}_{\bar\alpha\in \bar I},$$
$$D^C = \operatorname{span}\{Z_a\}_{a\in I\cup\bar I}, \quad T^C M = \operatorname{span}\{Z_A\}_{A\in I\cup\bar I\cup I_0},$$

where $\{e_\alpha; \varphi e_\alpha; \xi,\ \alpha\in I\}$ is an orthonormal φ – bases of TM. The index sets I, $\bar I$, I_0 are $\{1,\dots,n\}$, $\{\bar 1,\dots,\bar n\}$, $\{0\}$ respectively and $Z_0 = \xi$.

The essential complex Cristoffel's symbols D^C_{AB} of the Hermitian-like natural connection D are

$$D^\gamma_{\alpha\beta} = \Gamma^\gamma_{\alpha\beta} + \frac{3}{2}i\Phi^\gamma_{\alpha\beta}, \quad D^\gamma_{\bar\alpha\beta} = \Gamma^\gamma_{\bar\alpha\beta} - \frac{3}{2}i\Phi^\gamma_{\bar\alpha\beta}, \quad D^\gamma_{0\beta} = \Gamma^\gamma_{0\beta} - \frac{3}{2}i\Phi^\gamma_{0\beta},$$

where Γ^C_{AB} are Cristoffel's symbols of the Levi-Civita connection ∇ and

$$\Phi^C_{AB} = g^{SC}\Phi_{ABS}, \ \Phi_{ABS} = d\Phi(Z_A, Z_B, Z_S) \ [10].$$

The essential complex components of the Hermitian-like torsion

$$T : T_{ABC} = T(Z_A, Z_B, Z_C) = -T_{BAC} \quad \text{are} \quad [10]$$

$$\begin{aligned} &T_{\alpha\beta\gamma} = \frac{1}{4}N_{\alpha\beta\gamma}, \quad T_{\alpha\beta\bar\gamma} = 3i\Phi_{\alpha\beta\bar\gamma}, \qquad T_{0\beta\gamma} = \frac{1}{2}N_{0\beta\gamma}, \\ &T_{0\beta\bar\gamma} = \eta_{\beta\bar\gamma}, \qquad T_{\alpha\beta 0} = 2\eta_{\alpha\beta} = \frac{1}{2}N_{\alpha\beta 0}, \quad T_{\alpha 00} = 2\eta_{\alpha 0} = N_{\alpha 00}. \end{aligned} \tag{1}$$

where $N_{ABC} = g(N(Z_A, Z_B), Z_C)$ are components of the Nijenhuis tensor N and $\eta_{AB} = d\eta(Z_A, Z_B)$.

Let $K = [D, D] - D_{[,]}$ be the curvature tensor of type (1,3) formed with the Hermitian-like natural connection D (simply called *Hermitian-like curvature tensor*) on arbitrarry $M^{2n+1}(\varphi,\xi,\eta,g)$ and let by the same letter be denoted the corresponding to K tensor of type (0,4): $K(X,Y,Z,U) = g(K(X,Y)Z,U)$. The following equalities for K of type (1,3) are valid:

i) $K(Z,U)V = -K(U,Z)V$;

ii) $K(Z,U) \circ \varphi = \varphi \circ K(Z,U)$;

iii) $K(\varphi Z, \varphi U)V = K(Z,U)V, \quad Z, U \in D^{\mathbf{c}}$;

iv) $K(Z,U)\xi = 0$;

v) $\eta(K(Z,U)V) = 0$,

and the corresponding K of type (0,4) satisfies [10]

i) $K(Z,U,V,W) = -K(U,Z,V,W) = -K(Z,U,W,V)$;

ii) $K(Z,U,\varphi V,\varphi W) = K(Z,U,V,W), \quad V, W \in D^{\mathbf{c}}$;

iii) $K(\varphi Z,\varphi U,V,W) = K(Z,U,V,W), \quad Z, U \in D^{\mathbf{c}}$;

iv) $K(Z,U,V,\xi) = 0$.

So, we find that the essential complex components

$$\begin{aligned} K^S_{ABC} &= -K^S_{BAC} \\ &= Z_A D^S_{BC} - Z_B D^S_{AC} + D^D_{BC} D^S_{AD} - D^D_{AC} D^S_{BD} - C^D_{AB} D^S_{DC} \end{aligned}$$

for K of type (1,3) and $K_{ABCD} = K(Z_A, Z_B, Z_C, Z_D) = g_{DS} K^S_{ABC}$ for K of type (0,4) are $K^{\sigma}_{\alpha\bar{\beta}\gamma}$, $K^{\sigma}_{0\bar{\beta}\gamma}$, $K^{\sigma}_{0\beta\gamma}$ and $K_{\alpha\bar{\beta}\gamma\bar{\delta}}$, $K_{0\alpha\beta\bar{\gamma}}$ respectively.

Let $V^{2n+1}(\varphi,\xi,\eta,g)$ be an Euclidian vector space of real dimension $2n+1$, endowed with almost contact metric structure (φ,ξ,η,g) and let $V^{\mathbf{c}}$ be its complexification, as above. Let $\mathcal{K}$ be the space of all tensors $K \in \otimes^4 V^{\mathbf{c}*}$, having the symmetries of the Hermitian-like curvature tensor of type (0,4). Any $K \in \mathcal{K}$ generates the following complex Ricci-type contractions [9]:

$$k_{AB} = g^{\lambda\bar{\mu}} K_{AB\lambda\bar{\mu}}, \quad s_{AB} = g^{\bar{\lambda}\mu} K_{A\bar{\lambda}\mu B},$$
$$k^*_{AB} = g^{\lambda\bar{\mu}} K_{\lambda\bar{\mu}AB}, \quad s^*_{AB} = g^{\lambda\bar{\mu}} K_{A\lambda\bar{\mu}B},$$

and the corresponding scalars $\kappa = g^{\alpha\bar{\beta}} k_{\alpha\bar{\beta}}$, $\sigma = g^{\alpha\bar{\beta}} s_{\alpha\bar{\beta}}$. We have the relations

$$\overline{k_{\alpha\bar{\beta}}} = -k_{\bar{\alpha}\beta} = k_{\beta\bar{\alpha}}, \quad \overline{k^*_{\alpha\bar{\beta}}} = -k^*_{\bar{\alpha}\beta} = k^*_{\beta\bar{\alpha}}, \quad \overline{s_{\alpha\bar{\beta}}} = s^*_{\bar{\alpha}\beta},$$
$$k_{\bar{\alpha}0} = -\overline{k_{\alpha 0}} = -k_{0\bar{\alpha}}, \quad \overline{s_{0\bar{\beta}}} = s^*_{0\beta}.$$

Further the components k_{A0}, s_{0A} will be writen as k_A, s_A respectively.

We give the following

Theorem 3 (complete decomposition). *The space $\mathcal{K}$ is decomposable into 13 mutually orthogonal, $\mathcal{U}(n) \times 1$-invariant and irreducible factors:*

$$\mathcal{K} = \bigoplus_{i,j=1}^{3} h\mathcal{K}_{ij} \oplus \bigoplus_{i,j=1}^{2} v\mathcal{K}_{ij},$$

describing in the following table:

factor		*essential components and equalities for K*	
	$h\mathcal{K}_{11}$	$K_{\alpha\bar{\beta}\gamma\bar{\delta}} = \frac{\kappa}{n(n+1)} \underbrace{[g_{\alpha\bar{\beta}}g_{\gamma\bar{\delta}} + g_{\alpha\bar{\delta}}g_{\gamma\bar{\beta}}]}_{=(\pi_1+\pi_2)_{\alpha\bar{\beta}\gamma\bar{\delta}}}$	$K_{\alpha\bar{\beta}\gamma\bar{\delta}} = K_{\gamma\bar{\delta}\alpha\bar{\beta}}$ $= K_{\gamma\bar{\beta}\alpha\bar{\delta}};$
$h\mathcal{K}_1$	$h\mathcal{K}_{12}$	$K_{\alpha\bar{\beta}\gamma\bar{\delta}} = \frac{1}{n+2}[g_{\alpha\bar{\beta}}k_{\gamma\bar{\delta}} + g_{\alpha\bar{\delta}}k_{\gamma\bar{\beta}}$ $+k_{\alpha\bar{\beta}}g_{\gamma\bar{\delta}} + k_{\alpha\bar{\delta}}g_{\gamma\bar{\beta}}],\ \kappa = 0$	$k_{\alpha\bar{\beta}} = k^*_{\alpha\bar{\beta}} = s_{\alpha\bar{\beta}};$
	$h\mathcal{K}_{13}$	$k_{\alpha\bar{\beta}} = 0$	$\kappa = \sigma$
	$h\mathcal{K}_{21}$	$K_{\alpha\bar{\beta}\gamma\bar{\delta}} = \frac{\kappa}{n(n-1)} \underbrace{[g_{\alpha\bar{\beta}}g_{\gamma\bar{\delta}} - g_{\alpha\bar{\delta}}g_{\gamma\bar{\beta}}]}_{=-\frac{1}{2}(3\pi_1-\pi_2)_{\alpha\bar{\beta}\gamma\bar{\delta}}}$	$K_{\alpha\bar{\beta}\gamma\bar{\delta}} = K_{\gamma\bar{\delta}\alpha\bar{\beta}}$ $= -K_{\gamma\bar{\beta}\alpha\bar{\delta}};$
$h\mathcal{K}_2$	$h\mathcal{K}_{22}$	$K_{\alpha\bar{\beta}\gamma\bar{\delta}} = \frac{1}{n-2}[g_{\alpha\bar{\beta}}k_{\gamma\bar{\delta}} - g_{\alpha\bar{\delta}}k_{\gamma\bar{\beta}}+$ $+k_{\alpha\bar{\beta}}g_{\gamma\bar{\delta}} - k_{\alpha\bar{\delta}}g_{\gamma\bar{\beta}}],\ \kappa = 0$	$k_{\alpha\bar{\beta}} = k^*_{\alpha\bar{\beta}} = -s_{\alpha\bar{\beta}};$
	$h\mathcal{K}_{23}$	$k_{\alpha\bar{\beta}} = 0$	$\kappa = -\sigma$
	$h\mathcal{K}_{31}$	$K_{\alpha\bar{\beta}\gamma\bar{\delta}} = \frac{1}{n}[k_{\alpha\bar{\beta}}g_{\gamma\bar{\delta}} - g_{\alpha\bar{\beta}}k_{\gamma\bar{\delta}}],\ s_{\alpha\bar{\beta}} = 0$	$K_{\alpha\bar{\beta}\gamma\bar{\delta}} = -K_{\gamma\bar{\delta}\alpha\bar{\beta}};$
$h\mathcal{K}_3$	$h\mathcal{K}_{32}$	$K_{\alpha\bar{\beta}\gamma\bar{\delta}} = \frac{1}{n}[s_{\alpha\bar{\delta}}g_{\gamma\bar{\beta}} - g_{\alpha\bar{\delta}}s_{\gamma\bar{\beta}}],\ k_{\alpha\bar{\beta}} = 0$	$s_{\alpha\bar{\beta}} = -s_{\beta\bar{\alpha}};$ $k^*_{\alpha\bar{\beta}} = -k_{\alpha\bar{\beta}};$
	$h\mathcal{K}_{33}$	$k_{\alpha\bar{\beta}} = s_{\alpha\bar{\beta}} = 0$	$\kappa = \sigma = 0$
	$v\mathcal{K}_{11}$	$K_{0\bar{\beta}\gamma\bar{\delta}} = \frac{1}{n+1}[k_{\bar{\beta}}g_{\gamma\bar{\delta}} + k_{\bar{\delta}}g_{\gamma\bar{\beta}}]$	$K_{0\bar{\beta}\gamma\bar{\delta}} = K_{0\bar{\delta}\gamma\bar{\beta}};$
$v\mathcal{K}_1$	$v\mathcal{K}_{12}$	$k_{\beta} = 0$	$k_{\beta} = -s_{\beta}$
	$v\mathcal{K}_{21}$	$K_{0\bar{\beta}\gamma\bar{\delta}} = \frac{1}{n-1}[k_{\bar{\beta}}g_{\gamma\bar{\delta}} - k_{\bar{\delta}}g_{\gamma\bar{\beta}}]$	$K_{0\bar{\beta}\gamma\bar{\delta}} = -K_{0\bar{\delta}\gamma\bar{\beta}};$
$v\mathcal{K}_2$	$v\mathcal{K}_{22}$	$k_{\beta} = 0$	$k_{\beta} = s_{\beta}$

Proof. The projections $h = -\varphi^2$ and $v = \eta \otimes \xi$ act from V to $hV \equiv D = \ker \eta$ and $vV \equiv (\operatorname{Im} \eta).\xi$ respectively and give rise to the decomposition $\mathcal{K} = h\mathcal{K} \oplus v\mathcal{K}$ – orthogonal and $\mathcal{U}(n) \times 1$-invariant, where

$$\begin{aligned} h\mathcal{K} &= \{K \in \mathcal{K} :\ K(X,Y,Z,U) = hK(X,Y,Z,U) = K(hX,hY,hZ,hU)\}, \\ v\mathcal{K} &= \{K \in \mathcal{K} :\ K(X,Y,Z,U) = vK(X,Y,Z,U) \\ &= [\eta \wedge K(\xi, h*, hZ, hU)](X,Y)\}. \end{aligned}$$

The essential components for $K \in h\mathcal{K}$ are $K_{\alpha\bar{\beta}\gamma\bar{\delta}}$ and for $K \in v\mathcal{K}$ are $K_{0\bar{\beta}\gamma\bar{\delta}}$.

The complete decomposition $h\mathcal{K} = \bigoplus_{i,j=1}^{3} h\mathcal{K}_{ij}$ in [9] is obtained ($\pi_1 + \pi_2$ and $3\pi_1 - \pi_2$ are the basic invariant curvature tensors [12]). To decompose $v\mathcal{K}$, we consider as in [9] the operator

$$\mathcal{V} : v\mathcal{K} \to v\mathcal{K} : \quad (\mathcal{V}K)_{0\bar{\beta}\gamma\bar{\delta}} = \frac{1}{2}(K_{0\bar{\beta}\gamma\bar{\delta}} + K_{0\bar{\delta}\gamma\bar{\beta}}).$$

It's easy to check this operator is selfconjugated projection, commuting with $\mathcal{U}(n) \times 1$ and hence give rise to $v\mathcal{K} = v\mathcal{K}_1 \oplus v\mathcal{K}_2$, where $v\mathcal{K}_1(V^{\mathbf{c}}) = \operatorname{Im}\mathcal{V}$ and $v\mathcal{K}_2(V^{\mathbf{c}}) = \operatorname{Ker}\mathcal{V} = (\operatorname{Im}\mathcal{V})^{\perp}$ are $\mathcal{U}(n) \times 1$-invariant. By direct computations we check, that the operators

$$\mathcal{W}^{(j)} : v\mathcal{K}_j \to v\mathcal{K}_j; \; (\mathcal{W}^{(j)}K)_{0\bar{\beta}\gamma\bar{\delta}} = K_{0\bar{\beta}\gamma\bar{\delta}} - \frac{1}{n-(-1)^j}[k_{\bar{\beta}}g_{\gamma\bar{\delta}} - (-1)^j k_{\bar{\delta}}g_{\gamma\bar{\beta}}]$$

are also selfconjugated projections, commuting with $\mathcal{U}(n) \times 1$, $j = 1, 2$. Then for the subspaces $v\mathcal{K}_j$ the following complete, orthogonal and $\mathcal{U}(n) \times 1-$invariant decompositions are valid

$$v\mathcal{K}_j = v\mathcal{K}_{j1} \oplus v\mathcal{K}_{j2},$$

where $v\mathcal{K}_{j1} = \operatorname{Ker}\mathcal{W}^{(j)}$, $v\mathcal{K}_{j2} = \operatorname{Im}\mathcal{W}^{(j)} = (\operatorname{Ker}\mathcal{W}^{(j)})^{\perp}$, $j = 1, 2$, which ends the proof. □

We say an almost contact metric manifold $M^{2n+1}(\varphi, \xi, \eta, g)$ belongs to the class $\mathcal{K}_i$, formed from the mentioned 13 classes if its Hermitian-like curvature tensor K of type (0,4) satisfies $K_p \in \mathcal{K}_i(T_p^{\mathbf{c}}M)$ for any point $p \in M$.

3. Applications to the real hypersurfaces of a complex space form

Let $M^{2n+1}(\varphi, \xi, \eta, g)$ be real hypersurface of a Kaehler manifold $\overline{M}^{2n+2}(J, G)$, endowed with the naturally arised almost contact metric structure and let A and h be the shape operator and the second fundamental tensor on M resp. The following equalities are well known:

$$\begin{aligned} &\nabla\varphi = A \otimes \eta - h \otimes \xi, \quad \varphi A = \nabla\xi, \\ &\nabla\eta = -h(*, \varphi*), \quad d\Phi \equiv 0, \quad N = \eta \wedge (\varphi \circ A - A \circ \varphi), \\ &2d\eta = -\mathcal{S}^-(h(*, \varphi*)) = g((\varphi \circ A + A \circ \varphi)*, *), \\ &\mathcal{L}_\xi g = -\mathcal{S}^+(h(*, \varphi*)) = g((\varphi \circ A - A \circ \varphi)*, *), \end{aligned} \tag{2}$$

where $(\mathcal{S}^{\pm}h)(X,Y) = h(X,Y) \pm h(Y,X)$, $\mathcal{L}$-denotes the Lie-differentiation. Moreover

$$F(X,Y,Z) = [\eta \wedge h(X,*)](Y,Z),$$
$$F(X,Y,\xi) = -h(X,Y),$$
$$F(\xi,\xi,Z) = \omega(Z) = h(\xi,Z),$$

for $X, Y, Z \in \mathcal{X}M$. Then $M^{2n+1}(\varphi,\xi,\eta,g) \in W_1 \oplus W_2 \oplus W_4 \oplus W_6$ and the characterisations of the 16 possible classes in thermes of the second fundamental tensor in [7] are given.

Let (A_B^C) be the matrix of the shape operator A in thermes of a local complex frame fields $\{Z_A\}$: $AZ_B = A_B^C.Z_C$, $h_{CD} = g(AZ_C, Z_D) = A_C^S.g_{SD} = A_D^S.g_{SC} = h_{DC}$, $A_0^0 = h_{00} = h(\xi,\xi) = \nu$. The function ν is not expressible by F. Then the norm and the trace of h are

$$||h||^2 = A_B^C.A_C^B = \nu^2 + 2g^{\alpha\bar\beta}g^{\gamma\bar\delta}(h_{\gamma\bar\beta}h_{\alpha\bar\delta} + h_{\alpha\gamma}h_{\bar\beta\bar\delta}) + 4g^{\alpha\bar\beta}h_{\alpha0}h_{\bar\beta0}\,,$$
$$\operatorname{tr} h = A_B^B = \nu + 2g^{\alpha\bar\beta}h_{\alpha\bar\beta} = \nu - f(\xi)\,,$$

where $f(\xi) = g^{ij}F(e_i,e_j,\xi)$, $\{e_i\}$ – real basis. From Theorem 1, (1) and (2) we obtain

Theorem 4. *Let $M^{2n+1}(\varphi,\xi,\eta,g)$ be real hypersurface of a Kaehler manifold $\overline{M}^{2n+2}(J,G)$. Then:*

(i) the Hermitian-like natural connection $D \equiv \overset{0}{D} = \nabla - \varphi A \otimes \eta + \nabla\eta \otimes \xi$:

$$D_X Y = \overset{0}{D}_X Y = \nabla_X Y - \eta(Y)\varphi AX - h(X,\varphi Y)\xi;$$

(ii) the Hermitian-like torsion tensor field of type (1,2) is $T = 2d\eta \otimes \xi + \eta \wedge \varphi A$ and

$$T(X,Y,Z) = [\eta \wedge h(X,\varphi*)](Y,Z) - [\eta \wedge h(Y,\varphi*)](X,Z)$$
$$= 2d\eta(X,Y)\eta(Z) - [\eta \wedge h(*,\varphi Z)](X,Y);$$

(iii) for the four basic classes the concrete expressions and the essential components for T are as follows:

W_1: $T(X,Y,Z) = -[(\eta \wedge \omega \circ \varphi) \otimes \eta](X,Y,Z)$,

$$T_{\alpha00} = N_{\alpha00} = -(\mathcal{L}_\xi g)_{\alpha0} = 2\eta_{\alpha0} = iF_{00\alpha} = i\omega_\alpha = ih_{\alpha0},\ \operatorname{tr} h = \nu;$$

W_2: $T(X,Y,Z) = -\dfrac{f(\xi)}{2n}[\eta \wedge \Phi(h*,hZ)](X,Y)$,

$$T_{0\alpha\bar\beta} = \eta_{\alpha\bar\beta} = -iF_{\alpha\bar\beta0} = -i\frac{f(\xi)}{2n}g_{\alpha\bar\beta} = ih_{\alpha\bar\beta},\ \operatorname{tr} h = \nu - f(\xi),\ f(\xi) \neq 0;$$

$$W_4\colon T(X,Y,Z) = -\frac{f(\xi)}{2n}[\eta\wedge\Phi(h*,hZ)](X,Y) - [\eta\wedge d\eta(\varphi*,\varphi Z)](X,Y) + 2d\eta(X,Y)\eta(Z),$$

$$T_{0\alpha\bar\beta} = \eta_{\alpha\bar\beta} = -iF_{\alpha\bar\beta 0} = ih_{\alpha\bar\beta},\ \mathrm{tr}\, h = \nu,\ f(\xi) = 0;$$

$$W_6\colon T(X,Y,Z) = -\frac{1}{2}[\eta\wedge(\mathcal{L}_\xi g)(\varphi*,\varphi Z)](X,Y) - \eta(Z)(\mathcal{L}_\xi g)(\varphi X,\varphi Y),$$

$$T_{0\alpha\beta} = \frac{1}{2}N_{0\alpha\beta} = \frac{1}{2}(\mathcal{L}_\xi g)_{\alpha\beta} = iF_{\alpha\beta 0} = -ih_{\alpha\beta},\ \mathrm{tr}\, h = \nu.$$

The class $W_0 : h(X,Y) = \nu.\eta(X)\eta(Y)$ of the cosymplectic manifolds belongs to each of the basic classes and is torsion-free.

By direct computations, using Theorem 4 - i), ii) and (2), we get

Lemma 1. *For a real hypersurface of a Kaehler manifold the following equalities are valid:*

i) $(D_X T)(Y,Z) = [2D_X d\eta\otimes\xi + \eta\wedge\varphi D_X A](Y,Z)$
$$= 2\left\{[(\nabla_X d\eta) - \eta\wedge d\eta(*,\varphi AX)](Y,Z) - \begin{vmatrix} h(X,\varphi Y) & h(X,\varphi Z) \\ h(\xi,\varphi Y) & h(\xi,\varphi Z)\end{vmatrix}\right\}.\xi$$
$$+[\eta\wedge\varphi(\nabla_X A)](Y,Z) - [\eta\wedge h(*,\xi)(Y,Z)].\varphi^2 AX + [\eta\wedge h(X,\varphi*)(Y,Z)].\varphi A\xi;$$

ii) $T(T(X,Y),Z) = \{4[d\eta\otimes d\eta(\xi,*)](X,Y,Z) + 2[\eta\wedge d\eta(\varphi A*,Z)](X,Y)\}.\xi$
$+2[d\eta\otimes(\varphi A - \eta\otimes\varphi A\xi)](X,Y,Z) - [(\eta\wedge\varphi A\varphi A)\otimes\eta](X,Y,Z);$

iii) $R(X,Y)\xi = [\nabla_X,\nabla_Y]\xi - \nabla_{[X,Y]}\xi = (\nabla_X\varphi A)Y - (\nabla_Y\varphi A)X$
$= (\nabla_X\varphi)AY - (\nabla_Y\varphi)AX + \varphi[(\nabla_X A)Y - (\nabla_Y A)X]$
$= h(Y,\xi)AX - h(X,\xi)AY + \varphi[(\nabla_X A)Y - (\nabla_Y A)X].$

Theorem 5. *The Hermitian-like curvature tensor $K = [D,D] - D_{[\,,\,]}$ for a real hypersurface of a Kaehler manifold satisfies*

$$K(X,Y)Z = R(X,Y)Z + \begin{vmatrix} h(X,\varphi Z) & h(Y,\varphi Z) \\ \varphi AX & \varphi AY\end{vmatrix}$$
$$- [(\nabla_X h)(Y,\varphi Z) - (\nabla_Y h)(X,\varphi Z)].\xi - \begin{vmatrix} h(X,\xi) & h(Y,\xi) \\ h(X,Z) & h(Y,Z)\end{vmatrix}.\xi$$
$$- \eta(Z).\varphi[(\nabla_X A)Y - (\nabla_Y A)X] + \eta(Z).\begin{vmatrix} h(X,\xi) & h(Y,\xi) \\ AX & AY\end{vmatrix},$$

$$
\begin{aligned}
K(X,Y,Z,U) = R(X,Y,Z,U) &- \begin{vmatrix} h(X,\varphi Z) & h(Y,\varphi Z) \\ h(X,\varphi U) & h(Y,\varphi U) \end{vmatrix} \\
&+ \eta(Z)\,[(\nabla_X h)(Y,\varphi U) - (\nabla_Y h)(X,\varphi U)] \\
&- \eta(U)\,[(\nabla_X h)(Y,\varphi Z) - (\nabla_Y h)(X,\varphi Z)] \\
&+ \eta(Z). \begin{vmatrix} h(X,\xi) & h(Y,\xi) \\ h(X,U) & h(Y,U) \end{vmatrix} - \eta(U). \begin{vmatrix} h(X,\xi) & h(Y,\xi) \\ h(X,Z) & h(Y,Z) \end{vmatrix}.
\end{aligned}
$$

Further, let $\overline{M}^{2n+2}(c)$ be a complex space form, i.e. Kaehler manifold with constant holomorphic sectional curvature c. Then the equations of Gauss and Codazzi imply the well known equalities [4]

$$
\begin{aligned}
&R(X,Y,Z,U) \\
&= \frac{c}{4}[g(Y,*) \wedge g(X,*) + \Phi(Y,*) \wedge \Phi(X,*) - 2\Phi(X,Y).\Phi](Z,U) \\
&\quad + [h(Y,*) \wedge h(X,*)](Z,U),
\end{aligned}
$$

$$
\begin{aligned}
g((\nabla_X A)Y - (\nabla_Y A)X,\ Z) &= (\nabla_X h)(Y,Z) - (\nabla_Y h)(X,Z) \\
&= \frac{c}{4}[\eta \wedge \Phi(Z,*)](X,Y) + 2\eta(Z)\Phi(X,Y).
\end{aligned}
$$

Now, from Theorem 5 we get

Theorem 6. *The Hermitian-like curvature tensor K of type $(0,4)$ for a real hypersurface $M^{2n+1}(\varphi,\xi,\eta,g)$ of a complex space form $\overline{M}^{2n+2}(c)$ satisfies*

$$
\begin{aligned}
&K(X,Y,Z,U) \\
&= \frac{c}{4}\left[\begin{vmatrix} g(Y,Z) & g(Y,U) \\ g(X,Z) & g(X,U) \end{vmatrix} + \begin{vmatrix} \Phi(Y,Z) & \Phi(Y,U) \\ \Phi(X,Z) & \Phi(X,U) \end{vmatrix} - 2\Phi(X,Y)\Phi(Z,U)\right] \\
&+ \frac{c}{4}\left[\eta(Z)\begin{vmatrix} \eta(X) & \eta(Y) \\ g(U,X) & g(U,Y) \end{vmatrix} - \eta(U)\begin{vmatrix} \eta(X) & \eta(Y) \\ g(Z,X) & g(Z,Y) \end{vmatrix}\right] \\
&+ \begin{vmatrix} h(Y,Z) & h(Y,U) \\ h(X,Z) & h(X,U) \end{vmatrix} + \begin{vmatrix} h(Y,\varphi Z) & h(Y,\varphi U) \\ h(X,\varphi Z) & h(X,\varphi U) \end{vmatrix} \\
&+ \eta(Z). \begin{vmatrix} h(X,\xi) & h(Y,\xi) \\ h(X,U) & h(Y,U) \end{vmatrix} - \eta(U). \begin{vmatrix} h(X,\xi) & h(Y,\xi) \\ h(X,Z) & h(Y,Z) \end{vmatrix}.
\end{aligned}
$$

The essential complex components for K are

$$
K_{\alpha\overline{\beta}\gamma\overline{\delta}} = \frac{c}{2}(\pi_1 + \pi_2)_{\alpha\overline{\beta}\gamma\overline{\delta}} + 2\begin{vmatrix} h_{\overline{\beta}\gamma} & h_{\overline{\beta}\,\overline{\delta}} \\ h_{\alpha\gamma} & h_{\alpha\overline{\delta}} \end{vmatrix} = R_{\alpha\overline{\beta}\gamma\overline{\delta}} + \begin{vmatrix} h_{\overline{\beta}\gamma} & h_{\overline{\beta}\,\overline{\delta}} \\ h_{\alpha\gamma} & h_{\alpha\overline{\delta}} \end{vmatrix},
$$

$$
K_{0\alpha\beta\overline{\gamma}} = 2\begin{vmatrix} h_{\alpha\beta} & h_{\alpha\overline{\gamma}} \\ h_{0\beta} & h_{0\overline{\gamma}} \end{vmatrix} = 2R_{0\alpha\beta\overline{\gamma}},
$$

where $(\pi_1+\pi_2)_{\alpha\bar\beta\gamma\bar\delta} = g_{\alpha\bar\beta}g_{\gamma\bar\delta} + g_{\gamma\bar\beta}g_{\alpha\bar\delta}$.

Now we get the relations:

$$K_{\alpha\bar\beta\gamma\bar\delta} = K_{\gamma\bar\delta\alpha\bar\beta} = K_{\gamma\bar\beta\alpha\bar\delta} + 2\begin{vmatrix} h_{\bar\beta\gamma} & h_{\alpha\bar\beta} \\ h_{\bar\delta\gamma} & h_{\alpha\bar\delta}\end{vmatrix},$$

$$(3)\quad k_{\alpha\bar\beta} = k^*_{\alpha\bar\beta} = \frac{c}{2}(n+1)g_{\alpha\bar\beta} + 2g^{\gamma\bar\delta}\begin{vmatrix} h_{\bar\beta\gamma} & h_{\bar\beta\bar\delta} \\ h_{\alpha\gamma} & h_{\alpha\bar\delta}\end{vmatrix} = s_{\alpha\bar\beta} + 2g^{\gamma\bar\delta}\begin{vmatrix} h_{\bar\beta\gamma} & h_{\alpha\bar\beta} \\ h_{\bar\delta\gamma} & h_{\alpha\bar\delta}\end{vmatrix},$$

$$\kappa = \frac{c}{2}n(n+1) + 2g^{\alpha\bar\beta}g^{\gamma\bar\delta}\begin{vmatrix} h_{\bar\beta\gamma} & h_{\bar\beta\bar\delta} \\ h_{\alpha\gamma} & h_{\alpha\bar\delta}\end{vmatrix} = \sigma + 2g^{\alpha\bar\beta}g^{\gamma\bar\delta}\begin{vmatrix} h_{\bar\beta\gamma} & h_{\alpha\bar\beta} \\ h_{\bar\delta\gamma} & h_{\alpha\bar\delta}\end{vmatrix}$$

and

$$(4)\quad \begin{aligned} K_{0\alpha\beta\bar\gamma} &= K_{0\beta\alpha\bar\gamma} + 2\begin{vmatrix} h_{\beta\bar\gamma} & h_{\alpha\bar\gamma} \\ h_{0\beta} & h_{0\alpha}\end{vmatrix}, \\ k_\alpha = s_\alpha + 2g^{\beta\bar\gamma}\begin{vmatrix} h_{\alpha\bar\gamma} & h_{\beta\bar\gamma} \\ h_{0\alpha} & h_{0\beta}\end{vmatrix} &= -2g^{\beta\bar\gamma}\begin{vmatrix} h_{\alpha\beta} & h_{\alpha\bar\gamma} \\ h_{0\beta} & h_{0\bar\gamma}\end{vmatrix}.\end{aligned}$$

From Theorem 3, Theorem 4, (3) and (4) we get

Theorem 7. *Let* $M^{2n+1}(\varphi,\xi,\eta,g)$ *be real hypersurface of a complex space form* $\overline{M}^{2n+2}(c)$. *Then* $M^{2n+1}(\varphi,\xi,\eta,g) \in h\mathcal{K}_1 \oplus h\mathcal{K}_2 \oplus v\mathcal{K}_1 \oplus v\mathcal{K}_2$. *If* $M^{2n+1}(\varphi,\xi,\eta,g)$ *belongs to the basic class:*

W_1, *then* $||h||^2 = \nu^2 + ||\omega||^2$,

$$\begin{aligned}&K_{\alpha\bar\beta\gamma\bar\delta} = K_{\gamma\bar\delta\alpha\bar\beta} = K_{\gamma\bar\beta\alpha\bar\delta} = \frac{c}{2}(\pi_1+\pi_2)_{\alpha\bar\beta\gamma\bar\delta}, \quad K_{0\alpha\beta\bar\gamma} = 0,\\ &k_{\alpha\bar\beta} = k^*_{\alpha\bar\beta} = s_{\alpha\bar\beta} = \frac{c}{2}(n+1)g_{\alpha\bar\beta}, \quad \kappa = \sigma = \frac{c}{2}n(n+1),\\ &M^{2n+1}(\varphi,\xi,\eta,g) \in h\mathcal{K}_{11}\left(\kappa = \sigma = \frac{c}{2}n(n+1),\ c \neq 0\right),\\ &K \equiv 0 \iff c = 0;\end{aligned}$$

W_2, *then* $||h||^2 = \nu^2 + \dfrac{f^2(\xi)}{2n} > 0$,

$$\begin{aligned}&K_{\alpha\bar\beta\gamma\bar\delta} = K_{\gamma\bar\delta\alpha\bar\beta} = \frac{c}{2}(\pi_1+\pi_2)_{\alpha\bar\beta\gamma\bar\delta} + \frac{f^2(\xi)}{2n^2}g_{\gamma\bar\beta}g_{\alpha\bar\delta}\\ &\qquad = K_{\gamma\bar\beta\alpha\bar\delta} + \frac{f^2(\xi)}{4n^2}(3\pi_1-\pi_2)_{\alpha\bar\beta\gamma\bar\delta},\\ &K_{0\alpha\beta\bar\gamma} = 0,\\ &k_{\alpha\bar\beta} = k^*_{\alpha\bar\beta} = \left[\frac{c}{2}(n+1) + \frac{f^2(\xi)}{2n^2}\right]g_{\alpha\bar\beta} = s_{\alpha\bar\beta} - \frac{f^2(\xi)}{2n^2}(n-1)g_{\alpha\bar\beta},\\ &\kappa = \frac{c}{2}n(n+1) + \frac{f^2(\xi)}{2n} = \sigma - \frac{f^2(\xi)}{2n}(n-1),\end{aligned}$$

$M^{2n+1}(\varphi,\xi,\eta,g) \in h\mathcal{K}_1 \oplus h\mathcal{K}_2;$

W_4, *then* $||h||^2 = \nu^2 + 2g^{\alpha\bar{\beta}}g^{\gamma\bar{\delta}}h_{\gamma\bar{\beta}}h_{\alpha\bar{\delta}}$, $f(\xi) = g^{\alpha\bar{\beta}}h_{\alpha\bar{\beta}} = 0$,

$$K_{\alpha\bar{\beta}\gamma\bar{\delta}} = K_{\gamma\bar{\delta}\alpha\bar{\beta}} = \frac{c}{2}(\pi_1+\pi_2)_{\alpha\bar{\beta}\gamma\bar{\delta}} + 2h_{\gamma\bar{\beta}}h_{\alpha\bar{\delta}} = K_{\gamma\bar{\beta}\alpha\bar{\delta}} + 2\begin{vmatrix} h_{\bar{\beta}\gamma} & h_{\alpha\bar{\beta}} \\ h_{\bar{\delta}\gamma} & h_{\alpha\bar{\delta}} \end{vmatrix},$$

$$K_{0\alpha\beta\bar{\gamma}} = 0,$$

$$k_{\alpha\bar{\beta}} = k^*_{\alpha\bar{\beta}} = \frac{c}{2}(n+1)g_{\alpha\bar{\beta}} + 2g^{\gamma\bar{\delta}}h_{\gamma\bar{\beta}}h_{\alpha\bar{\delta}}, \quad s_{\alpha\bar{\beta}} = \frac{c}{2}(n+1)g_{\alpha\bar{\beta}},$$

$$\kappa = \frac{c}{2}n(n+1) + ||h||^2 - \nu^2, \quad \sigma = \frac{c}{2}n(n+1),$$

$M^{2n+1}(\varphi,\xi,\eta,g) \in h\mathcal{K}_1 \oplus h\mathcal{K}_2;$

W_6, *then* $||h||^2 = \nu^2 + 2g^{\alpha\bar{\beta}}g^{\gamma\bar{\delta}}h_{\alpha\gamma}h_{\bar{\beta}\bar{\delta}} = \nu^2 + 2g^{\alpha\bar{\beta}}g^{\gamma\bar{\delta}}(\mathcal{L}_\xi g)_{\alpha\gamma}(\mathcal{L}_\xi g)_{\bar{\beta}\bar{\delta}}$,

$$K_{\alpha\bar{\beta}\gamma\bar{\delta}} = K_{\gamma\bar{\delta}\alpha\bar{\beta}} = \frac{c}{2}(\pi_1+\pi_2)_{\alpha\bar{\beta}\gamma\bar{\delta}} - 2h_{\alpha\gamma}h_{\bar{\beta}\bar{\delta}} = K_{\gamma\bar{\beta}\alpha\bar{\delta}}, \quad K_{0\alpha\beta\bar{\gamma}} = 0,$$

$$k_{\alpha\bar{\beta}} = k^*_{\alpha\bar{\beta}} = \frac{c}{2}(n+1)g_{\alpha\bar{\beta}} - 2g^{\gamma\bar{\delta}}h_{\alpha\gamma}h_{\bar{\beta}\bar{\delta}} = s_{\alpha\bar{\beta}},$$

$$\kappa = \frac{c}{2}n(n+1) - ||h||^2 + \nu^2 = \sigma,$$

$M^{2n+1}(\varphi,\xi,\eta,g) \in h\mathcal{K}_1.$

Remark. The conditions a real hypersurface $M^{2n+1}(\varphi,\xi,\eta,g) \in W_i$, $i = 2, 4, 6$ of complex space form $\overline{M}^{2n+2}(c)$ to belongs in $h\mathcal{K}_{ij}$, $i, j = 1, 2$ depend on c, n, tr h, $||h||$ etc. and will be discussed in later papers.

References

1. V. Alexiev and G. Ganchev, Math. and Educ. in Math., Proc. of the XV Spring Conf. of UBM, Sunny Beach, 155 (1986).
2. ———, Comp. rend. Acad. bulg. Sci. 39, **9** (1986), 27.
3. ———, Ann. de Univ. de Sofia, **81** (1987).
4. Ch. Baikoussis, Canad. Math. Bull., **40** (1997), 3, 257.
5. D. Chinea and C. Gonzalez, Ann. Mat. Pura Appl., **156** (1990) 4, 15.
6. G. Ganchev and V. Alexiev, Math. and Educ. in Math., Proc. of the XV Spring Conf. of UBM, Sunny Beach, 186(1986).
7. G. Ganchev and M. Hristov, Perspectives of Complex Analysis, Differential Geometry and Mathematical Physics, Proc. 5th Int. Workshop on Complex Structures and Vector Fields, Varna, 2000, eds S. Dimiev and K. Sekigawa (World Scientific, Singapore, 147(2001)).
8. ———, Third Int. Conf. on Geometry, Integrability and Quantization, June 14–23, 2001, Varna, Bulgaria, eds. I. Mladenov and G. Naber, **Coral Press**, Sofia 2001, pp 238–248.
9. G. Ganchev, S. Ivanov, and V. Michova, IC/91/40, Int. Centre for Theoretical Physics, Miramare-Trieste, May, 1991.
10. M. Hristov, SERDICA-Bulg. math. publ., **20** (1994), 111.

11. ______, Aspects of Complex analysis, Differential Geometry and Mathematical Phisics, Proc 4th Int. Workshop on Complex Structures and Vector Fields, Varna, 1998, eds. S. Dimiev and K. Sekigawa, (World Scientific, Singapore, 134(1999)).
12. F. Tricerri and L. Vanhecke, Trans. Am. Math. Soc., **267** (1981), 365.

ON OPERATOR DIFFERENTIATION IN THE ACTION PRINCIPLE IN QUANTUM FIELD THEORY

BOZHIDAR Z. ILIEV

Department of Mathematical Modeling,
Institute for Nuclear Research and Nuclear Energy, Bulgarian Academy of Sciences,
Boul. Tzarigradsko chaussée 72, 1784 Sofia, Bulgaria
E-mail address: bozho@inrne.bas.bg
URL: http://theo.inrne.bas.bg/~bozho/

An analysis of the Schwinger's action principle in Lagrangian quantum field theory is presented. A solution of a problem contained in it is proposed via a suitable definition of a derivative with respect to operator variables. This results in a preservation of Euler-Lagrange equations and a change in the operator structure of conserved quantities. Besides, it entails certain relation between the field operators and their variations (which is identically valid for some fields, e.g. for the free ones). The general theory is illustrated on a number of particular examples.

1. Introduction

The paper deals with the following problems in Lagrangian (canonical) quantum field theory: meaning of derivatives with respect to operator argument, order of the operators in the structure of conserved quantities, and commutation of the variations of the field operators with these operators or their derivatives in Schwinger's action principle. These problems are reviewed, analyzed and their solution is proposed.

We should mention, the above problems exist only in the Lagrangian (canonical) quantum field theory in which the quantum fields are represented as operators, called field operators, acting on some Hilbert space, which in general is unknown if interacting fields are studied. These operators are supposed to satisfy some equations of motion, from them are constructed conserved quantities satisfying

---

Subject class: Quantum field theory
2000 MSC numbers: 81P99, 81Q99, 81T99
2001 PACS numbers : 02.90.+p, 03.70.+k, 11.10.Ef
Key-Words: Quantum field theory, Action principle, Schwinger's action principle, Action principles in quantum field theory, Conserved quantities, Operators of conserved quantities in quantum field theory, Energy-momentum operator, Current operator, Spin angular momentum density operator, Euler-Lagrange equations, Field equations, Derivative with respect to operator argument

conservation laws, etc. From the view- point of present-day quantum field theory, this approach is only a preliminary stage for more or less rigorous formulation of the theory in which the fields are represented via operator-valued distributions, a fact required even for description of free fields. Moreover, in non- perturbative directions, like constructive and conformal field theories, the main objects are the vacuum mean (expectation) values of the fields and from these are reconstructed the Hilbert space of states and the acting in it fields. Regardless of these facts, the Lagrangian (canonical) quantum field theory is an inherent component of the most of the ways of presentation of quantum field theory adopted explicitly or implicitly in books like [1, 2, 3, 4, 5, 6, 7, 8]. Besides the Lagrangian approach is a source of many ideas for other directions of research, like the axiomatic quantum field theory [3, 7, 8]. By these reasons, we devote the present paper to the improvement of some places of Lagrangian field theory which are non-rigorous or unsatisfactory form mathematical point of view.

The first two of the above-mentioned problems are discussed in Sect. 2. Sect. 3 reviews the Schwinger's action principle and some its consequences. Special attention is paid to the problem with the commutativity of the fields' variations and the field operators or/and their partial derivatives. It has been notice at first by Schwinger in his original work [9] but later, in serious books like [3], it has been forgotten. A suitable solution of that problem is proposed in Sect. 4 by giving a rigorous meaning of a derivative of operator-valued function of operator arguments with respect to such an argument. It entails preservation of Euler-Lagrange equations for the field operators and a unique definition of the operators of conserved quantities. A new moment is that the variations of the field operators cannot be completely arbitrary in the general case (e.g. for some interacting fields) as they should satisfy some conditions derived in this work. Sect. 5 illustrates the general theory of Sect. 4 with particular examples: free neutral or charged scalar field, (self-)interacting scalar fields, free spinor field, and system of fields described via quadratic Lagrangian. It is presented an example of a Lagrangian, describing free (or with some self-interaction) spinor field, for which the (classical operator) Euler-Lagrange equations do not exist in a sense that they are identities, like $0 = 0$. Regardless of that fact, this Lagrangian entails completely reasonable field equations. The main results of the work are summarized in Sect. 6.

In the Lagrangians we consider is not supposed normal ordering (of the products of creation and annihilation operators). Besides, no (anti)commutation (or paracommutation) relations are supposed to be fulfilled. But the results obtained are, of course, valid and if normal ordering of products is used and some kind of (anti)commutation (or paracommutation) relations are taken into account.

2. Problems with the equations of motion and with conserved quantities

Suppose a system of *classical* fields $\varphi_i(x)$, $i = 1, \dots, n \in \mathbb{N}$, over the Minkowski spacetime M, $x \in M$, is described via a Lagrangian L depending on them and their first partial derivatives $\partial_\mu \varphi_i(x) = \frac{\partial \varphi_i(x)}{\partial x^\mu}$, $\{x^\mu\}$ being the (local) coordinates of $x \in M$, i.e. $L = L(\varphi_j(x), \partial_\nu \varphi_i(x))$. Here and henceforth the Greek indices $\mu, \nu, \dots$ run from 0 to $\dim M - 1 = 3$ and the Latin indices $i, j, \dots$ run from 1 to some integer n. The equations of motion for $\varphi_i(x)$, known as the *Euler-Lagrange equations*, are[1]

$$\frac{\partial L}{\partial \varphi_i(x)} - \frac{\partial}{\partial x^\mu}\Big(\frac{\partial L}{\partial(\partial_\mu \varphi_i(x))}\Big) = 0 \tag{2.1}$$

and are derived from the variational principle of stationary action, known as the *action principle* (see, e.g, [1, § 1], [10, § 67], [3, pp. 19–20]).

The (first) Noether theorem [1, § 2] says that, if the action's variation is invariant under C^1 transformations

$$\begin{aligned} &x \mapsto x^\omega = x^\omega(x) \quad x^\omega|_{\omega=\mathbf{0}} = x \qquad \omega = (\omega^{(1)}, \dots, \omega^{(s)}) \\ &\varphi_i(x) \mapsto \varphi_i^\omega(x^\omega) \quad \varphi_i^\omega(x^\omega)|_{\omega=\mathbf{0}} = \varphi_i(x) \end{aligned} \tag{2.2}$$

depending on $s \in \mathbb{N}$ independent real parameters $\omega^{(1)}, \dots, \omega^{(s)}$, then the quantities

$$\begin{aligned} \theta^\mu_{(a)}(x) := -\pi^{i\mu}\Big\{ \frac{\partial \varphi_i^\omega(x^\omega)}{\partial \omega^{(a)}}\Big|_{\omega=0} - (\partial_\nu \varphi_i(x)) \frac{\partial x^{\omega\,\nu}}{\partial \omega^{(a)}}\Big|_{\omega=0} \Big\} \\ - L(x) \frac{\partial x^{\omega\,\mu}}{\partial \omega^{(a)}}\Big|_{\omega=0}, \end{aligned} \tag{2.3}$$

where $a = 1, \dots, s$ and

$$\pi^{i\mu} := \frac{\partial L}{\partial(\partial_\mu \varphi_i(x))}, \tag{2.4}$$

are conserved in a sense that their divergences vanish, viz.

$$\partial_\mu \theta^\mu_{(a)}(x) = 0. \tag{2.5}$$

In particular, the invariance with respect to spacetime translations, i.e. $x \mapsto x^b = x + b$, with $b \in M$, and $\varphi_i(x) \mapsto \varphi_i(x)$, leads to the conservation of the *energy-*

[1] In this paper the Einstein's summation convention over indices appearing twice on different levels is assumed over the whole range of their values.

momentum tensor:

$$T^{\mu\nu}(x) := \pi^{i\mu}(x)\partial^\nu \varphi_i(x) - L(x)\eta^{\mu\nu} \tag{2.6}$$

$$\partial_\nu T^{\mu\nu}(x) = 0, \tag{2.7}$$

where $\eta^{\mu\nu}$ is the Lorentz metric tensor of M with signature $(+\ -\ --)$ and the spacetime indices are raised (lowered) by $\eta^{\mu\nu}$ (by the inverse tensor $\eta_{\mu\nu}$ of $\eta^{\mu\nu}$). Analogously, the invariance relative to constant phase transformations, viz. $x \mapsto x$ and $\varphi_i(x) \mapsto \mathrm{e}^{\frac{q}{\mathrm{i}\hbar c}\lambda}\varphi_i(x)$ where $q = \text{const}$, λ is a real parameter, $\hbar$ is the Planck's constant (divided by 2π), and c is the velocity of light in vacuum, implies the conservation of the corresponding current:

$$J_\mu(x) := \frac{q}{\mathrm{i}\hbar c}\sum_i \varepsilon(\varphi_i)\pi^i_\mu(x)\varphi_i(x) \tag{2.8}$$

$$\partial^\mu J_\mu = 0 \tag{2.9}$$

where $\varepsilon(\varphi_i) = 0$ if $\varphi_i(x)$ is real, $\varepsilon(\varphi_i) = +1$ if $\varphi_i(x)$ is complex, and $\varepsilon(\varphi_j) = -1$ if $\varphi_j(x)$ is the complex conjugate to $\varphi_i(x)$.[2]

Below we shall be interested in the quantum case, when the fields $\varphi_i(x)$ become linear operators depending on $x \in M$ and acting on system's Hilbert space $\mathcal{F}$ of states. The above scheme is repeated *mutatis mutandis* in Heisenberg picture of motion in canonical quantum field theory, when the fields $\varphi_i(x)$ are spacetime dependent and the state vectors are spacetime independent. Details of this procedure will be given in Sect. 3 below. However, there are three related problems which should find suitable answers:

1. How the quantum Lagrangian L should be defined? For example, if a quantum system has a classical analogue, can we simply replace in the classical Lagrangian the classical fields with the corresponding quantum operators?

2. What is the meaning of the derivative operators $\frac{\partial}{\partial\varphi_i(x)}$ and $\frac{\partial}{\partial(\partial_\mu\varphi_i(x))}$, appearing in, e.g., the Euler- Lagrange equations, when $\varphi_i(x)$ are operator-valued, not classical, functions?

3. If the previous two problems are satisfactory (well) and uniquely solved, how should be defined the conserved quantities (2.3) in the quantum case? For instance, can we write the *energy-momentum operator* as

$$T^{\mu\nu}(x) := \pi^{i\mu}(x) \circ (\partial^\nu\varphi_i(x)) - L(x)\eta^{\mu\nu}? \tag{2.10}$$

Here $\circ$ is the sign of mappings/operators composition/product and all quantities are the operator analogues of the corresponding classical ones.

---

[2] It is a convention whether to a complex field or to its complex conjugate to be assigned the value $+1$ of the function ε.

Partially the nature of these problems is in the fact that, generally, the field operators $\varphi_i(x)$ do not commute. So, the order in which the field operators or functions of them appear in some composition (product) is significant, contrary to the classical case.

The solution of the first problem for the known fields, free or not, has been found a long time ago [1, 10, 3]. The basic requirements for L being that it should be a Hermitian operator which is invariant under Lorentz/Poincaré transformations and other symmetries of the system, if any. Besides, if a quantum system has a classical analogue, the classical Lagrangian should be equal to the quantum one when in the latter the field operators are replaced with the corresponding classical fields.

A posteriori there can be different solutions of the second problem. However, *a priori* there is a simple rule silently followed in the literature [10, 3]. According to it, one replaces in the quantum Lagrangian the field operators $\varphi_i(x)$ (and their partial derivatives) with classical fields $\varphi_i^{\text{cl}}(x)$ and the composition of mappings sign with multiplication sign in such a way that the order of the (quantum) fields to be preserved. Then, from the so-obtained Lagrangian function L^{cl} are calculated the derivatives $\frac{\partial L^{\text{cl}}}{\partial \varphi_i^{\text{cl}}(x)}$ and $\frac{\partial L^{\text{cl}}}{\partial(\partial_\mu \varphi_i^{\text{cl}}(x))}$ by preserving the order of all fields and their derivatives. At the end, one replaces in $\frac{\partial L^{\text{cl}}}{\partial \varphi_i^{\text{cl}}(x)}$ and $\frac{\partial L^{\text{cl}}}{\partial(\partial_\mu \varphi_i^{\text{cl}}(x))}$ the fields $\varphi_i^{\text{cl}}(x)$ with the field operators $\varphi_i(x)$ and the multiplication of fields with compositions of the corresponding operators. In short, all this means that we differentiate a quantum Lagrangian with respect to its operator arguments by the same rules as in the classical case with the only addition that one should always retain the initial order of all operators [11, § 2]. Following this procedure, one should keep in mind that a change of the order of some operators in the initial Lagrangian may result in different derivatives of it even if the Lagrangian is not changed as an operator.[3]

When one analyzes the third of the afore presented problems, there are two guiding principles: the conserved operators $\theta^\mu_{(a)}$ must be Hermitian and, if a system has a classical analogue, these operators should reduce to the corresponding classical conserved fields (2.3) when the field operators are replace with the corresponding to them classical fields and the composition of operators is replaced by the multiplication of (classical) fields. However, these guidelines are not enough for the explicit determination of the conserved operators and one should 'guess'

---

[3] If A and B are operators, the above rule implies $\frac{\partial}{\partial A}(A \circ B) = \frac{\partial}{\partial A}(B \circ A) = B$. So, if A and B anticommute, i.e. $A \circ B = -B \circ A$, we have $\frac{\partial}{\partial A}(A \circ B) = B$ and $\frac{\partial}{\partial A}(-B \circ A) = -B$. Consequently, the derivative relative to non-commuting operator argument has a 'memory' for the place (left or right in our example) where the arguments have been situated before the differentiation. This phenomenon will find natural explanation in Sect. 4; in particular, see remark 4.1.

their functional form; the result can be justified or rejected *a posteriori* by examining the consequences of the model hypothesis.[4]

For instance, a straightforward transferring of (2.6) into the quantum region results in (2.10) but this operator is, generally non-Hermitian. As a working hypothesis, one may assume a 'Hermitian symmetrization' of (2.10), viz.

$$T_{\mu\nu} = \frac{1}{2}\{\pi^i_\mu(x) \circ (\partial_\nu \varphi_i(x)) + (\partial_\nu \varphi^\dagger_i(x)) \circ (\pi^i_\mu(x))^\dagger\} - \eta_{\mu\nu} L(x) \qquad (2.11)$$

where the dagger, "†", denotes Hermitian conjugation of operators. As $L^\dagger = L$, the last expression for energy-momentum operator satisfies the above-written requirements.

Similar is the situation with the current operator. *Prima facie* one may write (cf. (2.8))

$$J_\mu = \frac{q}{\mathrm{i}\hbar c} \sum_i \varepsilon(\varphi_i) \pi^i_\mu \circ \varphi_i \qquad (2.12)$$

but, generally, this expression is not Hermitian, $J^\dagger_\mu \neq J_\mu$. As a working hypothesis, a 'Hermitian symmetrization' may be assumed (note, $\varepsilon(\varphi^\dagger_i) = -\varepsilon(\varphi_i)$):

$$J_\mu = \frac{q}{\mathrm{i}\hbar c} \sum_i \varepsilon(\varphi_i) \{\pi^i_\mu \circ \varphi_i - \varphi^\dagger_i \circ (\pi^i_\mu)^\dagger\}. \qquad (2.13)$$

A partial discussion of the above problems with (energy-)momentum and (current or) charge operator can be found in [12].

3. Schwinger's action principle (review and problems)

The particular variant of a variation action principle, adapted to the needs of quantum field theory, is known as the *Schwinger's action principle*. Its description can be found, for instance, in [3, sec. 2.1] or in the original paper [9] (see also [13]). The purpose of the present section is a concise summary of this method, some its consequences and problems it contains. For details, the reader is referred to [3, sec. 2.1], from where the below-presented resumé of Schwinger's action principle is extracted.

Let there be given a system of quantum fields represented via linear field operators $\varphi_i(x)$. Let $L = L(x) = L(\varphi_i(x), \partial_\mu \varphi_i(x))$ be the Lagrangian of the system. It is supposed to depend only on the field operators and their first partial derivatives. Let σ_1 and σ_2 be two 3-dimensional spacelike surfaces and R be

[4] Since in quantum field theory are important the constant operators $C_{(a)} := \int_\sigma \theta^\mu_{(a)}\, d\sigma_\mu$, the integration being along some 3-dimensional spacelike surface σ, sometimes different definitions of $\theta^\mu_{(a)}$ may result in identical operators $C_{(a)}$, even if different Lagrangians are employed.

the 4-dimensional region (submanifold) bounded by them. The *action operator* is then defined by

$$W := \frac{1}{c}\int\limits_{R} L(x)\,\mathrm{d}^4x =: \frac{1}{c}\int\limits_{\sigma_1}^{\sigma_2} L(x)\,\mathrm{d}^4x. \tag{3.1}$$

Consider the (infinitesimal) transformations

$$x^\mu \mapsto x'^{\,\mu} = x^\mu + \delta x^\mu \tag{3.2a}$$

$$\varphi_i(x) \mapsto \varphi_i'(x) + \delta_0\varphi_i(x) \tag{3.2b}$$

as a result of a change of a spacetime point x and field operator $\varphi_i(x)$ when a transition to a *new* reference frame is made; in (3.2b) the symbol x in $\varphi_i(x)$ refers to a point in the *new* frame. If $x \in \sigma$ for some spacelike surface σ, it is supposed that the infinitesimal change $\delta_0\varphi_i(x)$ to be generated by some generator $F[\sigma]$ which is operator-valued functional of σ, i.e.

$$\delta_0\varphi_i(x) = \mathrm{i}\hbar[F[\sigma], \varphi_i(x)]_{\text{-}} \tag{3.3}$$

where $[A, B]_\pm := A \circ B \pm B \circ A$ for operators A and B.

The Schwinger's action principle postulates that, if (3.2) induces the change $W \mapsto W + \delta W$ of the action integral (3.1), then the infinitesimal change δW of the action integral is a difference of two surface integrals and

$$\delta W = F[\sigma_2] - F[\sigma_1]. \tag{3.4}$$

To work out consequences of (3.4), we notice that the variation δW is due to independent effects of the variations $\delta_0\varphi_i(x)$ of the field operators and the change $R \mapsto R'$ of the integration region as a result of the change (3.2a) of the points of its boundary. So, neglecting second and higher order terms in the variations and applying (3.1), we can write

$$\delta W = \frac{1}{c}\int\limits_{R}\Big\{\delta_0 L + L\frac{\partial(\delta x^\mu)}{\partial x^\mu}\Big\}\,\mathrm{d}^4x \tag{3.5}$$

where

$$\delta_0 L := L\big(\varphi_i(x)+\delta_0\varphi_i(x), \partial_\mu\varphi_i(x)+\partial_\mu(\delta_0\varphi_i(x))\big) - L(\varphi_i(x), \partial_\mu\varphi_i(x)) \tag{3.6}$$

is the variation of the Lagrangian. *Expanding the first term in* (3.6) *into Taylor series and neglecting second and higher order terms, we get*

$$\delta_0 L = \frac{\partial L}{\partial\varphi_i(x)} \circ \delta_0\varphi_i(x) + \frac{\partial L}{\partial(\partial_\mu\varphi_i(x))} \circ \delta_0(\partial_\mu\varphi_i(x)). \tag{3.7}$$

Here the derivatives are understood as described in Sect. 2.

Remark 3.1. We have emphasized the last phrase because the transition from (3.6) to (3.7) is generally incorrect and wrong, if 'arbitrary' variations $\delta_0\varphi_i(x)$ are considered. This will be explained at length in Sect. 4. At this point, we recall only the Schwinger's remark [9, the comments after eq. (2.17)] that the expression (3.7) for $\delta_0 L$ should be considered as symbolic one because it must be taken into account the (anti)commutation properties of $\delta_0\varphi_i(x)$. To overcome this problem, he makes the hypothesis that these properties and the structure of L should be such that terms with different positions of $\delta_0\varphi_i(x)$ must lead to equal portions in the variation δW. As we shall see in Sect. 4, this is a severe restriction which, generally, cuts off part of the information, including the (anti)commutation properties of $\delta_0\varphi_i(x)$, the Schwinger's action principle contains. The above problem, in other form, is mentions in [14, p. 149] too.

Remark 3.2. Alternatively, one can put the variations in (3.7) to the left of the Lagrangian's derivatives. Such a modification does not change anything, except the order of some operators, in the following. This problem is partially mentioned in [13].

Substituting (3.7) into (3.5), noting that, by its definition, $\delta_0(\partial_\mu\varphi_i(x)) = \partial_\mu(\delta_0\varphi_i(x))$ and integrating by parts the term originating from the second term in (3.7), we obtain

$$\delta W = \frac{1}{c}\int\limits_R \Big\{\Big(\frac{\partial L}{\partial\varphi_i(x)} - \frac{\partial}{\partial x^\mu}\Big(\frac{\partial L}{\partial(\partial_\mu\varphi_i(x))}\Big)\Big)\circ\delta_0\varphi_i(x) + \frac{\partial}{\partial x^\mu}\Big(\frac{\partial L}{\partial(\partial_\mu\varphi_i(x))}\circ\delta_0\varphi_i(x) + L\delta x^\mu\Big)\Big\}\,\mathrm{d}^4x\,. \tag{3.8}$$

Introducing the local (functional) variation

$$\delta\varphi_i(x) := \varphi_i'(x') - \varphi_i(x) = \delta_0\varphi_i(x) + (\partial_\nu\varphi_i(x))\delta x^\nu, \tag{3.9}$$

where x and x' refer to the coordinates of one and the same geometric point with respect to the 'old' and 'new' frames, applying the Stokes' (Gauss') theorem, and repeating the steps in [3, p. 63], from (3.9) we get

$$\delta W = \frac{1}{c}\int\limits_R \Big\{\Big(\frac{\partial L}{\partial\varphi_i(x)} - \frac{\partial}{\partial x^\mu}\Big(\frac{\partial L}{\partial(\partial_\mu\varphi_i(x))}\Big)\Big)\circ\delta_0\varphi_i(x)\Big\}\,\mathrm{d}^4x + F[\sigma_2] - F[\sigma_1] \tag{3.10}$$

where

$$F[\sigma] := \frac{1}{c}\int\limits_{\sigma} \{\pi^{i\mu}(x) \circ \delta_0\varphi_i(x) + L(x)\delta x^\mu\}\, \mathrm{d}\sigma_\mu$$

$$= \frac{1}{c}\int\limits_{\sigma} \{\pi^{i\mu}(x) \circ \delta\varphi_i(x) - \big(\pi^{i\mu}(x) \circ (\partial_\nu\varphi_i(x)) - \delta^\mu_\nu L(x)\big)\delta x^\nu\}\, \mathrm{d}\sigma_\mu. \quad (3.11)$$

Here σ is a spacelike surface with surface element $\mathrm{d}\sigma_\mu$, the notation (2.4) has been used, and δ^ν_μ is the (mixed) Kroneker δ-symbol, i.e. $\delta^\nu_\mu = 1$ for $\mu = \nu$ and $\delta^\nu_\mu = 0$ for $\mu \neq \nu$.

Following the known argumentation (see [3, sec. 2.1] and [9, sec. 2]), from (3.10) and (3.11) a number of fundamental consequences can be derived. For example, we shall mention three of them.

Since (3.4) demands δW to be a difference of two surface integrals and R and $\delta_0\varphi_i(x)$ are completely arbitrary, from (3.10) the Euler-Lagrange equations (2.1) for the field operators $\varphi_i(x)$ follow.

Identifying (3.11) with $\delta x^\nu = 0$ with the generator of the corresponding transformation (3.2) (with $\delta x^\nu = 0$), equation (3.3) implies

$$\delta\varphi_i(x) = \mathrm{i}\hbar\Big[\int\limits_{\sigma} \pi^{j\nu}(x') \circ \delta\varphi_j(x')\, \mathrm{d}\sigma_\mu(x'), \varphi_i(x)\Big]_{-} \quad (3.12)$$

as, by (3.9), $\delta\varphi_i(x) = \delta_0\varphi_i(x)$ for $\delta x^\mu = 0$. Similarly

$$\delta\pi^{i\mu}(x) = \mathrm{i}\hbar\Big[\int\limits_{\sigma} \pi^{j\nu}(x') \circ \delta\varphi_j(x')\, \mathrm{d}\sigma_\mu(x'), \pi^{i\mu}(x)\Big]_{-}. \quad (3.13)$$

The last two equalities should be identities for arbitrary $\delta\varphi_i(x)$ as long as $(x' - x)^2 := \eta_{\mu\nu}(x'^{\,\mu} - x^\mu)(x'^{\,\nu} - x^\nu) < 0$. They can be satisfied if one assumes (as additional postulate) the famous equal-time (anti)commutation relations, as it is proved in [3, pp. 65–67].

Consider transformations (3.2) leaving the action operator unchanged, i.e. such that

$$\delta W = 0. \quad (3.14)$$

Then, by (3.10) and (2.1),

$$F[\sigma_1] = F[\sigma_2] \quad (3.15)$$

for any 3-dimensional spacetime surfaces σ_1 and σ_2, i.e. the operators (3.11) are surface-independent:

$$\frac{\delta F[\sigma]}{\delta\sigma(y)} = 0 \qquad y \in \sigma \quad (3.15')$$

where $\frac{\delta}{\delta\sigma(y)}$ means the derivative of a functional of σ relative to σ at $y \in \sigma$ [3, p. 10]. From here follows that the 'current (density)'

$$f^{\mu}(x) := \pi^{i\mu}(x) \circ \delta\varphi_i(x) - \big(\pi^{i\mu}(x) \circ (\partial_\nu\varphi_i(x)) - \delta^{\mu}_{\nu} L(x)\big)\delta x^{\nu} \tag{3.16}$$

is conserved, viz.

$$\partial_\mu f^{\mu}(x) = 0. \tag{3.17}$$

It is a simple exercise to be verified, if we take (2.2) as a particular realization of (3.2), then (3.16) and (3.17) will respectively read

$$f^{\mu}(x) = -\sum_{a=1}^{s} \theta^{\mu}_{(a)}(x)\delta\omega^{(a)} \tag{3.18}$$

$$\partial_\mu \theta^{\mu}_{(a)}(x) = 0, \tag{3.19}$$

where

$$\theta^{\mu}_{(a)}(x) := -\pi^{i\mu} \circ \Big\{ \frac{\partial\varphi_i^{\omega}(x^{\omega})}{\partial\omega^{(a)}}\Big|_{\omega=0} - (\partial_\nu\varphi_i(x)) \frac{\partial x^{\omega\,\nu}}{\partial\omega^{(a)}}\Big|_{\omega=0} \Big\} - L(x)\frac{\partial x^{\omega\,\mu}}{\partial\omega^{(u)}}\Big|_{\omega=0} \tag{3.20}$$

is the quantum version of (2.3). In this way, we arrive to the quantum variant of the (first) Noether theorem saying that, if the action (3.1) is invariant under finite parameter transformation (2.2), the current operators (3.20) are conserved in a sense of (3.19) or, equivalently, in a sense that the integrals

$$C_{(a)}(\sigma) := \int_{\sigma} \theta^{\mu}_{(a)}\, \mathrm{d}\sigma_\mu \tag{3.21}$$

are surface-independent, i.e.

$$\frac{\delta C_{(a)}(\sigma)}{\delta\sigma(y)} = 0 \qquad y \in \sigma. \tag{3.22}$$

In particular, the choice $\sigma = \{x : x^0 = ct = \text{const}\}$ results in

$$C_{(a)}(t) := \int_{x^0=ct} \theta^{\mu}_{(a)}(x)\, \mathrm{d}^3\boldsymbol{x} \tag{3.21$'$}$$

$$\frac{\mathrm{d}C_{(a)}(t)}{\mathrm{d}t} = 0. \tag{3.22$'$}$$

We shall end the review of the Schwinger's action principle and its consequences with the remark that the particular transformations (3.2) with the choices

$$\delta x^{\mu} = a^{\mu} \qquad \delta_0\varphi_i(x) = 0 \tag{3.23}$$

$$\delta x^{\mu} = 0 \qquad \delta_0\varphi_i(x) = \varepsilon(\varphi_i)\frac{q}{\mathrm{i}\hbar c}\lambda\varphi_i(x), \tag{3.24}$$

where a^μ and λ are real parameters, lead to the canonical energy-momentum operator (2.10) and current operator (2.12), respectively, and, consequently, to the accompanying them problems, as discussed in Sect. 2.

In remark 3.1, we mentioned that the representation (3.7) for the r.h.s of (3.6) is, in the general case, incorrect and wrong. Since the Lagrangian L is supposed to be polynomial or convergent power series in $\varphi_i(x)$ and $\partial_\mu\varphi_i(x)$, it must be a sum of terms like $\alpha\psi_1(x)\circ\cdots\circ\psi_a(x)$, where α is real or complex non-zero number, $a\in\mathbb{N}$ and $\psi_b(x)$, $b=1,\dots,a$, is a field operator of a partial derivative of a field operator relative to some coordinate. The variation of such a term, under the transformation (3.2b), is

$$\alpha\sum_{b=1}^{a}\psi_1(x)\circ\cdots\circ\psi_{b-1}(x)\circ\delta_0\psi_b(x)\circ\psi_{b+1}(x)\circ\cdots\circ\psi_a(x)$$

and can be put in the form (3.7), i.e.

$$\alpha\Big(\sum_{b=1}^{a}\psi_1(x)\circ\cdots\circ\psi_{b-1}(x)\circ\psi_{b+1}(x)\circ\cdots\circ\psi_a(x)\Big)\circ\delta_0\psi_b(x),$$

if and only if

$$[\delta_0\psi_b(x),\psi_{b+1}(x)\circ\cdots\circ\psi_a(x)]_{\_}=0. \tag{3.25}$$

These are the conditions Schwinger assumed to hold in [9, the comments after eq. (2.17)] (see also [13]). The particular form of these conditions depends, of course, on the concrete Lagrangian under consideration. Generally, they say that the Lagrangian and variations of the field operators cannot be completely independent and arbitrary.

Since, usually, the Lagrangian is considered as a basic object in the theory, one may ask: can the conditions (3.25) hold for and arbitrary Lagrangian? The answer is positive. For example, if the variations $\delta_0\varphi_i(x)$ are chosen as multiples of the identity mapping $\mathrm{id}_{\mathcal{F}}$ of the system's Hilbert space $\mathcal{F}$ of states, i.e.

$$\delta_0\varphi_i(x)=f_i(x)\,\mathrm{id}_{\mathcal{F}} \tag{3.26}$$

for *completely arbitrary functions* $f_i\colon M\to\mathbb{C}$, then

$$\delta_0(\partial_\mu\varphi_i(x))=(\partial_\mu f_i(x))\,\mathrm{id}_{\mathcal{F}} \tag{3.27}$$

and, hence, the conditions (3.25) are identically satisfied. It is a simple exercise to be proved that the choices (3.26) are sufficient for a rigorous derivation of all the results concerning Schwinger's action principle reviewed above, as well as the ones in the literature [3, 9, 13, 15, 14].[5] However, the choices (3.26) entail also the problems with the conserved quantities mentioned in Sect. 2.

As the purpose of the present paper is not the investigation of the conditions under which (3.7) holds, we shall end this section with a simple example illustrating the problem with (3.7). Consider a free neutral scalar field $\varphi(x)$ with mass parameter m. Its Lagrangian is [10, 3]

$$L = -\frac{1}{2}m^2c^4\varphi\circ\varphi + \frac{1}{2}c^2\hbar^2(\partial_\mu\varphi)\circ(\partial^\mu\varphi) \tag{3.28}$$

so that

$$\begin{aligned} \delta L = -\frac{1}{2}m^2c^4\{\varphi\circ(\delta_0\varphi) + (\delta_0\varphi)\circ\varphi\} \\ +\frac{1}{2}c^2\hbar^2\{(\partial_\mu\varphi)\circ(\delta_0(\partial^\mu\varphi)) + (\delta_0(\partial_\mu\varphi))\circ(\partial^\mu\varphi)\}. \end{aligned} \tag{3.29}$$

Obviously, we can write the last expression in the form (3.7), i.e. as

$$\delta L = \frac{\partial L}{\partial\varphi}\circ\delta_0\varphi + \frac{\partial L}{\partial(\partial_\mu\varphi)}\circ\delta_0(\partial_\mu\varphi), \tag{3.30}$$

where

$$\frac{\partial L}{\partial\varphi} = -m^2c^4\varphi \qquad \frac{\partial L}{\partial(\partial_\mu\varphi)} = c^2\hbar^2(\partial^\mu\varphi), \tag{3.31}$$

if and only if

$$(\delta_0\varphi)\circ\varphi = \varphi\circ(\delta_0\varphi) \quad (\delta_0(\partial_\mu\varphi))\circ(\partial^\mu\varphi) = (\partial_\mu\varphi)\circ(\delta_0(\partial^\mu\varphi)). \tag{3.32}$$

If one assumes these equalities (or (3.30)) to hold for *completely arbitrary* $\delta_0\varphi$, as it is done everywhere in the literature, (the Schur's lemma (see, e.g., [16, sec. 8.2] or [17, ch. 5, sec. 3]) implies that) the field φ must be proportional to the identity mapping of system's Hilbert space $\mathcal{F}$, $\varphi(x) = g(x)\,\mathsf{id}_\mathcal{F}$ with $g\colon M\to\mathbb{C}$ of class C^2 such that $m^2c^2g+\hbar^2(\partial_\mu\circ\partial^\mu)g = 0$, due to (2.1) (i.e. due to the equation (4.2″) below). This is equivalent to a consideration of a classical free real scalar field g. However, if we restrict the variety of variations $\delta_0\varphi$ to the choice (3.26), i.e.

$$\delta_0\varphi(x) = f(x)\,\mathsf{id}_\mathcal{F} \tag{3.33}$$

with completely arbitrary $f\colon M\to\mathbb{C}$, we recover the standard quantum field theory of a free neutral scalar field φ [3, 10, 1], accompanied with the mentioned problems concerning the conserved quantities, i.e. the energy-momentum tensorial operator in this particular case.

---

[5] The choices (3.26) explain also the 'usual' meaning of the derivatives with respect to operators, as the ones in (2.1) and (3.8).

4. A solution of the problems

As we pointed in Sect. 3, a possible solution of the problem with the representation (3.7) is to restrict the field variations to multiples of the identity mapping $\mathsf{id}_{\mathcal{F}}$ (see (3.26)) which rigorously reproduces the known results and problems following from Schwinger's action principle. However, it is our opinion, the representation (3.7), as well as all efforts to ensure its validity, is not inherent to the Schwinger's variational principle of quantum field theory. By imposing it, one restricts the possible Lagrangians and/or the variety of possible variations of the field operators by a purely technical reason, which is not in harmony with the other principles of quantum field theory. Moreover, by demanding the validity of (3.7), one 'changes the rules of the game' after it has been started, viz. after the variational principle is formulated and an extraction of consequences of it has began, one suddenly imposes the equality (3.7) only because it is valid in the classical case. We find such a situation unsatisfactory and propose the below-described solution of all problems mentions until now. But, to illustrate the method we intend to apply, we first consider the example Lagrangian (3.28).

Looking over (3.29) and (3.30), we see that, for a free neutral scalar field, the derivative $\frac{\partial L}{\partial \varphi}$ should be regarded as a mapping acting on the operators on $\mathcal{F}$, not on vectors in $\mathcal{F}$, such that $\frac{\partial L}{\partial \varphi}\colon v \mapsto \frac{\partial L}{\partial \varphi}(v) = -\frac{1}{2}m^2c^4(\varphi \circ v + v \circ \varphi)$ for any $v\colon \mathcal{F} \to \mathcal{F}$. Defining similarly $\frac{\partial L}{\partial(\partial_\mu \varphi)}$ by $\frac{\partial L}{\partial(\partial_\mu \varphi)}\colon v \mapsto \frac{1}{2}c^2\hbar^2((\partial^\mu \varphi) \circ v + v \circ (\partial^\mu \varphi))$, we can replace (3.30) with the equality

$$\delta L = \frac{\partial L}{\partial \varphi}(\delta_0 \varphi) + \frac{\partial L}{\partial(\partial_\mu \varphi)}(\delta_0(\partial_\mu \varphi)) \tag{4.1}$$

where no additional conditions, like (3.32), on φ and $\delta_0\varphi$ have been imposed. Further, repeating the derivation of Euler-Lagrange equations, with (4.1) for (3.7), we get

$$\Big(\frac{\partial L}{\partial \varphi} - \frac{\partial}{\partial x^\mu}\Big(\frac{\partial L}{\partial(\partial_\mu \varphi)}\Big)\Big)(\delta_0 \varphi) = 0. \tag{4.2}$$

Substituting here the just-obtained derivatives, we find

$$\frac{1}{2}(-m^2c^4\varphi - c^2\hbar^2\Box(\varphi)) \circ \delta_0\varphi + \delta_0\varphi \circ \frac{1}{2}(-m^2c^4\varphi - c^2\hbar^2\Box(\varphi)) = 0 \tag{4.2$'$}$$

where $\Box := \partial_\mu \partial^\mu$ is the D'Alembert operator and no additional condition have been imposed. The choice $\delta_0\varphi = \mathsf{id}_{\mathcal{F}}$ reduces the last equality to the standard Klein-Gordon equation for φ,

$$m^2c^2\varphi + \hbar^2\Box(\varphi) = 0 \tag{4.2$''$}$$

which, in turn, converts (4.2′) into identity. So, without the *additional* conditions (3.32), we derived from the action principle the 'right' Klein-Gordon equation describing free neutral scalar field. For the energy-momentum operator $T_{\mu\nu}$ of such a field, one can repeat its derivation, as described in Sect. 3, but with the new definition of the Lagrangian's derivatives. The result reads (cf. (2.10))

$$T_{\mu\nu} = \pi_\mu(\partial_\nu\varphi) - L\eta_{\mu\nu} \tag{4.3}$$

where $\pi^\mu := \frac{\partial L}{\partial(\partial_\mu\varphi)} : v \mapsto \frac{1}{2}c^2\hbar^2((\partial^\mu\varphi)\circ v + v\circ(\partial^\mu\varphi))$ for $v\colon \mathcal{F}\to\mathcal{F}$, or, equivalently

$$\begin{aligned} T_{\mu\nu} = {} & \frac{1}{2}c^2\hbar^2\{(\partial_\mu\varphi)\circ(\partial_\nu\varphi) + (\partial_\nu\varphi)\circ(\partial^\mu\varphi)\} \\ & + \frac{1}{2}\{m^2c^4\varphi\circ\varphi - c^2\hbar^2(\partial_\mu\varphi)\circ(\partial^\mu\varphi)\}\eta_{\mu\nu}. \end{aligned} \tag{4.3′}$$

Thus, without any additional hypotheses, the action principle (3.4) leads to the 'Hermitian symmetrized' energy-momentum tensor (2.11) (with the 'usual' meaning of the $\pi^{i\mu}$, i.c. $\pi_\mu = c^2\hbar^2\partial_\mu\varphi$). It can be proved that, after imposing the commutation relations and normal ordering, the quantum field φ described via (4.2″) and (4.3′) is identical with the one described via (4.2″) and $T_{\mu\nu} = c^2\hbar^2(\partial_\mu\varphi)\circ(\partial_\nu\varphi) - \eta_{\mu\nu}L$, corresponding to (2.10)) and usually considered in the literature [1, 10].

Having in mind the above example, we turn now our attention to the general case of arbitrary Lagrangian, which is supposed to be polynomial or convergent power series in the field operators and their first partial derivatives.

The main idea of the following is the derivatives of the Lagrangian with respect to the field operators and/or their partial derivatives to be defined as *mappings acting on operators* such that (cf. (3.7) and (4.1))

$$\delta L = \sum_i \frac{\partial L}{\partial\varphi_i(x)}(\delta_0\varphi_i(x)) + \sum_{i,\mu} \frac{\partial L}{\partial(\partial_\mu\varphi_i(x))}(\delta_0(\partial_\mu\varphi_i(x))). \tag{4.4}$$

If one elaborates this definition, one will come the following definition of a derivative of operator-valued function of operator arguments with respect to an operator variable.

Definition 4.1. Let $\mathcal{F}$ be a $\mathbb{C}$-vector space, $\omega \subseteq \{\mathcal{F}\to\mathcal{F}\}$ be a subset of the space of operators acting on $\mathcal{F}$, $n\in\mathbb{N}$, $u_1, \dots, u_n \in \omega$, $u \in \{u_1,\dots,u_n\}$, and $f\colon (u_1,\dots,u_n) \mapsto f(u_1,\dots,u_n)\colon \mathcal{F}\to\mathcal{F}$ be operator-valued function of $u_1,\dots,u_n$ which is polynomial (or convergent power series) in its operator arguments. The derivative of f with respect to u is an n-argument mapping with domain $\omega\times\cdots\times\omega$ (n-times), denoted by $\frac{\partial f}{\partial u}$, such that:

(i) Its value at $(u_1, \dots, u_n)$, denoted by $\frac{\partial f}{\partial u}(u_1, \dots, u_n) := \frac{\partial f(u_1,\dots,u_n)}{\partial u} := \frac{\partial f}{\partial u}\big|_{(u_1,\dots,u_n)}$, is a mapping $\omega \to \{\mathcal{F} \to \mathcal{F}\}$ from the subset ω on the space of operators on $\mathcal{F}$.

(ii) The mapping $\frac{\partial}{\partial u}: f \mapsto \frac{\partial f}{\partial u}$ is linear relative to complex-valued functions on M. In particular, it is $\mathbb{C}$-linear.

(iii) Let $v: \mathcal{F} \to \mathcal{F}$ be such that $u + v \in \omega$, $a \in \mathbb{N}$, $i_1, \dots, i_a \in \{1, \dots, n\}$ and $I := \{i \in \{i_1, \dots, i_a\} : u_i = u\}$ be the set of indices which label all operators among $u_{i_1}, \dots, u_{i_a}$ equal to u. Then

$$\Big(\frac{\partial}{\partial u}(u_{i_1} \circ \dots \circ u_{i_a})\Big)(v) := \sum_{i \in I} \Big\{ (u_{i_1} \circ \dots \circ u_{i_a})\big|_{u_i = v} \Big\}. \tag{4.5}$$

In particular, if I is empty, $I = \varnothing$, the r.h.s. of (4.5) is set equal to the zero operator of $\mathcal{F}$.

Remark 4.1. The restriction $u + v \in \omega$ is essential one in quantum field theory, in which the field operators, usually, satisfy some (anti)commutation relations and, hence, in it $\omega \neq \{\mathcal{F} \to \mathcal{F}\}$; for some general remarks on that item, see [18, sec. 21.1]. For instance, let us find the derivative of $A \circ B$ with respect to A, where $A, B: \mathcal{F} \to \mathcal{F}$ are anticommuting operators, $A \circ B = -B \circ A$. In this particular case $\frac{\partial}{\partial A}(A \circ B)$ is defined only on those $v: \mathcal{F} \to \mathcal{F}$ for which $(A + v) \circ B = -B \circ (A + v)$, i.e. such that $v \circ B = -B \circ v$, and hence $\omega = \{z: \mathcal{F} \to \mathcal{F} : z \circ B = -B \circ z\}$. In accord with (4.5), we have $\frac{\partial (A \circ B)}{\partial A}(v) = v \circ B = -B \circ v = \frac{\partial(-B \circ A)}{\partial A}(v)$; the evaluation of the derivative on element $w \in \{\mathcal{F} \to \mathcal{F}\} \backslash \omega$ leads to a contradiction, $\frac{\partial (A \circ B)}{\partial A}(w) \neq \frac{\partial(-B \circ A)}{\partial A}(w)$.

Remark 4.2. From a view-point of functional analysis, the definition 4.1 defines the notion of partial Fréchet derivative of particular kind of functionals employed in quantum field theory. From this position, the r.h.s. of (4.4) is nothing else, but the Fréchet differential of the Lagrangian considered as a mapping between some operator spaces.

In short, definition 4.1 means that a derivative of operator-valued function, polynomial or convergent power series, of operator arguments with respect to some of its operator arguments is calculated by differentiating each its term according to (4.5). In particular, this is valid for Lagrangians of the type we consider in this work.

It is a trivial to be verified, the derivatives introduce via definition 4.1 possess all 'standard' derivative properties; in particular, they satisfy the Leibnitz rule for differentiation of compositions (products) of functions and the rule for differentiation of composite functions.

If $c\colon M \to \mathbb{C}$ and $u + c(x)\,\mathsf{id}_{\mathcal{F}} \in \omega$, $x \in M$, it is a trivial corollary of definition 4.1 that

$$\frac{\partial f(u_1, \dots, u_n)}{\partial u}(c(x)\,\mathsf{id}_{\mathcal{F}}) = c(x)\frac{\partial^{\mathrm{cl}} f(u_1, \dots, u_n)}{\partial u}, \tag{4.6}$$

where all operators are supposed to be linear and $\frac{\partial^{\mathrm{cl}}}{\partial u}$ means the 'classical' derivative with respect to u as it was defined in Sect. 2, i.e. the derivative in the r.h.s. of (4.6) should be calculated as if $u_1, \dots, u_n$ were classical fields over M with a preservation of the relative order of all operators. This is exactly the definition of a derivative of a function of non-commuting arguments accepted, e.g., in [11, § 2]. For example, we have $\frac{\partial \varphi^3}{\partial \varphi}(v) = \varphi^2 \circ v + \varphi \circ v \circ \varphi + v \circ \varphi^2$ and $\frac{\partial \varphi^3}{\partial \varphi}(c(x)\,\mathsf{id}_{\mathcal{F}}) = 3c(x)\varphi^2$ where $\varphi^a := \varphi \circ \dots \varphi$ (a-times) for $a \in \mathbb{N}$ and $\varphi\colon \mathcal{F} \to \mathcal{F}$.

Equipped with the new definition of a derivative of a Lagrangian with respect to a field operator or its partial derivative, it is trivial to verify that (4.4) is an equivalent version of (3.6), up to second and higher order terms, without making any additional hypotheses with respect to the Lagrangian or/and variations of the field operators. Moreover, if we consider the variations (3.26), then, in view of (4.6) and (4.4), we get

$$\begin{aligned}
&\delta_0 L|_{\delta_0\varphi_i(x) = f_i(x)\,\mathsf{id}_{\mathcal{F}}} \\
&= \left\{ \frac{\partial^{\mathrm{cl}} L}{\partial \varphi_i(x)} \circ \delta_0\varphi_i(x) + \frac{\partial^{\mathrm{cl}} L}{\partial(\partial_\mu \varphi_i(x))} \circ \delta_0(\partial_\mu \varphi_i(x)) \right\}\Big|_{\delta_0\varphi_i(x) = f_i(x)\,\mathsf{id}_{\mathcal{F}}} \\
&= f_i(x)\frac{\partial^{\mathrm{cl}} L}{\partial \varphi_i(x)} + (\partial_\mu f_i(x))\frac{\partial^{\mathrm{cl}} L}{\partial(\partial_\mu \varphi_i(x))}
\end{aligned} \tag{4.7}$$

from where all standard consequences (and problems) of the Schwinger's action principle can be derived (recovered).[6] It is natural to be expected that the restriction of the field variations to ones given via (3.26) should lead to a broadening of the consequences of the variational principle (3.4). Below we shall examine them without accepting any additional conditions, like (3.26) or, in the general case, (3.25). In this case, one can expect new consequences of the Schwinger's action principle, which otherwise are 'swallowed' by (3.26) (or (3.25)) in its standard presentations.

We shall now work out the explicit form of the action variation (3.5) with (3.7) replaced by (4.4) with the new meaning of the derivatives in it. Inserting (4.4) into (3.5) and integrating by parts the term coming from the second one in (4.4),[7]

---

[6] See the remark in [14, p. 149] on the above topic.

[7] By virtue of properties (ii) and (iii) in definition 4.1, the integration by parts is a rigorous operation; in general, we have $\frac{\partial f}{\partial u}(\partial_\mu v) = \partial_\mu\big(\frac{\partial f}{\partial u}(v)\big) - \big(\partial_\mu\big(\frac{\partial f}{\partial u}\big)\big)(v)$.

we get (cf (3.8))

$$\delta W = \frac{1}{c} \int\limits_R \Big\{ \sum_i \Big(\frac{\partial L}{\partial \varphi_i(x)} - \frac{\partial}{\partial x^\mu} \Big(\frac{\partial L}{\partial(\partial_\mu \varphi_i(x))} \Big) \Big) (\delta_0 \varphi_i(x)) \\ + \sum_\mu \frac{\partial}{\partial x^\mu} \Big(\sum_i \frac{\partial L}{\partial(\partial_\mu \varphi_i(x))} (\delta_0 \varphi_i(x)) + L(x) \delta x^\mu \Big) \Big\} \, \mathrm{d}^4 x. \tag{4.8}$$

From here, repeating *mutatis mutandis* the derivation of (3.10), as given in [3], we obtain

$$\delta W = \frac{1}{c} \int\limits_R \Big\{ \sum_i \Big(\frac{\partial L}{\partial \varphi_i(x)} - \frac{\partial}{\partial x^\mu} \Big(\frac{\partial L}{\partial(\partial_\mu \varphi_i(x))} \Big) \Big) (\delta_0 \varphi_i(x)) \Big\} \, \mathrm{d}^4 x \\ + F[\sigma_2] - F[\sigma_1] \tag{4.9}$$

where

$$F[\sigma] := \frac{1}{c} \int\limits_\sigma \sum_\mu \Big\{ \sum_i \pi^{i\mu}(x) (\delta \varphi_i(x)) \\ - \sum_\nu \Big(\sum_i \pi^{i\mu}(x) (\partial_\nu \varphi_i(x)) - \delta^\mu_\nu L(x) \Big) \delta x^\nu \Big\} \, \mathrm{d}\sigma_\mu \tag{4.10}$$

with

$$\pi^{i\mu}(x) := \frac{\partial L}{\partial(\partial_\mu \varphi_i(x))} : \{\mathcal{F} \to \mathcal{F}\} \to \{\mathcal{F} \to \mathcal{F}\} \tag{4.11}$$

being the derivative of L relative to $\partial_\mu \varphi_i(x)$ according to definition 4.1. Formally, (4.9) and (4.10) can be obtained from (3.10) and (3.11), respectively, via the replacements:

$$\frac{\partial L}{\partial \varphi_i(x)} \circ v \mapsto \frac{\partial L}{\partial \varphi_i(x)} (v) \\ \Big\{ \frac{\partial}{\partial x^\mu} \Big(\frac{\partial L}{\partial \varphi_i(x)} \Big) \Big\} \circ v \mapsto \Big\{ \frac{\partial}{\partial x^\mu} \Big(\frac{\partial L}{\partial \varphi_i(x)} \Big) \Big\} (v) \\ \pi^{i\mu} \circ v = \frac{\partial L}{\partial(\partial_\mu \varphi_i(x))} \circ v \mapsto \pi^{i\mu}(v) = \frac{\partial L}{\partial(\partial_\mu \varphi_i(x))} (v) \tag{4.12}$$

where $v \colon \mathcal{F} \to \mathcal{F}$ (is some variation, i.e. $\delta_0 \varphi_i(x)$ or $\delta_0 \varphi_i$). As we shall see further, these changes can be used for 'repairing' the standard consequences of Schwinger's action principle. Evidently, the replacements opposite to (4.12) transform (4.9) and (4.10) to (3.10) and (3.11), respectively.

Evidently, in view of (4.6), it is clear that the 'old' variation (3.10) (with the standard meaning of the derivatives in it) and the 'new' variation (4.9) (with the derivatives in it given via definition 4.1) are identical if variations like (3.26), i.e. multipliers of the identity mapping, are employed.

Let us proceed with extraction of consequences of the variational principle (3.4) on a base of the representation (4.9).

Since (3.4) states that δW must be a difference of two surface integrals, (4.9) implies the vanishment of the volume integral in it, which, due to the arbitrariness of the integration region R, is equivalent to

$$\sum_i \Big\{ \Big(\frac{\partial L}{\partial \varphi_i(x)} - \frac{\partial}{\partial x^\mu} \Big(\frac{\partial L}{\partial(\partial_\mu \varphi_i(x))} \Big) \Big)(v_i) \Big\} = 0 \tag{4.13}$$

where, for brevity, we have denoted by v_i the variation $\delta_0\varphi_i(x)$ of $\varphi_i(x)$, $v_i := \delta_0\varphi_i(x)$. We shall refer to (4.13) as the *Euler-Lagrange relation.* This is the prototype of the *Euler-Lagrange equations for the field operators* $\varphi_i(x)$ *and their variations* $v_i := \delta_0\varphi_i(x)$. We should emphasize, now this is an equation both for $\varphi_i(x)$ and v_i, contrary to the standard procedure where one gets, due to the arbitrariness of v_i, equations only for $\varphi_i(x)$.[8] So, the new moment with respect to the 'old' variational principle is that (4.13) puts, in general, restrictions both on the field operators and on their variations, i.e. the variations cannot be considered as completely arbitrary (if one does not wont to deal with trivial fields in some cases). In Sect. 5, examples will be considered when completely arbitrary and not such variations are admissible; the particular situation depends on the concrete Lagrangian employed.[9] However, from the derivation of (4.9) (and therefore of (4.13)), it is clear that the variations (3.26), i.e. ones proportional to the identity mapping $\mathsf{id}_{\mathcal{F}}$ of the system's Hilbert space $\mathcal{F}$ of states, are always admissible. For them, in view of (4.6) and the complete arbitrariness of the functions $f_i \colon M \to \mathbb{C}$ in (3.26), we derive from (4.13) the 'classical' Euler-Lagrange equations for the field operators as

$$\frac{\partial^{\mathrm{cl}} L}{\partial \varphi_i(x)} - \frac{\partial}{\partial x^\mu} \Big(\frac{\partial^{\mathrm{cl}} L}{\partial(\partial_\mu \varphi_i(x))} \Big) = 0 \tag{4.14}$$

which, due to the meaning of the operator derivatives in them, coincide with the ones obtained form the 'old' variations (3.10). But it should clearly be understood, if one requires (4.13) to be valid for *completely arbitrary* variations v_i, other restrictions on the field operators may arise.[10] The alternative point of view

[8] This does not exclude the coincidence of the final equations for φ_i; in particular, such is the case with the free fields — see Sect. 5.

[9] As a rule, the Lagrangians describing free fields admit arbitrary variations, while those describing (self-)interacting fields require some restrictions on the variety of field variations.

[10] In some cases, these new restrictions imply the field operators to be proportional to $\mathsf{id}_{\mathcal{F}}$, i.e., in a sense, leading to classical, not quantum, fields.

is to look on (4.14) as on field equations for the field operators and on the remaining consequences of (4.13), if any, as on restrictions on the admissible variations v_i. We shall discuss this topic in Sect. 5 on concrete examples; in particular, in Subsect. 5.6 it will be presented an example of a Lagrangian which leads to completely reasonable field equations which are *not* the Euler-Lagrange equations for it (the latter being simply identities with respect to the fields and their variations).

Let us turn now to the problem with conserved quantities. Consider a transformation (3.2), in which the field operators $\varphi_i(x)$ and their variations $v_i = \delta_0\varphi_i(x)$ satisfy (4.13), leaving the action (3.1) unchanged. For these operators equations (3.1)–(3.15′) hold (see (4.9) and (4.13)) with $F[\sigma]$ defined, now, by (4.10), not by (3.11). Therefore, the 'current (density)' (cf. (3.16))

$$f^\mu(x) := \sum_i \pi^{i\mu}(x)(\delta\varphi_i(x)) - \sum_\nu \Big(\sum_i \pi^{i\mu}(x)(\partial_\nu\varphi_i(x)) - \delta^\mu_\nu L(x)\Big)\delta x^\nu \tag{4.15}$$

is conserved, i.e. satisfies the continuity equation (3.17). In particular, if an s-parameter transformations (2.2) satisfy the above conditions, the equalities (4.15) and (3.17) will reduce respectively to (3.18) and (3.19) with the 'Noether currents' $\theta^\mu_{(a)}(x)$, $a = 1, \dots, s$, given by (cf (3.20))

$$\begin{aligned}\theta^\mu_{(a)}(x) := &-\sum_i \pi^{i\mu}(x)\Big(\frac{\partial\varphi^\omega_i(x^\omega)}{\partial\omega^{(a)}}\Big|_{\omega=0}\Big)\\ &+\sum_{i,\nu}\pi^{i\mu}(x)\big(\partial_\nu\varphi_i(x)\big)\frac{\partial x^{\omega\,\nu}}{\partial\omega^{(a)}}\Big|_{\omega=0} - L(x)\frac{\partial x^{\omega\,\mu}}{\partial\omega^{(a)}}\Big|_{\omega=0}\end{aligned} \tag{4.16}$$

where $\pi^{i\mu(x)}$ is defined via (4.11). The equations (3.21)–(3.22′), of course, remain valid with the new definition (4.16) of $\theta^\mu_{(a)}(x)$.

To feel better the difference between (4.15) and (3.16) (or between (4.16) and (3.20)), let us consider the transformations (3.23) and (3.24), generating in the classical case the energy-momentum tensor (2.6) and current vector (2.8), respectively. For them the operator (4.15) reduces respectively to $f^\mu(x) = -T^\mu{}_\nu(x)a^\nu$ and $f^\mu(x) = J^\mu(x)\lambda$, where

$$T^{\mu\nu}(x) = \sum_i \pi^{i\mu}(x)\big(\partial^\nu\varphi_i(x)\big) - \eta^{\mu\nu}L(x) \tag{4.17}$$

$$J^\mu(x) = \frac{q}{\mathrm{i}\hbar c}\sum_i \varepsilon(\varphi_i)\pi^{i\mu}(x)\big(\varphi_i(x)\big) \tag{4.18}$$

are the energy-momentum and (charge) current operators, respectively. In some cases, these expressions may differ significantly from (2.10) and (2.12), respectively, which will be illustrated on concrete examples in Sect. 5.

As a last example of a conserved quantity, we consider the angular momentum operator $M^{\lambda}_{\mu\nu}(x)$. Suppose the action operator is invariant under 4-rotations, i.e. under the changes $x^{\mu} \mapsto x^{\varepsilon\,\mu} = x^{\mu} + \varepsilon^{\mu\nu} x_{\nu}$, with $x_{\nu} := \eta_{\nu\mu} x^{\mu}$ and antisymmetric real parameters $\varepsilon^{\mu\nu} = -\varepsilon^{\nu\mu}$, and $\varphi_i(x) \mapsto \varphi_i^{\varepsilon}(x^{\varepsilon})$ with $\varphi_i^{\varepsilon}(x^{\varepsilon}) = \varphi_i(x) + \sum_{\mu<\nu} I^{j}_{i\mu\nu} \varphi_j(x) \varepsilon^{\mu\nu} + \cdots$, where the dots stand for second and higher order terms in $\varepsilon^{\mu\nu}$ and $I^{j}_{i\mu\nu} = -I^{j}_{i\nu\mu}$ are numbers characterizing the behaviour of the field operators under 4-rotations. Since $x^{\varepsilon\,\rho} = x^{\rho} + \sum_{\mu<\nu} (\delta^{\rho}_{\mu} x_{\nu} - \delta^{\rho}_{\nu} x_{\mu}) \varepsilon^{\mu\nu}$, from (4.16) (with changed sign), in view of (4.17), we obtain

$$M^{\lambda}_{\mu\nu}(x) = \big(x_{\mu} T^{\lambda}{}_{\nu}(x) - x_{\nu} T^{\lambda}{}_{\mu}(x)\big) + S^{\lambda}_{\mu\nu}(x), \tag{4.19}$$

where $T^{\lambda}{}_{\nu}(x)$ are the mixed components of the energy-momentum operator (4.17) and

$$S^{\lambda}_{\mu\nu}(x) := \sum_{i,j} \pi^{i\lambda}(x)(\varphi_j(x)) I^{j}_{i\mu\nu} \tag{4.20}$$

is the spin angular momentum operator.

As for the quantization rules, equation (3.12) should now be replaced by

$$\delta\varphi_i(x) = \mathrm{i}\hbar \int_{\sigma} \Big[\sum_j \pi^{j\nu}(x') \big(\delta\varphi_j(x')\big), \varphi_i(x) \Big]_{-} \mathrm{d}\sigma_{\nu}(x') \tag{4.21}$$

and similarly for (3.13). These equations and the variations (3.26), combined with the know argumentation [3, sect 2.1 (ii)], produce the canonical (anti)commutation relations. However, a different choice of the field variations, if such are admissible, may result in new restrictions on the field operators.

5. Examples

The main purpose of this section is an illustration of the general theory of Sect. 4 for particular Lagrangians. As we shall see, known results are reproduce with some corrections. The 'quadratic' Lagrangians will be pointed as the 'best' ones selected by the Schwinger's action principle.

5.1. *Free neutral scalar field*

The Lagrangian of a free neutral scalar field $\varphi = \varphi^{\dagger}$ with mass parameter m is (3.28). In accord with definition 4.1, the action of its operator derivatives on an operator v are (cf. (3.31))

$$\begin{aligned} \frac{\partial L}{\partial \varphi}(v) &= -\frac{1}{2} m^2 c^4 (\varphi \circ v + v \circ \varphi) \\ \pi^{\mu}(v) &= \frac{\partial L}{\partial(\partial_{\mu}\varphi)}(v) = \frac{1}{2} c^2 \hbar^2 \big((\partial^{\mu}\varphi) \circ v + v \circ (\partial^{\mu}\varphi)\big). \end{aligned} \tag{5.1}$$

Hence the Euler-Lagrange relation (4.13) for it reads

$$\frac{1}{2}(-m^2c^4\varphi - c^2\hbar^2\Box(\varphi)) \circ v + v \circ \frac{1}{2}(-m^2c^4\varphi - c^2\hbar^2\Box(\varphi)) = 0 \tag{5.2}$$

where $\Box = \partial_\mu\partial^\mu$ is the D'Alembert operator. The choice $v = \mathrm{id}_{\mathcal{F}}$, $\mathcal{F}$ being the field's (system's) Hilbert space of states, results in the Klein-Gordon equation

$$m^2c^2\varphi + \hbar^2\Box(\varphi) = 0 \tag{5.3}$$

which corresponds to (4.14) with Lagrangian (3.28). Evidently, (5.3) converts (5.2) into identity relative to v. Thus the equations (4.14) do not impose any restrictions for the variations v in a case of the Lagrangian (3.28). In view of (5.1), the energy-momentum operator (4.17) now reads

$$\begin{aligned} T_{\mu\nu} = {} & \frac{1}{2}c^2\hbar^2\{(\partial_\mu\varphi)\circ(\partial_\nu\varphi) + (\partial_\nu\varphi)\circ(\partial_\mu\varphi)\} \\ & - \eta_{\mu\nu}\{-\frac{1}{2}m^2c^4\varphi\circ\varphi + \frac{1}{2}c^2\hbar^2(\partial_\mu\varphi)\circ(\partial^\mu\varphi)\}, \end{aligned} \tag{5.4}$$

which corresponds to (2.11) with Lagrangian (3.28), not to (2.10), when the additional conditions (3.25), i.e. (3.32) in the particular case, in Schwinger's action principle are imposed.

In almost the same way, the reader may wish to consider an electromagnetic field with 4-potential operators A_μ and, e.g., gauge invariant Lagrangian $L = -\frac{1}{4}c^2\hbar^2 F_{\mu\nu}F^{\mu\nu}$ with $F_{\mu\nu} := \partial_\mu A_\nu - \partial_\nu A_\mu$. In this case, we have $\frac{\partial L}{\partial A^\mu} = 0$ and (do not sum over μ!) $\pi^\nu_\mu(v_\mu) = \frac{\partial L}{\partial(\partial_\nu A^\mu)}(v_\mu) = \frac{1}{2}c^2\hbar^2\{v_\mu\circ F_{\nu\mu} + F_{\nu\mu}\circ v_\mu\}$ for a variation v_μ of A_μ.

5.2. *Free charged scalar field*

The standard choice of a Lagrangian of a free charged scalar field $\varphi \neq \varphi^\dagger$ with mass parameter m is [1, 10, 3][11]

$$L = -m^2c^4\varphi^\dagger\circ\varphi + c^2\hbar^2(\partial_\mu\varphi^\dagger)\circ(\partial^\mu\varphi). \tag{5.5}$$

So, we have:

$$\begin{aligned} &\frac{\partial L}{\partial\varphi}(v) = -m^2c^4\varphi^\dagger\circ v \qquad \pi^\mu(v) = \frac{\partial L}{\partial(\partial_\mu\varphi)}(v) = c^2\hbar^2(\partial^\mu\varphi^\dagger)\circ v \\ &\frac{\partial L}{\partial\varphi^\dagger}(w) = -m^2c^4 w\circ\varphi \qquad \pi^{\dagger\,\mu}(w) = \frac{\partial L}{\partial(\partial_\mu\varphi^\dagger)}(w) = c^2\hbar^2 w\circ(\partial^\mu\varphi) \end{aligned} \tag{5.6}$$

---

[11] In some sense, the Lagrangian $L = -\frac{1}{2}m^2c^4(\varphi^\dagger\circ\varphi + \varphi\circ\varphi^\dagger) + \frac{1}{2}c^2\hbar^2\big((\partial_\mu\varphi^\dagger)\circ(\partial^\mu\varphi) + (\partial_\mu\varphi)\circ(\partial^\mu\varphi^\dagger)\big)$ is better than (5.5). But, after imposing the commutation relations and normal ordering, the quantum field theory arising from both Lagrangians turns to be one and the same.

for operators v and w having a meaning of variations of φ and $\varphi^\dagger$, respectively. Therefore the relation (4.13) now reads:

$$(-m^2c^4\varphi^\dagger - c^2\hbar^2\Box(\varphi^\dagger)) \circ v + w \circ (-m^2c^4\varphi - c^2\hbar^2\Box(\varphi)) = 0. \tag{5.7}$$

From here, using the standard choices $(v, w) = (0, \mathsf{id}_{\mathcal{F}}), (\mathsf{id}_{\mathcal{F}}, 0)$, we derive the Klein-Gordon equations

$$m^2c^2\varphi + \hbar^2\Box(\varphi) = 0 \qquad m^2c^2\varphi^\dagger + \hbar^2\Box(\varphi^\dagger) = 0, \tag{5.8}$$

which convert (5.7) into identity relative to v and w. Thus, the variations v and w in (5.7) can be completely arbitrary. Substituting (5.6) into (4.17) and (4.18), we get the energy-momentum and current operators respectively as

$$\begin{aligned} T_{\mu\nu} &= c^2\hbar^2\{(\partial_\mu\varphi^\dagger) \circ (\partial_\nu\varphi) + (\partial_\nu\varphi^\dagger) \circ (\partial_\mu\varphi)\} \\ &\quad - \eta_{\mu\nu}\{-m^2c^4\varphi^\dagger \circ \varphi + c^2\hbar^2(\partial_\mu\varphi^\dagger) \circ (\partial^\mu\varphi)\} \qquad (5.9) \\ J_\mu &= -\mathrm{i}\hbar c q\{(\partial_\mu\varphi^\dagger) \circ \varphi - \varphi^\dagger \circ (\partial_\mu\varphi)\}, \qquad (5.10) \end{aligned}$$

where, for definiteness, we have chosen $\varepsilon(\varphi) = +1$ and $\varepsilon(\varphi^\dagger) = -1$. These expressions correspond to (2.11) and (2.13), not to (2.10) and (2.12), respectively. Evidently, $T_{\mu\nu} = T_{\nu\mu}$, $T^\dagger_{\mu\nu} = T_{\mu\nu}$, and $J^\dagger_\mu = J_\mu$. Consequently, our formalism produces the known results form the literature [1, 10], where expressions, like (5.9) and (5.10), are more a matter of convention/postulate than a one of rigorous derivation.

5.3. *Self-interacting neutral scalar field*

Consider a neutral scalar field $\varphi = \varphi^\dagger$ with Lagrangian

$$L = -\frac{1}{2}m^2c^4\varphi \circ \varphi + \frac{1}{2}c^2\hbar^2(\partial_\mu\varphi) \circ (\partial^\mu\varphi) + a\varphi^k \tag{5.11}$$

where a is a real non-zero parameter, $k \in \mathbb{N}$, and $\varphi^k := \varphi \circ \ldots \varphi$ (k-times). Applying definition 4.1, we get:

$$\begin{aligned} \frac{\partial L}{\partial\varphi}(v) &= -\frac{1}{2}m^2c^4(\varphi \circ v + v \circ \varphi) \\ &\quad + a(v \circ \varphi^{k-1} + \varphi \circ v \circ \varphi^{k-2} + \cdots + \varphi^{k-2} \circ v \circ \varphi + \varphi^{k-1} \circ v) \\ \pi^\mu(v) &= \frac{\partial L}{\partial(\partial_\mu\varphi)}(v) = \frac{1}{2}c^2\hbar^2\big((\partial^\mu\varphi) \circ v + v \circ (\partial^\mu\varphi)\big). \end{aligned} \tag{5.12}$$

So, the energy-momentum operator is given by (5.4), with the term in the braces replaced by the r.h.s. of (5.11), and the relation (4.13) reads (cf. (5.2))

$$\frac{1}{2}(-m^2c^4\varphi - c^2\hbar^2\Box(\varphi)) \circ v + v \circ \frac{1}{2}(-m^2c^4\varphi - c^2\hbar^2\Box(\varphi)) \\ + a(v \circ \varphi^{k-1} + \varphi \circ v \circ \varphi^{k-2} + \cdots + \varphi^{k-2} \circ v \circ \varphi + \varphi^{k-1} \circ v) = 0. \quad (5.13)$$

Choosing $v = \mathrm{id}_{\mathcal{F}}$, we obtain a 'classical' Euler-Lagrange equation for φ:

$$m^2c^4\varphi + c^2\hbar^2\Box(\varphi) = ak\varphi^{k-1}. \quad (5.14)$$

Combining (5.14) with (5.13), we see that the variation $v = \delta_0\varphi$ *cannot be arbitrary* (for $a \neq 0$) as it must satisfy the equation

$$(1 - k/2)(\varphi^{k-1} \circ v + v \circ \varphi^{k-1}) \\ + \varphi \circ v \circ \varphi^{k-2} + \varphi^2 \circ v \circ \varphi^{k-3} + \cdots + \varphi^{k-2} \circ v \circ \varphi = 0 \quad (5.15)$$

which always has solutions of the form $v = f(x)\,\mathrm{id}_{\mathcal{F}}$ with $f\colon M \to \mathbb{C}$. In particular, for $k = 1, 2, 3, 4$, this equation respectively reads:

$$\begin{aligned} &0 = 0 \quad (k = 1) \qquad 0 = 0 \quad (k = 2) \\ &[\varphi, [\varphi, v]_{\text{-}}]_{\text{-}} = 0 \quad (k = 3) \qquad [\varphi^2, [\varphi, v]_{\text{-}}]_{\text{-}} = 0 \quad (k = 4). \end{aligned} \quad (5.16)$$

Consequently, for $k \geq 3$ either the set of the variation v should be restricted to the ones satisfying (5.15) with φ being a solution (5.14), or to the field equation (5.14) should be added the condition (5.15) with arbitrary operator/variation v.

5.4. *Interacting neutral scalar fields*

Consider a system of two neutral scalar fields φ_1 and φ_2 with Lagrangian

$$L = -\frac{1}{2}m^2c^4\varphi_1 \circ \varphi_1 + \frac{1}{2}c^2\hbar^2(\partial_\mu\varphi_1) \circ (\partial^\mu\varphi_1) \\ - \frac{1}{2}m^2c^4\varphi_2 \circ \varphi_2 + \frac{1}{2}c^2\hbar^2(\partial_\mu\varphi_2) \circ (\partial^\mu\varphi_2) + a\varphi_1\varphi_2 \quad (5.17)$$

where a is a non-vanishing real parameter. Performing a procedure similar to the ones in the previous subsections, we obtain the field equations

$$m^2c^4\varphi_1 + c^2\hbar^2\Box(\varphi_1) = a\varphi_2 \qquad m^2c^4\varphi_2 + c^2\hbar^2\Box(\varphi_2) = a\varphi_1 \quad (5.18)$$

and the conditions

$$[\varphi_1, v_2]_{\text{-}} = 0 \qquad [\varphi_2, v_1]_{\text{-}} = 0 \quad (5.19)$$

which must satisfy the solutions of (5.18) and the variations v_1 and v_2 of φ_1 and φ_2, respectively. The last conditions are quite natural from physical view-point because they mean that we can make a variation of φ_1 independently of φ_2 and *vice versa*. The energy-momentum operator of the system under consideration turns to be

$$T_{\mu\nu} = \frac{1}{2}c^2\hbar^2 \sum_{i=1,2} \{(\partial_\mu\varphi_i)\circ(\partial_\nu\varphi_i) + (\partial_\nu\varphi_i)\circ(\partial_\mu\varphi_i)\} - \eta_{\mu\nu}L. \tag{5.20}$$

5.5. *Free Dirac (spinor) field 1. Standard Lagrangian*

As a standard Lagrangian of a spin $\frac{1}{2}$ Dirac field ψ, we take [1]

$$L = \frac{1}{2}\mathrm{i}\hbar c\{\overline{\psi}\gamma^\mu \odot (\partial_\mu\psi) - (\partial_\mu\overline{\psi})\gamma^\mu \odot \psi\} - mc^2\overline{\psi}\odot\psi. \tag{5.21}$$

Here: γ^μ are the Dirac's γ-matrices, $\overline{\psi} := \psi^\dagger\gamma^0$ is the Dirac conjugate of $\psi = (\psi_0, \psi_1, \psi_2, \psi_3)^\top$ with $\top$ being the matrix transposition sign, in products like $\overline{\psi}\gamma^\mu$ the matrix multiplication sign is dropped, and $\odot$ denotes a composition combined with matrix multiplication, e.g. $\overline{\psi}\odot\psi = \sum_\mu \overline{\psi}_\mu \circ \psi_\mu = (\gamma^0)^{\mu\nu}\psi_\mu^\dagger \circ \psi_\nu$. If v and $\overline{v}$ denote variations of ψ and $\overline{\psi}$, respectively, we, in view of definition 4.1, obtain

$$\begin{aligned} &\frac{\partial L}{\partial\psi}(v) = -\frac{1}{2}\mathrm{i}\hbar c(\partial_\mu\overline{\psi})\gamma^\mu \odot v - mc^2\overline{\psi}\odot v & &\frac{\partial L}{\partial(\partial_\mu\psi)}(v) = \frac{1}{2}\mathrm{i}\hbar c\overline{\psi}\gamma^\mu \odot v \\ &\frac{\partial L}{\partial\overline{\psi}}(\overline{v}) = \frac{1}{2}\mathrm{i}\hbar c\overline{v}\odot\gamma^\mu(\partial_\mu\psi) - mc^2\overline{v}\odot\psi & &\frac{\partial L}{\partial(\partial_\mu\overline{\psi})}(\overline{v}) = -\frac{1}{2}\mathrm{i}\hbar c\overline{v}\odot\gamma^\mu\psi. \end{aligned} \tag{5.22}$$

Thus, the Euler-Lagrange relation (4.13) takes the form

$$-\{\mathrm{i}\hbar c(\partial_\mu\overline{\psi})\gamma^\mu + mc^2\overline{\psi}\}\odot v + \overline{v}\odot\{\mathrm{i}\hbar c\gamma^\mu(\partial_\mu\psi) - mc^2\psi\} = 0. \tag{5.23}$$

The choices when v or $\overline{v}$ is the zero operator and $\overline{v}$ or v, respectively, is arbitrary, result in the Dirac equation and its conjugate,[12] viz.

$$\mathrm{i}\hbar\gamma^\mu(\partial_\mu\psi) - mc\psi = 0 \qquad \mathrm{i}\hbar(\partial_\mu\overline{\psi})\gamma^\mu + mc\overline{\psi} = 0. \tag{5.24}$$

These equations convert (5.23) into identity relative to v and $\overline{v}$. So, the variations of a free spinor field are completely arbitrary. Combining (5.22) with (4.17)–(4.20), we get the energy-momentum, current and spin angular momentum oper-

[12] More precisely, the choice when all but one of the components v_α and $\overline{v}_\alpha$ vanish, results in the (conjugate) Dirac equation for this component; the substitution of these results into (5.23) converts it into identity relative to v_α and $\overline{v}_\alpha$.

ators as:

$$T_{\mu\nu} = \frac{1}{2}\mathrm{i}\hbar c\{\overline{\psi}\gamma_\mu \odot (\partial_\nu \psi) - (\partial_\nu \overline{\psi})\gamma_\mu \odot \psi\} \tag{5.25}$$

$$J_\mu = qc\overline{\psi}\gamma_\mu \odot \psi \tag{5.26}$$

$$S^\lambda_{\mu\nu} = -\frac{1}{4}\hbar c\overline{\psi}(\gamma^\lambda \sigma_{\mu\nu} + \sigma_{\mu\nu}\gamma^\lambda) \odot \psi \qquad \sigma_{\mu\nu} := \frac{\mathrm{i}}{2}(\gamma^\mu\gamma^\nu - \gamma^\nu\gamma^\mu), \tag{5.27}$$

where we have used that a 4-rotation $x^\mu \mapsto x^\mu + \varepsilon^{\mu\nu}x_\nu$ implies $\psi \mapsto \mathrm{e}^{-\frac{\mathrm{i}}{4}\sigma_{\mu\nu}\varepsilon^{\mu\nu}}\psi$ and $\overline{\psi} \mapsto \mathrm{e}^{+\frac{\mathrm{i}}{4}\sigma_{\mu\nu}\varepsilon^{\mu\nu}}\overline{\psi}$. These expressions are identical with the ones in [1].

5.6. *Free Dirac (spinor) field 2. Charge symmetric Lagrangian*

In the present subsection, we shall present an example of a Lagrangian for which the Euler-Lagrange equations are identities, like $0 = 0$, but which, regardless of this 'strange' fact, entails completely reasonable field equations. In a classical sense, this Lagrangian is a 'completely singular' one, as all its classical derivatives are zero, but it is not a constant operator.

Let ψ be a Dirac 4-spinor and (see, e.g., [1, 2, 3])

$$\breve{\psi} := C\overline{\psi}^\top = (\overline{\psi}C^\top)^\top \tag{5.28}$$

be its *charge conjugate* one, where the matrix C satisfies the conditions

$$C^{-1}\gamma^\mu C = -\gamma^{\mu\,\top} := -(\gamma^\mu)^\top \qquad C^\top = -C. \tag{5.29}$$

Let us consider ψ and $\breve{\psi}$ as independent field variables. In their terms, the Lagrangian (5.21) reads

$$\begin{aligned} L' = &-\frac{1}{2}\mathrm{i}\hbar c\{\breve{\psi}^\top(x)C^{-1}\gamma^\mu \odot (\partial_\mu\psi(x)) - (\partial_\mu\breve{\psi}^\top(x))C^{-1}\gamma^\mu \odot \psi(x)\} \\ &+ mc^2\breve{\psi}^\top(x)C^{-1} \odot \psi(x). \end{aligned} \tag{5.30}$$

We would like to emphasize on the change of the signs and the appearance of the matrix C in (5.30) with respect to (5.21). An alternative to this Lagrangian is a one with changed positions of ψ and $\breve{\psi}$, viz.

$$\begin{aligned} L'' = &-\frac{1}{2}\mathrm{i}\hbar c\{\psi^\top(x)C^{-1}\gamma^\mu \odot (\partial_\mu\breve{\psi}(x)) - (\partial_\mu\psi^\top(x))C^{-1}\gamma^\mu \odot \breve{\psi}(x)\} \\ &+ mc^2\psi^\top(x)C^{-1} \odot \breve{\psi}(x). \end{aligned} \tag{5.31}$$

Evidently, the variables ψ and $\breve{\psi}$ do not enter in (5.30) and (5.31) on equal footing. We shall try to 'symmetrize' the situation by considering a Lagrangian which is the half sum of the last two ones, i.e.

$$
\begin{aligned}
L''' = \frac{1}{4}\mathrm{i}\hbar c\{ &-\breve{\psi}^\top(x)C^{-1}\gamma^\mu \odot (\partial_\mu\psi(x)) + (\partial_\mu\breve{\psi}^\top(x))C^{-1}\gamma^\mu \odot \psi(x) \\
&- \psi^\top(x)C^{-1}\gamma^\mu \odot (\partial_\mu\breve{\psi}(x)) + (\partial_\mu\psi^\top(x))C^{-1}\gamma^\mu \odot \breve{\psi}(x)\} \\
&+ \frac{1}{2}mc^2\{\breve{\psi}^\top(x)C^{-1} \odot \psi(x) + \psi^\top(x)C^{-1} \odot \breve{\psi}(x)\}. \qquad (5.32)
\end{aligned}
$$

Let v and $\breve{v}$ denote variations of ψ and $\breve{\psi}$, respectively. Applying definition 4.1, we can calculate the derivatives of the Lagrangians (5.30)–(5.32). They are as follows:

$$
\begin{aligned}
\frac{\partial L'}{\partial\psi^\top}(v) &= +\frac{1}{2}\mathrm{i}\hbar c\{C^{-1}\gamma^\mu(\partial_\mu\breve{\psi})\}^\top \odot v - mc^2\{C^{-1}\breve{\psi}\}^\top \odot v \\
\frac{\partial L'}{\partial\breve{\psi}^\top}(\breve{v}) &= -\frac{1}{2}\mathrm{i}\hbar c\breve{v}^\top \odot C^{-1}\gamma^\mu(\partial_\mu\psi) + mc^2 v^\top \odot C^{-1}\psi \\
\frac{\partial L'}{\partial(\partial_\mu\psi^\top)}(v) &= -\frac{1}{2}\mathrm{i}\hbar c\{C^{-1}\gamma^\mu\breve{\psi}\}^\top \odot v \\
\frac{\partial L'}{\partial(\partial_\mu\breve{\psi}^\top)}(\breve{v}) &= +\frac{1}{2}\mathrm{i}\hbar c\breve{v}^\top \odot C^{-1}\gamma^\mu\psi \\
\frac{\partial L''}{\partial\psi^\top}(v) &= -\frac{1}{2}\mathrm{i}\hbar c v^\top \odot C^{-1}\gamma^\mu(\partial_\mu\breve{\psi}) + mc^2 v^\top \odot C^{-1}\breve{\psi} \\
\frac{\partial L''}{\partial\breve{\psi}^\top}(\breve{v}) &= +\frac{1}{2}\mathrm{i}\hbar c\{C^{-1}\gamma^\mu(\partial_\mu\psi)\}^\top \odot \breve{v} - mc^2\{C^{-1}\psi\}^\top \odot \breve{v} \\
\frac{\partial L''}{\partial(\partial_\mu\psi^\top)}(v) &= +\frac{1}{2}\mathrm{i}\hbar c v^\top \odot C^{-1}\gamma^\mu\breve{\psi} \\
\frac{\partial L''}{\partial(\partial_\mu\breve{\psi}^\top)}(\breve{v}) &= -\frac{1}{2}\mathrm{i}\hbar c\{C^{-1}\gamma^\mu\psi\}^\top \odot \breve{v} \\
2\frac{\partial L'''}{\partial\psi^\top}(v) &= +\frac{1}{2}\mathrm{i}\hbar c\{C^{-1}\gamma^\mu(\partial_\mu\breve{\psi})\}^\top \odot v - mc^2\{C^{-1}\breve{\psi}\}^\top \odot v \\
&\quad - \frac{1}{2}\mathrm{i}\hbar c v^\top \odot C^{-1}\gamma^\mu(\partial_\mu\breve{\psi}) + mc^2 v^\top \odot C^{-1}\breve{\psi} \\
2\frac{\partial L'''}{\partial\breve{\psi}^\top}(\breve{v}) &= -\frac{1}{2}\mathrm{i}\hbar c\breve{v}^\top \odot C^{-1}\gamma^\mu(\partial_\mu\psi) + mc^2 v^\top \odot C^{-1}\psi \\
&\quad + \frac{1}{2}\mathrm{i}\hbar c\{C^{-1}\gamma^\mu(\partial_\mu\psi)\}^\top \odot \breve{v} - mc^2\{C^{-1}\psi\}^\top \odot \breve{v} \\
2\frac{\partial L'''}{\partial(\partial_\mu\psi^\top)}(v) &= -\frac{1}{2}\mathrm{i}\hbar c\{C^{-1}\gamma^\mu\breve{\psi}\}^\top \odot v + \frac{1}{2}\mathrm{i}\hbar c v^\top \odot C^{-1}\gamma^\mu\breve{\psi} \\
2\frac{\partial L'''}{\partial(\partial_\mu\breve{\psi}^\top)}(\breve{v}) &= +\frac{1}{2}\mathrm{i}\hbar c\breve{v}^\top \odot C^{-1}\gamma^\mu\psi - \frac{1}{2}\mathrm{i}\hbar c\{C^{-1}\gamma^\mu\psi\}^\top \odot \breve{v}.
\end{aligned}
$$

Notice, in these equalities the matrix transposition serves only to ensure proper matrix multiplication.

Substituting the just calculated derivatives in the Euler- Lagrange equation (4.13), we see that for the Lagrangians (5.30)–(5.32) it reduces respectively to

$$+\big(A(\breve{\psi})\big)^\top \odot v - \breve{v}^\top \odot A(\psi) = 0 \qquad (5.33)$$

$$-v^\top \odot A(\breve{\psi}) + \big(A(\psi)\big)^\top \odot \breve{v} = 0 \qquad (5.34)$$

$$+\big(A(\breve{\psi})\big)^\top \odot v - v^\top \odot A(\breve{\psi}) + \big(A(\psi)\big)^\top \odot \breve{v} - \breve{v}^\top \odot A(\psi) = 0 \qquad (5.35)$$

where

$$A\colon \psi \mapsto A(\psi) := \mathrm{i}\hbar c C^{-1}\gamma^\mu \partial_\mu \psi - mc^2 C^{-1}\psi \qquad (5.36)$$

Let $\alpha \in \{0,1,2,3\}$ be arbitrarily fixed. Choosing all but the αth component of v (resp. $\breve{v}$) to vanish and setting this component to be equal to the identity mapping $\mathsf{id}_{\mathcal{F}}$, we see that (5.33) or (5.34) is equivalent to the equations $A(\psi_\alpha) = 0$ and $A(\breve{\psi}_\alpha) = 0$ for all $\alpha = 0, 1, 2, 3$, which, by virtue of (5.36), are equivalent to the system of Dirac equations

$$\mathrm{i}\hbar\gamma^\mu\partial_\mu\psi - mc^2\psi = 0 \qquad \mathrm{i}\hbar\gamma^\mu\partial_\mu\breve{\psi} - mc^2\breve{\psi} = 0. \qquad (5.37)$$

However, if one substitutes in (5.35) the just-described choices of v and $\breve{v}$, one will get the identity $0 = 0$ for any one of them, instead of some equations for the components of ψ and $\breve{\psi}$.[13] In this sense,*classical Euler-Lagrange equations for the Lagrangian* (5.32) *do not exist*. In this case, which cannot be handled by the standard methods, we shall proceed as follows. Let us make the same as the above selection of the variations v and $\breve{v}$ without setting the non-vanishing components to be equal to $\mathsf{id}_{\mathcal{F}}$. Such choices reduces (5.35) to the following system of relations (do not sum over α!)

$$[A(\breve{\psi}_\alpha), v_\alpha]_{\_} = 0 \quad [A(\psi_\alpha), \breve{v}_\alpha]_{\_} = 0 \qquad (5.38)$$

for the variations v_α and $\breve{v}_\alpha$. If we now choose the operators v_α and $\breve{v}_\alpha$ to range in a unitary representation in $\mathcal{F}$ of some group, then the Schur's lemma[14] implies the existence of classical functions, not operators, $f_{\psi,\alpha}(x)$ and $f_{\breve{\psi},\alpha}(x)$, such that

$$A(\psi_\alpha(x)) = f_{\psi,\alpha}(x)\,\mathsf{id}_{\mathcal{F}} \quad A(\breve{\psi}_\alpha(x)) = f_{\breve{\psi},\alpha}(x)\,\mathsf{id}_{\mathcal{F}}. \qquad (5.39)$$

[13] Moreover, the derivatives of L''' vanish for the choices mentioned, i.e. the classical derivatives (see (4.6)) of the Lagrangian (5.32) identically vanish.

[14] See, e.g, [19, appendix II], or [16, sec. 8.2], or [17, ch. 5, sec. 3].

In view of (5.36), these equations are equivalent to the system

$$\mathrm{i}\hbar\gamma^\mu\partial_\mu\psi - mc^2\psi = \chi_\psi \,\mathsf{id}_{\mathcal{F}} \qquad \mathrm{i}\hbar\gamma^\mu\partial_\mu\breve{\psi} - mc^2\breve{\psi} = \breve{\chi}_{\breve{\psi}}\,\mathsf{id}_{\mathcal{F}}. \tag{5.40}$$

where χ_ψ and $\breve{\chi}_{\breve{\psi}}$ are some *classical*, not operator-valued, spinors. It is trivial to be checked that (5.40) converts (5.35) into identity with respect to v and $\breve{v}$. Therefore the system of (Dirac equations with, generally, non-vanishing r.h.s.) (5.40) plays a role of a system of field equations for the Lagrangian (5.32).

Ending this example, we note that if one wants the Lagrangian (5.32) to describe a free spinor field, the choices

$$\chi_\psi = 0 \quad \breve{\chi}_{\breve{\psi}} = 0 \tag{5.41}$$

should be made; otherwise, the equations (5.40) describe a spin $\frac{1}{2}$ quantum field with some selfinteraction. Elsewhere we shall demonstrate that the Lagrangians (5.30)–(5.32) and the additional conditions (5.41) lead to one and the same quantum field theory of free spinor fields.

5.7. *General quadratic Lagrangian*

Let us consider a system of quantum fields φ_i with a Lagrangian

$$\begin{aligned} L = a^i\varphi_i + m^{ij}\varphi_i\circ\varphi_j + b^{i\mu}\partial_\mu\varphi_i + g^{i\mu j\nu}(\partial_\mu\varphi_i)\circ(\partial_\nu\varphi_j) \\ + c^{ij\mu}\varphi_i\circ(\partial_\mu\varphi_j) + d^{ij\mu}(\partial_\mu\varphi_j)\circ\varphi_i, \end{aligned} \tag{5.42}$$

where a^i, m^{ij}, $b^{i\mu}$, $g^{i\mu j\nu}$, $c^{ij\mu}$, and $d^{ij\mu}$ are some (dimensional) constants. Calculating $\frac{\partial L}{\partial\varphi_i}$ and $\pi^{i\mu} = \frac{\partial L}{\partial(\partial_\mu\varphi_i)}$, according to definition 4.1, and substituting them into (4.13), we get the Euler-Lagrange relation

$$\alpha^i\circ v_i + v_i\circ\beta^i + a^i v_i = 0 \tag{5.43}$$

where

$$\begin{aligned} \alpha^i &:= m^{ij}\varphi_j + (c^{ij\mu} - d^{ji\mu})\partial_\mu\varphi_j - g^{i\mu j\nu}\partial_\mu\partial_\nu\varphi_j \\ \beta^i &:= m^{ji}\varphi_j + (d^{ij\mu} - c^{ji\mu})\partial_\mu\varphi_j - g^{j\nu i\mu}\partial_\mu\partial_\nu\varphi_j \end{aligned} \tag{5.44}$$

and $v_i = \delta_0\varphi_i$ is a variation of φ_i. Choosing $v_i = f_i(x)\,\mathsf{id}_{\mathcal{F}}$ with arbitrary $f\colon M \to \mathbb{C}$, from (5.43), we obtain the Euler-Lagrange equations

$$\alpha^i + \beta^i + a^i\,\mathsf{id}_{\mathcal{F}} = 0. \tag{5.45}$$

The combination of (5.45) and (5.43) results in the condition

$$\sum_i [\alpha^i, v_i]_{\_} = 0 \tag{5.46}$$

(which is equivalent to $\sum_i[\beta^i, v_i]_{\_} = 0$) for the field operators φ_i and their variations v_i. These conditions can be satisfied identically relative to v_i if all α^i (and, hence, all β^i) happen to be proportional to the identity operator $\mathsf{id}_{\mathcal{F}}$. In particular, this will be the case when

$$m^{ij} = m^{ji} \quad c^{ij\mu} - d^{ji\mu} = d^{ij\mu} - c^{ji\mu} \quad g^{i\mu j\nu} = g^{j\nu i\mu}, \tag{5.47}$$

so that $\alpha^i = \beta^i$ (see (5.44)) and the field equations (5.45) read

$$\alpha^i + \frac{1}{2} a^i \,\mathsf{id}_{\mathcal{F}} = 0. \tag{5.48}$$

As a special case of (5.47), the one of free fields should be single out. In this important case, we have (do not sum over i!)

$$a^i = b^i = c^{ij\mu} = d^{ij\mu} = 0 \quad m^{ij} = m_i^2 c^4 \delta^{ij} \quad g^{i\mu j\nu} = g^{\mu\nu} \delta^{ij} \tag{5.49}$$

where m_i and $g^{\mu\nu} = g^{\nu\mu}$ are some (dimensional) constants.

However, having in mind the considerations in Subsect. 5.6, one should take into account the possibility that the Euler-Lagrange equations (5.45) may turn to be identities, in which case the field equations should be derived from (5.43) in a different way.

The energy-momentum operator corresponding to the Lagrangian (5.42), in view of (4.17), is

$$\begin{aligned} T^{\mu\nu} = b^{i\mu}(\partial^\nu \varphi_i) + g^{i\mu j\lambda}(\partial^\nu \varphi_i) \circ (\partial_\lambda \varphi_j) + g^{j\lambda i\mu}(\partial_\lambda \varphi_j) \circ (\partial^\nu \varphi_i) \\ + c^{ji\mu} \varphi_j \circ (\partial^\nu \varphi_i) + d^{ji\mu}(\partial^\nu \varphi_i) \circ \varphi_j - \eta^{\mu\nu} L. \end{aligned} \tag{5.50}$$

If the system possesses a charge, the corresponding current operator can be obtained formally from the r.h.s. of (5.50) by deleting the last term and replacing $\partial^\nu \varphi_i$ with $\frac{q}{i\hbar c}\varepsilon(\varphi_i)\varphi_i$ (see (4.18)).

6. Conclusion

In this paper we have given an analysis of some aspects and corollaries of the Schwinger's action principle in Lagrangian (canonical) quantum field theory. As it was demonstrated, in the 'standard' presentation, this variational principle contains an additional hypothesis, which does not follow logically from the rest of the theory and modifies the conserved quantities so that they do not always have the required properties. We have removed the mentioned hypothesis by giving a suitable meaning of derivatives of operator-valued functions of operator arguments with respect to such an argument. After that modification is done, the following important consequences of the Schwinger's action principle were derived:

i. The classical (standard) Euler-Lagrange equation of motion for the field operators remain the same as before the change.

ii. The conserved quantities (operators) are changed so that they are unique and have the required properties, at least in the examples considered.

iii. In the general case (of (self-)interacting fields), the variations of the field operators cannot be completely arbitrary as they should satisfy some equations in which the field operators, satisfying the Euler-Lagrange equations, enter.

iv. The variations of the field operators which are proportional to the identity mapping/operator, like (3.26), are always admissible.

v. The variations of item iv above are sufficient for the derivation of the Euler-Lagrange equations and all spacetime conserved quantities. However, for other variations, such as the ones connected with internal symmetries (e.g. like (constant) phase transformations leading to charge conservation), one should always check whether they are admissible in a sense that they must satisfy the equations mentions in point iii above.

vi. If one insists on keeping the field variations completely arbitrary, it is quite likely that, for interacting fields, field operators which are multiples of the identity operator will be the only solutions of the variational problem determining them.[15]

Since the (classical) Euler-Lagrange equations remain unchanged after the described correction of Schwinger's action principle, the modification of the operator ordering in the structure of conserved quantities and, possibly, other equations for the fields (and their variations) should be regarded as the main outcome of the present investigation.

Since definition 4.1 is valid for any $\mathbb{C}$-vector space $\mathcal{F}$, non only for Hilbert one, and operators acting on it, the results of this paper are applicable to more general situations than the one of Lagrangian quantum field theory. For instance, one may consider the variational problem for operator-valued functional (3.1) in which L is a polynomial or convergent power series in some operators $\varphi_i(x)\colon \mathcal{F} \to \mathcal{F}$. We can assert that $\varphi_i(x)$ and their variations v_i must satisfy the Euler-Lagrange relation (4.13); in particular, the operators $\varphi_i(x)$ should be solutions of the 'classical' Euler-Lagrange equations (4.14).

References

1. N. N. Bogolyubov and D. V. Shirkov. *Introduction to the theory of quantized fields.* Nauka, Moscow, third edition, 1976. In Russian. English translation: Wiley, New York, 1980.

[15] For instance, such is the case considered in Subsect. 5.4: if (5.19) holds for any v_1 and v_2, then $\varphi_a(x) = f_a(x)\,\mathsf{id}_{\mathcal{F}}$, $a = 1, 2$, for some $f_a\colon M \to \mathbb{C}$ (as a result of, e.g., the Schur's lemma —see [16, sec. 8.2] or [17, ch. 5, sec. 3]).

2. J. D. Bjorken and S. D. Drell. *Relativistic quantum mechanics*, volume 1 and 2. McGraw-Hill Book Company, New York, 1964, 1965. Russian translation: Nauka, Moscow, 1978.
3. Paul Roman. *Introduction to quantum field theory*. John Wiley&Sons, Inc., New York-London-Sydney-Toronto, 1969.
4. Lewis H. Ryder. *Quantum field theory*. Cambridge Univ. Press, Cambridge, 1985. Russian translation: Mir, Moscow, 1987.
5. A. I. Akhiezer and V. B. Berestetskii. *Quantum electrodynamics*. Nauka, Moscow, 1969. In Russian. English translation: Authorized English ed., rev. and enl. by the author, Translated from the 2d Russian ed. by G.M. Volkoff, New York, Interscience Publishers, 1965. Other English translations: New York, Consultants Bureau, 1957; London, Oldbourne Press, 1964, 1962.
6. Pierre Ramond. *Field theory: a modern primer*, volume 51 of *Frontiers in physics*. Reading, MA Benjamin-Cummings, London-Amsterdam-Don Mills, Ontario-Sidney-Tokio, 1 edition, 1981. 2nd rev. print, Frontiers in physics vol. 74, Adison Wesley Publ. Co., Redwood city, CA, 1989; Russian translation from the first ed.: Moscow, Mir 1984.
7. N. N. Bogolubov, A. A. Logunov, and I. T. Todorov. *Introduction to axiomatic quantum field theory*. W. A. Benjamin, Inc., London, 1975. Translation from Russian: Nauka, Moscow, 1969.
8. N. N. Bogolubov, A. A. Logunov, A. I. Oksak, and I. T. Todorov. *General principles of quantum field theory*. Nauka, Moscow, 1987. In Russian. English translation: Kluwer Academic Publishers, Dordrecht, 1989.
9. J. Schwinger. The theory of quantized fields. I. *Physical Review*, 82(6):914, 1951. Russian translation in [20, pp. 115–137].
10. J. D. Bjorken and S. D. Drell. *Relativistic quantum fields*, volume 2. McGraw-Hill Book Company, New York, 1965. Russian translation: Nauka, Moscow, 1978.
11. W. Pauli and W. Heisenberg. Zur Quantendynamik der wellenfelder (On the quantum dynamics of wavefields). *Z. Phys.*, 56(1):1–61, 1929. In German. Russian translation in [21, pp. 30–88].
12. W. Pauli and V. Weisskopf. Über die Quantizierung der skalaren relativistischen Wellengleichung (On the quantization of the scalar relativistic wave equation). *Helv. Phys. Acta*, 7:709–731, 1934. In German. Russian translation in [21, pp. 194–214].
13. J. Schwinger. The theory of quantized fields. II. *Physical Review*, 91(3):713, 1953.
14. R. E. Peierls. The commutation laws of relativistic field theory. *Proceedings of the Royal society (London)*, 214(117):143–157, August 1952.
15. J. Schwinger. The theory of quantized fields. III. *Physical Review*, 91(3):728, 1953.
16. A. A. Kirillov. *Elements of the theory of representations*. Springer, Berlin, 1976. Translation from Russian (second ed., Nauka, Moscow, 1978).
17. Asim Barut and Ryszard Roczka. *Theory of group representations and applications*. PWN — Polish Scientific Publishers, Waszawa, 1977. Russian translation: Mir, Moscow, 1980.
18. Y. Ohnuki and S. Kamefuchi. *Quantum field theory and parafields*. University of Tokyo Press, Tokyo, 1982.
19. Yu. B. Rumer and A. I. Fet. *Group theory and quantized fields*. Nauka, Moscow, 1977. In Russian.

20. *Novejshee razvitie kvantovoj electrodinamiki. Sbornik statej (The newest development of quantum electrodynamics. Collection of papers),* editor Ivanenko D. D., Moscow, 1954. IL. In Russian.
21. Wolfgang Pauli. *Works on quantum theory. Articles 1928–1958.* Nauka Publ., Moscow, 1977. In Russian.

TIME-LIKE KILLING VECTOR FIELDS ON COMPACT INDEFINITE KÄHLER SURFACES

HIROYUKI KAMADA*

Numazu College of Technology,
3600, Ooka, Numazu,
Shizuoka 410-8501, Japan
E-mail: kamada@numazu-ct.ac.jp

Several results for time-like Killing vector fields on a compact indefinite Kähler surface are introduced. For example, under certain assumptions, the existence of an indefinite Kähler metric and a time-like Killing vector field implies that the underlying complex surface is biholomorphic to a Hirzebruch surface F_d. Moreover, for any indefinite Kähler metric g on a Hirzebruch surface F_d, every time-like Killing vector field associated with an S^1-symmetry on (F_d, g) is induced from the standard $\mathbb{C}^*$-action along the fibers of $F_d = \mathbb{P}(\mathcal{O}(d) \oplus \mathcal{O}) \to \mathbb{P}^1$. Furthermore, several remarks related to the existence of indefinite Kähler metrics are given.

1. Introduction

An indefinite Kähler metric g on a compact complex manifold is said to be scalar-flat if its scalar curvature s is identically zero. Let M be a compact complex surface admitting a scalar-flat indefinite Kähler metric g. Then we call (M, g) a scalar-flat indefinite Kähler surface. The standard indefinite product metric on the product $\mathbb{P}^1 \times \mathbb{P}^1$ of two complex projective lines with unit round metric is a typical example of a scalar-flat (and conformally-flat) indefinite Kähler metric. Furthermore, many explicit examples of (non-conformally-flat) scalar-flat indefinite Kähler metrics are constructed on $\mathbb{P}^1 \times \mathbb{P}^1$, with the help of an indefinite analogue of LeBrun's hyperbolic ansatz (see Kamada [5, 6]). By construction, with respect to each metric of these examples, there exists a semi-free S^1-symmetry (i.e., a nontrivial isometric S^1-action) such that the Killing vector field ξ generating the symmetry satisfies the following properties:

- ξ is time-like, that is, $g(\xi, \xi) < 0$ outside Zero(ξ),
- Zero(ξ) contains two two-dimensional connected components,

*Supported by JSPS Grant-in-Aid for Scientific Research "Encouragement of Young Scientists (B)#13740053"

where Zero(ξ) denotes the set of zeros of ξ. In this article, we study how the existence of an indefinite Kähler metric and a time-like Killing vector field affects the underlying compact complex surface. In fact, we see that, under certain assumptions, the existence of such structures implies that the underlying compact complex surface is biholomorphic to a Hirzebruch surface F_d ($d = 0, 1, 2, \dots$). Furthermore, we introduce properties of time-like Killing vector fields on F_d with an indefinite Kähler metric g.

2. Kähler surfaces

Let $M = (M, J)$ be a compact complex surface with complex structure J, and g a pseudo-Riemannian metric on M satisfying Hermitian property $g(J\cdot, J\cdot) = g$. We call (M, g) a Hermitian surface and $\Omega := g(J\cdot, \cdot)$ the fundamental form of (M, g). Then, g is indefinite (resp. definite) if and only if $\Omega^2 < 0$ (resp. $\Omega^2 > 0$) with respect to the orientation defined by J. If a Hermitian surface (M, g) is indefinite, then g is an indefinite metric of type $(2, 2)$. We sometimes call a metric of type $(2, 2)$ on a real four-manifold simply a neutral metric. A Hermitian surface (M, g) is called a Kähler surface if the fundamental form $\Omega := g(J\cdot, \cdot)$ is closed, and the fundamental form Ω is then called the Kähler form of (M, g). Throughout this article, we endow a Hermitian (or Kähler) surface (M, J, g) with the complex orientation determined by J. For a Kähler surface (M, g), the Kähler form Ω is a symplectic $(1, 1)$-form on M. Thus, if g is indefinite (resp. definite), then $b_2^-(M) \geq 1$ (resp. $b_2^+(M) \geq 1$), where $b_2^-(M)$ (resp. $b_2^+(M)$) denotes the dimension of a maximal subspace of the second cohomology group $H^2(M; \mathbb{R})$ where the intersection form is negative (resp. positive).

The existence of a positive-definite Kähler metric on a compact complex manifold implies that all of its odd Betti numbers are even (by the Serre duality). It is well-known (cf. Barth et al. [1]) that the converse is also true for a compact complex surface M. Indeed, M admits a positive-definite Kähler metric if and only if its first Betti number $b_1(M)$ is even.

In the indefinite case, the existence of an indefinite Kähler metric does not necessarily imply that its first Betti number is even. Perhaps no simple criterion, such as in the positive-definite case, has been known for the existence of an indefinite Kähler metric on a compact complex surface. Concerning the existence of indefinite Kähler metrics, Petean [10] obtained the following

Proposition 2.1. *Let M be a compact complex surface. Suppose that M admits an indefinite Kähler metric. Then M should be biholomorphic to a surface in the following list*:

M	$\kappa(M)$	$b_1(M)$	$b_2^+(M)$	$b_2^-(M)$
ruled surface of genus g	$-\infty$	2g	1	1
minimal surface of class VII_0 with no global spherical shell		1	0	even (>0)
hyperelliptic surface	0	2	1	1
complex torus		4	3	3
primary Kodaira surface		3	2	2
minimal properly elliptic surface with $\tau(M) = b_2^+(M) - b_2^-(M) = 0$	1	even odd	>0	>0
minimal surface of general type with $\tau(M) = b_2^+(M) - b_2^-(M) \geq 0$	2	even	odd (>0)	odd (>0)

where $\kappa(M)$ denotes the Kodaira dimension of M.

As in Petean [10], ruled surfaces and surfaces with $\kappa(M) = 1$ and $b_1(M)$ even admit indefinite Kähler metrics, and hyperelliptic surfaces, complex tori and primary Kodaira surfaces admit Ricci-flat indefinite Kähler metrics. Note that any primary Kodaira surface admits indefinite Kähler metrics, although its first Betti number equals three. There are examples of minimal surfaces of general type that admit indefinite Kähler metrics. The existence of indefinite Kähler metrics in the other cases still remains unknown. In fact, no surface of class VII_0 in the list above is known at present. He also studied indefinite Kähler Einstein metrics on compact complex surfaces.

Let (M, g) be a compact (possibly indefinite) Kähler surface and γ denote the Ricci form of (M, g). Then g is scalar-flat if and only if $\gamma \wedge \Omega \equiv 0$. Therefore, as a necessary condition, if (M, g) is scalar-flat Kähler, then the Kähler form Ω satisfies $c_1(M) \cdot [\Omega] = 0$, where $c_1(M)$ is the real first Chern class of M and $[\Omega]$ denotes the Kähler class, that is, the cohomology class of Ω. Note that scalar-flat indefinite (resp. positive-definite) Kähler surfaces form a particular class of oriented four-manifolds with self-dual neutral metrics (resp. anti-self-dual Riemannian metrics) (cf. Itoh [4]). Here, a (neutral or Riemannian) metric g on an oriented four-manifold M is said to be self-dual if the Weyl conformal tensor W, thought of an endomorphism on the space of two-forms, satisfies $*W = W$, where $*$ denotes the Hodge star operator of (M, g). In the positive-definite case, the possible candidates of compact scalar-flat positive-definite Kähler surfaces are the following (cf. Itoh [4]):

- surfaces covered by complex tori or $K3$ surfaces,

• (possibly non-minimal) ruled surfaces.

It is well-known that a surface covered by a complex torus or a $K3$ surface admits a Ricci-flat positive-definite Kähler metric (and vice versa), which is the unique scalar-flat Kähler metric in its Kähler class (cf. Barth et al. [1]). Regarding non-Ricci-flat case, the Riemannian product $\mathbb{P}^1 \times \Sigma_g$ of the complex projective line $\mathbb{P}^1$ with unit round metric $h_{\mathbb{P}^1}$ and a Riemann surface Σ_g of genus $g \geq 2$ with metric h_{Σ_g} of constant curvature -1 is a typical example of a scalar-flat (and moreover conformally-flat) Kähler surface. Furthermore, it is also known that the blow-up of any ruled surface at sufficiently many points, in appropriate position, admits a scalar-flat Kähler metric (see Kim-LeBrun-Pontecorvo [7], LeBrun [9]).

In the indefinite case, as mentioned before, there are many scalar-flat indefinite Kähler metrics on the product $\mathbb{P}^1 \times \mathbb{P}^1$ of complex projective lines. For other possible candidates of compact scalar-flat indefinite Kähler surfaces, see Kamada [5]. However, except for $\mathbb{P}^1 \times \mathbb{P}^1$ and the Ricci-flat case, only a few examples of such surfaces are known.

3. Time-like Killing vector fields

Let (M, g) be a (positive-definite or indefinite) Kähler surface. A Killing vector field $\xi \not\equiv 0$ on (M, g) is said to be time-like if $g(\xi, \xi) < 0$ outside the set Zero(ξ) of zeros of ξ. Note that, when g is positive-definite, there exists no time-like vector field on (M, g), but every vector field on M is time-like on $(M, -g)$. It is natural to ask which compact complex surfaces admit time-like S^1-symmetry. In regard to this problem, we first recall the following (cf. Kamada [6]):

Proposition 3.1. *Let (M, J, g) be a (possibly non-compact, indefinite) Kähler surface and ξ a Killing vector field on (M, g). Then $\xi - \sqrt{-1}\, J\xi$ is a holomorphic vector field on $M = (M, J)$.*

This can be proved by using the fact that (M, J, g) is indefinite (resp. positive-definite) Kähler if and only if the holonomy group is contained in the indefinite unitary group $U(1, 1)$ (resp. the unitary group $U(2)$). Note that, when M is compact and g is positive-definite, Proposition 3.1 and the same statement in any dimension are well-known (cf. Goldberg [3]).

In what follows, we focus our attention on the indefinite case.

Concerning the existence of time-like S^1-symmetry, we can obtain the following result (cf. Kamada [5]):

Theorem 3.1. *Let (M, g) be a compact indefinite Kähler surface whose Kähler form Ω satisfies $c_1(M) \cdot [\Omega] = 0$, and let ξ be a time-like Killing vector field on*

(M, g) associated with a semi-free time-like S^1-symmetry. Suppose that the set Zero(ξ) contains at least two two-dimensional connected components. Then M should be biholomorphic to a Hirzebruch surface F_d.

Furthermore, we can also show a similar result under assumptions weaker than those in Theorem 3.1 (see Kamada [6]):

Theorem 3.2. *Let (M, g) be as in Theorem* 3.1, *and let ξ be a Killing vector field on (M, g) associated with an S^1-symmetry. Suppose that Zero$(\xi) \neq \emptyset$. Then M should be biholomorphic to F_d.*

Note that a result due to Taubes [11] regarding the Seiberg-Witten invariant for compact symplectic four-manifolds plays an essential role in the proofs of these theorems.

In Theorem 3.1, we can also prove that Zero$(\xi) = \mathbb{P}^1 \coprod \mathbb{P}^1$. However, under the assumptions in Theorem 3.2, we do not know whether Zero(ξ) consists of two holomorphic two-spheres. If ξ is further assumed to be time-like, then we can conclude Zero$(\xi) = \mathbb{P}^1 \coprod \mathbb{P}^1$, as in Theorems 3.3 and 3.4.

In the reminder of this section, we study time-like Killing vector fields with respect to an indefinite Kähler metric on a Hirzebruch surface F_d. First of all, we recall the standard realization of F_d as a complex submanifold in the product $\mathbb{P}^2 \times \mathbb{P}^1$ of complex projective plane $\mathbb{P}^2$ and complex projective line $\mathbb{P}^1$:

$$F_d = \left\{ (w_0 : w_1 : w_2, z_0 : z_1) \in \mathbb{P}^2 \times \mathbb{P}^1 \;\middle|\; z_0^d w_2 = z_1^d w_1 \right\}$$

$(d = 0, 1, 2, \dots)$. Then, F_d has the following numerical characters:

$$\kappa = -\infty, \quad b_1 = h^{1,0} = q = p_g = 0, \quad b_2^+ = b_2^- = 1, \quad c_1^2 = 2c_2 = 8,$$

where $h^{1,0}$, q, p_g, c_1^2 and c_2 are the complex dimension of the space of holomorphic one-forms, the irregularity, the geometric genus, the squared first Chern class and the second Chern class, respectively. Furthermore, via the projection F_d to the second factor $\mathbb{P}^1$, we can regard F_d as a $\mathbb{P}^1$-bundle $\mathbb{P}(\mathcal{O}(d) \oplus \mathcal{O}) \to \mathbb{P}^1$, where $\mathcal{O}(d)$ and $\mathcal{O}$ denote the holomorphic line bundle of degree d and the trivial line bundle over $\mathbb{P}^1$, respectively. Note that F_d is biholomorphic to $\mathbb{P}^1 \times \mathbb{P}^1$ when $d = 0$, and is the one-point blow-up $\mathbb{P}^2 \# \overline{\mathbb{P}^2}$ of $\mathbb{P}^2$ when $d = 1$. We can also regard F_d as a toric surface, by the following $T_2 = (\mathbb{C}^*)^2$-action:

$$(\lambda, \mu) \cdot (w_0 : w_1 : w_2, z_0 : z_1) := (w_0 : \mu w_1 : \lambda^d \mu w_2, z_0 : \lambda z_1).$$

Let τ_1 and τ_2 be the holomorphic vector fields on F_d given by $\tau_1 = \partial/\partial\lambda$ and $\tau_2 = \partial/\partial\mu$, respectively. In particular, τ_2 is induced from the standard $\mathbb{C}^*$-action along the fibers of $F_d = \mathbb{P}(\mathcal{O}(d) \oplus \mathcal{O}) \to \mathbb{P}^1$.

To describe holomorphic vector fields on F_d, we set $U_i = \{ (w_0 : w_1 : w_2) \in \mathbb{P}^2 \mid w_i \neq 0 \}$, $V_j = \{ (z_0 : z_1) \in \mathbb{P}^1 \mid z_j \neq 0 \}$ and $W_{ij} = (U_i \times V_j) \cap F_d$ ($i = 0, 1, 2$; $j = 0, 1$). For simplicity, we also set $W_0 := W_{00}$, $W_1 := W_{01}$, $W_2 := W_{21}$ and $W_3 := W_{10}$, or equivalently,

$$\begin{aligned}
W_0 &:= \{ (1 : y^0 : (x^0)^d y^0, 1 : x^0) \} \cong \mathbb{C}^2 = \{ (x^0, y^0) \}, \\
W_1 &:= \{ (1 : (x^1)^d y^1 : y^1, x^1 : 1) \} \cong \mathbb{C}^2 = \{ (x^1, y^1) \}, \\
W_2 &:= \{ (y^2 : (x^2)^d : 1, \quad x^2 : 1) \} \cong \mathbb{C}^2 = \{ (x^2, y^2) \}, \\
W_3 &:= \{ (y^3 : 1 : (x^3)^d, \quad 1 : x^3) \} \cong \mathbb{C}^2 = \{ (x^3, y^3) \}.
\end{aligned}$$

Then $F_d = \bigcup_{a=0}^{3} W_a$, and, on $W_0 \cap W_1$, the following relation holds:

$$x^1 = 1/x^0, \quad y^1 = (x^0)^d y^0.$$

Let V be a holomorphic vector field on F_d. Then it is well-known (cf. Kodaira [8]) that V is expressed in terms of $(x, y) := (x^0, y^0)$ as

$$V = \alpha(x)\partial_x + \beta(x, y)\partial_y,$$

where

$$\alpha(x) = a_0 + a_1 x + a_2 x^2, \quad \beta(x, y) = b_0 + b_1(x) y + b_2(x) y^2$$

for constants $a_0, a_1, a_2, b_0 \in \mathbb{C}$ and polynomials $b_1(x)$, $b_2(x)$ in x. If $d > 0$, then

$$b_0 = 0, \quad b_1(x) = -d \cdot a_2 x + b_1, \quad b_2(x) = \sum_{k=0}^{d} c_k x^k$$

for $b_1, c_0, \ldots, c_d \in \mathbb{C}$. Here, we can express the holomorphic vector fields τ_1 and τ_2 in terms of these coordinates (x, y) as $\tau_1 = x\partial_x$ and $\tau_2 = y\partial_y$, respectively. Since V has $3 + 1 + (d + 1) = d + 5$ parameters, the complex dimension $\dim_{\mathbb{C}} H^0(F_d; \mathcal{O}(T^{1,0}F_d))$ of the space $H^0(F_d; \mathcal{O}(T^{1,0}F_d))$ of all holomorphic vector fields on F_d equals $d + 5$.

If $d = 0$, then $b_1(x)$ and $b_2(x)$ are constant, so we set $b_1(x) = b_1, b_2(x) = b_2$. Then V has 6 parameters a_0, a_1, a_2, b_0, b_1, $b_2 \in \mathbb{C}$, which implies $\dim_{\mathbb{C}} H^0(F_0; \mathcal{O}(T^{1,0}F_0)) = 6$.

In the case of $d = 0$, the following result is shown in Kamada [5, 6]:

Theorem 3.3. *Let g be an indefinite Kähler metric on $F_0 \cong \mathbb{P}^1 \times \mathbb{P}^1$, and let ξ be a time-like Killing vector field on (F_0, g). Then, in terms of suitable holomorphic coordinates (x, y) of $F_0 = \mathbb{P}^1 \times \mathbb{P}^1$, ξ may be expressed as either $\xi = \mathrm{Re}\,(\sqrt{-1}\, ay\partial_y)$ or $\xi = \mathrm{Re}\,(\sqrt{-1}\, bx\partial_x)$ $(a, b \in \mathbb{R})$. In either case, Zero$(\xi) \cong \mathbb{P}^1 \coprod \mathbb{P}^1$.*

Note that Theorem 3.3 plays an important role to distinguish isometry classes of explicit scalar-flat indefinite Kähler metrics on $\mathbb{P}^1 \times \mathbb{P}^1$ constructed by an indefinite analogue of LeBrun's hyperbolic ansatz.

In the case of $d > 0$, we can also prove the following result.

Theorem 3.4. *Let g be any indefinite Kähler metric on F_d $(d > 0)$. Suppose that there exists a time-like Killing vector field ξ on (F_d, g). Then the corresponding holomorphic vector field $\xi - \sqrt{-1}\,J\xi$ is tangent to the fibers of $F_d = \mathbb{P}(\mathcal{O}(d) \oplus \mathcal{O}) \to \mathbb{P}^1$. Furthermore, if ξ also generates an S^1-symmetry on (F_d, g), then Zero$(\xi) = \Delta_0 \coprod \Delta_\infty (\cong \mathbb{P}^1 \coprod \mathbb{P}^1)$, where*

$$\Delta_0 = \{(0 : z_0^d : z_1^d, z_0 : z_1) \in F_d\}, \quad \Delta_\infty = \{(1 : 0 : 0, z_0 : z_1) \in F_d\}.$$

In particular, $\xi - \sqrt{-1}\,J\xi = \sqrt{-1}\,a\tau_2$ holds for some $a \in \mathbb{R}$.

This result can be proved by making use of an argument similar to that in the case of $d = 0$ (cf. Kamada [6]).

4. Remarks

Let M be a compact complex manifold and $H^0(M; \mathcal{O}(T^{1,0}M))$ denote the Lie algebra of all holomorphic vector fields on M. For a positive-definite Kähler class $[\Omega]$ on M, a Lie algebra character $F_{[\Omega]} : H^0(M; \mathcal{O}(T^{1,0}M)) \to \mathbb{C}$, called the Bando-Calabi-Futaki character, can be defined. This is an obstruction to the existence of Kähler metrics of constant scalar curvature in the given class $[\Omega]$. Namely, if there exists a Kähler metric of constant scalar curvature whose Kähler class coincides with $[\Omega]$, then $F_{[\Omega]} \equiv 0$. Futaki-Mabuchi[2] generalizes the Bando-Calabi-Futaki character to the indefinite case, under certain restriction. By using this generalization, we can prove that no scalar-flat indefinite Kähler metric exists on F_d, if $d \geq 1$. For details, see Kamada [6].

On the other hand, for the existence of positive-definite Kähler metrics of constant scalar curvature, another obstruction, called Matsushima-Lichnerowicz obstruction, is known. Namely, if a compact complex manifold M admits a positive-definite Kähler metric of constant scalar curvature, then $H^0(M; \mathcal{O}(T^{1,0}M))$ is reductive (cf. Goldberg [3]). The author does not know whether a similar result holds in the indefinite case.

It should be noted that we can find an (explicit) indefinite Kähler metric g on F_d whose Kähler form Ω satisfies $c_1(F_d) \cdot [\Omega] = 0$, when $d = 1$. Namely, $F_1(\cong \mathbb{P}^2 \# \overline{\mathbb{P}^2})$ admits an indefinite Kähler metric with vanishing total scalar curvature, although it admits no scalar-flat indefinite Kähler metric. Thus it is natural to ask whether such a metric exists on F_d when $d \geq 2$. At present, the author has no answer to this question.

References

1. W. Barth, C. Peters and A. Van de Ven, Compact complex surfaces, Springer-Verlag, Berlin, Heidelberg, New York, Tokyo, 1984.
2. A. Futaki and T. Mabuchi, *Moment maps and symmetric multilinear forms associated with symplectic classes*, Asian J. Math. **6**, 349–371 (2002).
3. S.I. Goldberg, Curvature and Homology, Dover Publications, Inc., New York, 1982.
4. M. Itoh, *Self-duality of Kähler surfaces*, Compositio Math. **51**, 265-273 (1984).
5. H. Kamada, *Self-dual Kähler metrics of neutral signature on complex surfaces*, Tohoku Math. Publ. **24**, Tohoku University (Thesis), Sendai, 2002.
6. ———, Compact scalar-flat indefinite Kähler surfaces with Hamiltonian S^1-symmetry, preprint (2002).
7. J. Kim, C. LeBrun and M. Pontecorvo, *Scalar-flat Kähler surfaces of all genera*, J. Reine Angew. Math. **486**, 69–95 (1997).
8. K. Kodaira, Complex manifolds and deformation of complex structures, Grundlehren Math. Wiss. **283**, Springer-Verlag, New York, 1986.
9. C. LeBrun, *Scalar-flat Kähler metrics on blown-up ruled surfaces*, J. Reine Angew. Math. **420**, 161–177 (1991).
10. J. Petean, *Indefinite Kähler-Einstein metrics on compact complex surfaces*, Commun. Math. Phys. **189**, 227–235 (1997).
11. C.H. Taubes, *The Seiberg-Witten invariants and symplectic forms*, Math. Res. Lett. **1**, 809–822 (1994).

THREE THEOREMS ON ISOTROPIC IMMERSIONS

SADAHIRO MAEDA

Department of Mathematics, Shimane University
Matsue, Shimane, 690-8504, Japan
E-mail: smaeda@math.shimane-u.ac.jp

In this paper we survey three theorems on isotropic immersions related to results in [3], [4].

1. Introduction

An isometric immersion $f : M \longrightarrow \widetilde{M}$ of a Riemannian manifold M into an ambient Riemannian manifold $\widetilde{M}$ (with metric $\langle\ ,\ \rangle$) is said to be *isotropic* at $x \in M$ if $\|\sigma(X,X)\|/\|X\|^2 (= \lambda(x))$ does not depend on the choice of $X (\neq 0) \in T_xM$. If the immersion is isotropic at every point, then the immersion is said to be *isotropic*. When the function $\lambda = \lambda(x)$ is constant on M, M is called a constant (λ-)isotropic submanifold.

We note that a totally umbilic immersion is isotropic, but not *vice versa*. Motivated by this fact, we first give a necessary and sufficient condition for an isotropic immersion to be totally umbilic in an arbitrary Riemannian manifold (Theorem 1).

We next consider isotropic submanifolds in a standard sphere $S^N(c)$ of curvature c. Every G-equivariant isometric immersion f of a compact rank one symmetric space $M(= G/K)$ into $S^N(c)$ is such an example. We investigate the second fundamental form of isotropic submanifolds with low codimension in $S^N(c)$. We show that these submanifolds have parallel second fundamental form (Theorem 2).

Finally we pay particular attention to constant isotropic submanifolds M in an arbitrary Riemannian manifold $\widetilde{M}$. We characterize these submanifolds M by investigate the extrinsic shape of circles of the submanifold M (Theorem 3). This result is a generalization of the following fact due to Nomizu and Yano [5]: Let M^n be a Riemannian submanifold of $\widetilde{M}^{n+p}$ through an isometric immersion f. Then M^n is an extrinsic sphere of $\widetilde{M}^{n+p}$ (that is, M^n is a totally umbilic submanifold with parallel mean curvature vector of $\widetilde{M}^{n+p}$) if and only if, for some positive constant κ and for every circle $\gamma = \gamma(s)$ of curvature κ on M^n, the curve $f \circ \gamma$ is a circle in $\widetilde{M}^{n+p}$.

2. Characterization of totally umbilic immersions

In this section we investigate the second fundamental form at one point. Let V and W be Euclidean vector spaces with inner products $\langle\ ,\ \rangle$, whose dimensions are n and k, respectively. We abstract the second fundamental form at one point to a symmetric bilinear form $\sigma : V \times V \to W$. We adopt for σ the usual notation and terminology of isometric immersions. Let $S^2(V)$ be the space of all symmetric endomorphisms of V. Then we define the linear map $A : W \to S^2(V)$ by $\langle A_\xi x, y\rangle = \langle\sigma(x,y),\xi\rangle$ for $x, y \in V$ and $\xi \in W$. We say that λ-isotropic if there exists a real constant λ such that $\|\sigma(x,x)\| = \lambda$ for every unit vector $x \in V$. We prepare the following without proof for the later use ([6]):

Lemma 1. *Let $\sigma : V \times V \longrightarrow W$ be a symmetric bilinear form. Then the following are equivalent*:

(1) *σ is λ isotropic.*

(2) $\langle\sigma(x,x),\sigma(x,y)\rangle = 0$ *for each orthogonal pair of vectors $x, y \in V$.*

(3) $\langle\sigma(x,y),\sigma(z,w)\rangle+\langle\sigma(x,z),\sigma(y,w)\rangle+\langle\sigma(x,w),\sigma(y,z)\rangle = \lambda^2(\langle x,y\rangle\langle z,w\rangle$ $+\langle x,z\rangle\langle y,w\rangle + \langle x,w\rangle\langle y,z\rangle)$ *for any vectors $x, y, z, w \in V$.*

We here provide a characterization of umbilic bilinear forms ([4]):

Proposition 1. *Let $\sigma : V \times V \longrightarrow W$ be a symmetric bilinear form. Then the following are equivalet*:

(i) *σ is umbilic.*

(ii) *σ is isotropic and for any $\xi, \eta \in W$ $A_\xi A_\eta = A_\eta A_\xi$.*

Proof. (i)$\Longrightarrow$(ii): By assumption we have $A_\xi = \langle\xi, \mathfrak{h}\rangle Id$ for any $\xi \in W$, where $\mathfrak{h}$ is the mean curvature vector. Hence A_ξ and A_η commute for any $\xi, \eta \in W$.

(ii)$\Longrightarrow$(i): Suppose that σ is λ-isotropic. We take a unit vector $x \in V$ and put $\xi = \sigma(x,x) \in W$. It follows from the statement (2) in Lemma 1 that x is an eigenvector of A_ξ with eigenvalue λ^2. We choose an orthonormal basis $\{e_1, \ldots, e_n\}$ of V satisfying that $A_\xi e_i = \lambda_i e_i$ $(i = 1, \ldots, n)$, where $e_1 = x$ and $\lambda_1 = \lambda^2$. We fix $i(\geqq 2)$. By the statement (3) in Lemma 1 we find that

$$\langle\sigma(x,e_i),\sigma(x,e_j)\rangle = (\lambda^2 - \lambda_i)/2 \cdot \delta_{ij} \quad \text{for any } j,$$

and hence $A_{\sigma(x,e_i)}x = (\lambda^2 - \lambda_i)/2 \cdot e_i$. Since $A_\xi A_{\sigma(x,e_i)}x = A_{\sigma(x,e_i)}A_\xi x$, we obtain

$$(\lambda^2 - \lambda_i)\lambda_i/2 = (\lambda^2 - \lambda_i)\lambda^2/2$$

so that $\lambda_i = \lambda^2$ for $i \geqq 2$, that is, V is the eigenspace of A_ξ with eigenvalue λ^2. Therefore we find that σ is umbilic. □

As an immediate consequence of this proposition we have the following:

Theorem 1. *Let M be a Riemannian submanifold in a Riemannian manifold $\widetilde{M}$. Then the following are equivalent:*

(i) *M is totally umbilic in $\widetilde{M}$.*

(ii) *M is isotropic in $\widetilde{M}$ and for any normal vectors ξ, η on M $A_\xi A_\eta = A_\eta A_\xi$ holds.*

Let the ambient space $\widetilde{M}$ be a real space form $\widetilde{M}(c)$ of curvature c, which is either one of a standard sphere, a Euclidean space and a hyperbolic space. Then it follows from Theorem 1 that

Corollary 1. *Let M be a Riemannian submanifold in a real space form $\widetilde{M}(c)$. Then the following are equivalent:*

(i) *M is totally umbilic in $\widetilde{M}(c)$.*

(ii) *M is an isotropic submanifold with flat normal connection in $\widetilde{M}(c)$.*

3. Isotropic immersions with low codimensions

We shall show that an isotropic submanifold with low codimension in a real space form has parallel second fundamental form. For this purpose we prepare the following algebraic lemma without proof ([2]):

Lemma 2. *Let V and W be Euclidean vector spaces with inner products $\langle\ ,\ \rangle$, where $\dim V = n(\geqq 3)$. Let $\sigma : V \times V \to W$ be a λ-isotropic symmetric bilinear form. Suppose that $\dim W \leqq (n+2)/2$. Then σ satisfies one of the following:*

(i) *σ is umbilic, that is, there exists $\xi \in W$ satisfying $\sigma(x, y) = \langle x, y\rangle\xi$.*

(ii) *$n = 4, 8, 16$ and $\dim W = (n+2)/2$.*

In addition, in case of (ii), *the following statements* (a) *and* (b) *hold:*

(a) *For each unit vector $x \in V$ there exists an orthonormal basis $\{e_1, \ldots, e_n\}$ for V satisfying $\sigma(e_i, e_i) = \sigma(x, x)$ for $1 \leqq i \leqq n/2$ and $\sigma(e_j, e_j) = -\sigma(x, x)$ for $(n+2)/2 \leqq j \leqq n$.*

(b) *For each unit vector $\xi \in W$ there exists an orthonormal basis $\{e_1, \ldots, e_n\}$ for V satisfying $A_\xi e_i = \lambda e_i$ for $1 \leqq i \leqq n/2$ and $A_\xi e_j = -\lambda e_j$ for $(n+2)/2 \leqq j \leqq n$.*

Our aime here is to prove the following ([4]):

Theorem 2. *Let f be an isotropic immersion of an $n(\geqq 3)$-dimensional connected Riemannian manifold M^n into an $(n+p)$-dimensional real space form $\widetilde{M}^{n+p}(c)$ of curvature c. Suppose that $p \leqq (n+2)/2$. Then M has parallel second fundamental form, so that either f is totally umbilic or f is locally congruent to one of the first minial standard minimal immersion of M^n into $\widetilde{M}^{n+p}(c)$:*

(1) $M^n = \mathbb{C}P^2(4c/3)$, $\widetilde{M}^{n+p}(c) = S^7(c)$,

(2) $M^n = \mathbb{H}P^2(4c/3)$, $\widetilde{M}^{n+p}(c) = S^{13}(c)$,

(3) $M^n = \mathbb{C}ayP^2(4c/3)$, $\widetilde{M}^{n+p}(c) = S^{25}(c)$,

where $S^m(c)$ denotes an m-dimensional sphere of constant sectional curvature c; $\mathbb{C}P^2(4c/3)$, $\mathbb{H}P^2(4c/3)$ and $\mathbb{C}ayP^2(4c/3)$ denote a projective planes of maximal sectional curvature $4c/3$ over the complex, quaternions and Cayley numbers, respectively.

Proof. First we note that our manifold is totally umbilic in $\widetilde{M}^{n+p}(c)$ in the case of $p < (n+2)/2$ (for details, see pp. 130–131 in [2]). In the following, we consider the case that $p = (n+2)/2$ and f is not totally umbilic. Let $U = \{x \in M \mid x \text{ is an umbilic point of the isometric immersion } f\}$ and $U^c = M - U$. Note that $U^c \neq \phi$. We shall show that $U^c = M$ and the second fundamental form of f is parallel. We denote by R and S the curvature tensor and the Ricci tensor of M, respectively. Then it follows from the following Gauss equation

$$\begin{aligned}\langle\sigma(X,Y),\sigma(Z,W)\rangle &- \langle\sigma(Z,Y),\sigma(X,W)\rangle \\ &= \langle R(Z,X)Y,W\rangle - c(\langle X,Y\rangle\langle Z,W\rangle - \langle Z,Y\rangle\langle X,W\rangle)\end{aligned} \tag{3.1}$$

that $S(X,Y) = (n-1)(c+\lambda^2)\langle X,Y\rangle$ on U. On the other hand, by virtue of (3.1), the statement (3) in Lemma 1 and the fact that the immersion f is minimal on the open set U^c (see the statement (ii) in Lemma 2) we know that $S(X,Y) = \{(n-1)c - (n+2)/2 \cdot \lambda^2\}\langle X,Y\rangle$ on U^c. These, together with the hypothesis $\dim M \geqq 3$, imply that M is Einstein. Hence $M = U^c$. In addition, λ is constant so that

$$\langle(\bar{\nabla}_Y\sigma)(X,X),\sigma(X,X)\rangle = 0 \quad \text{for any } X, Y. \tag{3.2}$$

Here the covariant differentiation $\bar{\nabla}$ of the second fundamental form σ with respect to the connection in (tangent bundle)+(normal bundle) is defined as follows:

$$(\bar{\nabla}_X\sigma)(Y,Z) = D_X(\sigma(Y,Z)) - \sigma(\nabla_X Y, Z) - \sigma(Y, \nabla_X Z).$$

Similarly, from the statement (3) in Lemma 1 we have

$$\langle(\bar{\nabla}_X\sigma)(X,X),\sigma(X,Y)\rangle + \langle\sigma(X,X),(\bar{\nabla}_X\sigma)(X,Y)\rangle = 0 \quad \text{for any } X, Y. \tag{3.3}$$

It follows from (3.2), (3.3) and the Codazzi equation $(\bar{\nabla}_X\sigma)(Y,Z) = (\bar{\nabla}_Y\sigma)(X,Z)$ that $\langle(\bar{\nabla}_X\sigma)(X,X),\sigma(X,Y)\rangle = 0$ for all X, Y. We here recall the fact that the first normal space $N_1 := \{\sigma(X,Y) \mid X, Y \in TM\}_{\mathbb{R}}$ of f spans the normal space of M at its each point (see page 131 in [2]). Hence, $(\bar{\nabla}_X\sigma)(X,X) = 0$ for any X, so that $\bar{\nabla}\sigma = 0$. Therefore, by virtue of the works [1, 7] we can get the conclusion. □

4. Characterization of constant isotropic immersions

A smooth curve $\gamma = \gamma(s)$ parametrized by its arclength s in a Riemannian manifold M is called a *circle* if there exist a field $Y = Y_s$ of unit vectors along γ and a constant $\kappa(\geqq 0)$ satisfying

$$\begin{cases} \nabla_{\dot\gamma}\dot\gamma = \kappa Y \\ \nabla_{\dot\gamma}Y = -\kappa\dot\gamma, \end{cases} \tag{4.1}$$

where $\dot\gamma$ denotes the unit tangent vector of γ and $\nabla_{\dot\gamma}$ the covariant differentiation along γ with respect to the Riemannian connection ∇ of M. The constant κ is called the *curvature* of the circle. A circle of null curvature is nothing but a geodesic. For each point $x \in M$, each orthonormal pair (u, v) of vectors at x and each positive constant κ, there exists locally a unique circle $\gamma = \gamma(s)$ on M with initial condition that $\gamma(0) = x$, $\dot\gamma(0) = u$ and $\nabla_{\dot\gamma}\dot\gamma(0) = \kappa v$. The purpose of this section is to prove the following ([3]):

Theorem 3. *Let M^n be an n-dimensional connected Riemannian submanifold of an $(n+p)$-dimensional Riemannian manifold $\widetilde{M}^{n+p}$ through an isometric immersion f. Then the following are equivalent:*

(i) *M^n is a constant (λ-) isotropic submanifold of $\widetilde{M}^{n+p}$.*

(ii) *There exists $\kappa > 0$ satisfying that for each circle γ of curvature κ on the submanifold M^n the curve $f \circ \gamma$ in $\widetilde{M}^{n+p}$ has constant first curvature κ_1 along this curve.*

Proof. (i) ⇒ (ii): Let $f : M^n \to \widetilde{M}^{n+p}$ be a constant λ-isotropic immersion. In the following, for simplicity we also denote $f \circ \gamma$ by γ. It follows from equation (4.1) and the formula of Gauss $\widetilde{\nabla}_X Z = \nabla_X Z + \sigma(X,Z)$ that

$$\widetilde{\nabla}_{\dot\gamma(s)}\dot\gamma(s) = \kappa Y_s + \sigma(\dot\gamma(s),\dot\gamma(s)),\ s \in I. \tag{4.2}$$

Here $\widetilde{\nabla}$ is the Riemannian connection of the ambient space $\widetilde{M}^{n+p}$ and I is some open interval on $\mathbb{R}$. Then from (4.2) we can see that the first curvature $\kappa_1 = \|\widetilde{\nabla}_{\dot{\gamma}}\dot{\gamma}\|$ of the curve $f \circ \gamma$ is equal to $\sqrt{\kappa^2 + \lambda^2}$, which is constant on I.

(ii) $\Rightarrow$ (i): Let $f : M^n \to \widetilde{M}^{n+p}$ be an isometric immersion satisfying the condition (ii). We take a point $x \in M$ and choose an arbitrary orthonormal pair of vectors $u, v \in T_xM$. Let $\gamma = \gamma(s), s \in I$ be a circle of curvature κ on the submanifold M^n with initial condition that $\gamma(0) = x$, $\dot{\gamma}(0) = u$ and $\nabla_{\dot{\gamma}}\dot{\gamma}(0) = \kappa v$. By condition (ii) the first curvature $\kappa_1 = \|\widetilde{\nabla}_{\dot{\gamma}}\dot{\gamma}\|$ of the curve $f \circ \gamma$ is constant, so that equation (4.2) implies $\|\sigma(\dot{\gamma}, \dot{\gamma})\|$ is constant on I. Hence, denoting by D the connection of the normal bundle of M^n in $\widetilde{M}^{n+p}$, from (4.1) we obtain

$$\begin{aligned} 0 = \frac{d}{ds}\|\sigma(\dot{\gamma}, \dot{\gamma})\|^2 &= 2\langle D_{\dot{\gamma}}(\sigma(\dot{\gamma}, \dot{\gamma})), \sigma(\dot{\gamma}, \dot{\gamma})\rangle \\ &= 2\langle(\bar{\nabla}_{\dot{\gamma}}\sigma)(\dot{\gamma}, \dot{\gamma}) + 2\sigma(\nabla_{\dot{\gamma}}\dot{\gamma}, \dot{\gamma}), \sigma(\dot{\gamma}, \dot{\gamma})\rangle \\ &= 2\langle(\bar{\nabla}_{\dot{\gamma}}\sigma)(\dot{\gamma}, \dot{\gamma}), \sigma(\dot{\gamma}, \dot{\gamma})\rangle + 4\kappa\langle\sigma(\dot{\gamma}, \dot{\gamma}), \sigma(\dot{\gamma}, Y)\rangle. \end{aligned} \tag{4.3}$$

Evaluating equation (4.3) at $s = 0$, we get

$$\langle(\bar{\nabla}_u\sigma)(u, u), \sigma(u, u)\rangle + 2\kappa\langle\sigma(u, u), \sigma(u, v)\rangle = 0. \tag{4.4}$$

On the other hand, for another circle $\rho = \rho(s)$ of the same curvature κ on the submanifold M^n with initial condition that $\rho(0) = x$, $\dot{\rho}(0) = u$ and $\nabla_{\dot{\rho}}\dot{\rho}(0) = -\kappa v$, we have

$$\langle(\bar{\nabla}_u\sigma)(u, u), \sigma(u, u)\rangle - 2\kappa\langle\sigma(u, u), \sigma(u, v)\rangle = 0 \tag{4.5}$$

which corresponds to equation (4.4). Thus, from (4.4) and (4.5) we can see that $\langle\sigma(u, u), \sigma(u, v)\rangle = 0$ for any orthonormal pair of vectors u, v at each point x of M, so that the submanifold M^n is (λ-)isotropic in M^{n+p} through the isometric immersion f by Lemma 1.

Next, we shall show that $\lambda : M \to \mathbb{R}$ is constant. It follows from (4.4) and (4.5) that

$$\langle(\bar{\nabla}_u\sigma)(u, u), \sigma(u, u)\rangle = 0 \quad \text{for every unit vecor } u \text{ at each point } x \text{ of } M^n.$$

Then, for every geodesic $\tau = \tau(s)$ on the submanifold M^n we see that $\lambda = \lambda(s)$ is constant along τ. Therefore we can conclude that λ is constant on M. □

References

1. D. Ferus, *Immersions with parallel second fundamental form*, Math. Z. **140**(1974), 87–92.
2. S. Maeda, *Isotropic immersions with low codimensions*, Bull. Nagoya Inst. Tech. **46**(1994), 127–138.

3. ———, *A characterization of constant isotropic immersions by circles*, to appear in Archiv der Math.
4. S. Maeda and K. Tsukada, *Isotropic immersions into a real space form*, Canadian Math. Bull. **37**(1994), 245–253.
5. K. Nomizu and K. Yano, *On circles and spheres in Riemannian geometry*, Math. Ann. **219**(1974), 163–170.
6. B. O'Neill, *Isotropic and Kaehler immersions*, Canadian J. Math. (1965), 905–915.
7. M. Takeuchi, *Parallel submanifolds of space forms*, Manifolds and Lie groups in honor of Y. Matsushima, Birkhäuser, Boston. (1981), 429–447.

CIRCLES AND KÄHLER IMMERSIONS

SADAHIRO MAEDA

Department of Mathematics, Shimane University
Matsue, Shimane, 690-8504, Japan
E-mail: smaeda@math.shimane-u.ac.jp

TOSHIAKI ADACHI

Department of Mathematics, Nagoya Institute of Technology
Nagoya 466-8555, Japan
E-mail: adachi@math.kyy.nitech.ac.jp

In this paper we study Kähler immersions by observing the extrinsic shape of circles on Kähler submanifolds.

1. Introduction

As a crossing of complex analysis and differential geometry the theory of Kähler immersions is one of the most interesting objects in geometry of Riemannian submanifolds (cf. [12]). In this paper we study Kähler immersions from the viewpoint of circles on Kähler submanifolds.

A smooth curve $\gamma : \mathbb{R} \longrightarrow M$ parametrized by its arclength s in a complete Riemannian manifold M is called a *circle* of curvature $\kappa(\geqq 0)$ if it satisfies the following equation:

$$\nabla_{\dot{\gamma}}\nabla_{\dot{\gamma}}\dot{\gamma} = -\kappa^2\dot{\gamma}(s),$$

where κ is constant and $\nabla_{\dot{\gamma}}$ denotes the covariant differentiation along γ with respect to the Riemannian connection ∇ of M. Since $\|\nabla_{\dot{\gamma}}\dot{\gamma}\| = \kappa$, this equation is equivalent to the equation of geodesics when $\kappa = 0$. So we treat a geodesic as a circle of null curvature.

This paper, which is an expository paper related to our recent results ([14], [15], [5], [6]), consists of four parts. The first is to characterize totally geodesic Kähler immersions into an arbitrary Kähler manifold (Theorem 1). The second is to characterize all Veronese imbeddings of complex projective spaces into complex projective spaces which are typical examples of non-totally geodesic Kähler immersions (Theorem 2). The third is to characterize parallel complex hypersurfaces in a Kähler manifold of constant holomorphic sectional curvature (Theorems 3 and 4). The last is to characterize the Veronese imbedding of degree 2 (Theorem 5).

2. Isotropic immersions

An isometric immersion $f : M \to \widetilde{M}$ is said to be *isotropic* at $x \in M$ if $\|\sigma(u,u)\|/\|u\|^2$ does not depend on the choice of $u(\neq 0) \in T_xM$. In this case we denote this value by $\lambda(x)$. If the immersion is isotropic at every point, then the immersion is called isotropic. When the function $\lambda = \lambda(x)$ is constant on M, we call M a constant (λ-)isotropic submanifold. Note that a totally umbilic immersion is isotropic, but not *vice versa*. The following is well known ([11]).

Lemma A. *An isometric immersion f of M into $(\widetilde{M}, \langle\ ,\ \rangle)$ is isotropic at $x \in M$ if and only if the second fundamental form σ of f satisfies $\langle \sigma(u,u), \sigma(u,v)\rangle = 0$ for an arbitrary orthogonal pair u, $v \in T_xM$.*

3. Veronese imbeddings

We recall a Kähler isometric full immersion of a complex projective space $\mathbb{C}P^n(c)$ of constant holomorphic sectional curvature c into a complex projective space $\mathbb{C}P^N(\tilde{c})$ of constant holomorphic sectional curvature $\tilde{c}$. By virtue of the classification theorem ([4], [9]) this Kähler immersion is nothing but a Kähler imbedding $f_k : \mathbb{C}P^n(c/k) \to \mathbb{C}P^N(c)$ given by

$$[z_i]_{0 \leqq i \leqq n} \mapsto \left[\sqrt{\frac{k!}{k_0! \cdots k_n!}} z_0^{k_0} \cdots z_n^{k_n}\right]_{k_0 + \cdots + k_n = k},$$

where $[*]$ means the point of the projective space with the homogeneous coordinates $*$ and $N = (n+k)!/(n!\,k!) - 1$. We usually call f_k the k-th *Veronese imbedding*. The imbedding f_k has various geometric properties. For example, for each geodesic γ on $\mathbb{C}P^n(c/k)$ the curve $f_k \circ \gamma$ fully lies on a totally real totally geodesic $\mathbb{R}P^k(c/4)$ in $\mathbb{C}P^N(c)$ (cf. [13]). We here recall the work of [4], [9] for the later use. We denote by $M_N(c)$ a complex space form, which is locally complex analytically isometric to a complex projective space $\mathbb{C}P^N(c)$ when $c > 0$, a complex Euclidean space $\mathbb{C}^N$ when $c = 0$ or a complex hyperbolic space $\mathbb{C}H^N(c)$ when $c < 0$.

Theorem B ([4], [9]). *Let $f : M_n(c) \to M_N(\tilde{c})$ be a Kähler isometric immersion of a complex space form of constant holomorphic sectional curvature c into* *another complex space form of constant holomorphic sectional curvature* $\tilde{c}$. *Then the following hold*:

1) *If* $\tilde{c} > 0$ *and* f *is full, then* $\tilde{c} = kc$ *and* $N = (n+k)!/(n!\,k!) - 1$ *for some positive integer* k.

2) *If* $\tilde{c} \leqq 0$, *then* f *is totally geodesic.*

4. Frenet curves of order 2

A smooth curve $\gamma = \gamma(s)$ parametrized by its arclength s on a Riemannian manifold $(M, \langle\ ,\ \rangle)$ with Riemannian metric $\langle\ ,\ \rangle$ is called a *Frenet curve of proper order* 2 if there exist a field $Y = Y_s$ of unit vectors along γ and a positive smooth function $\kappa = \kappa(s)$ satisfying

$$\begin{cases} \nabla_{\dot{\gamma}}\dot{\gamma} = \kappa Y \\ \nabla_{\dot{\gamma}} Y = -\kappa\dot{\gamma}, \end{cases} \tag{4.1}$$

where $\dot{\gamma}$ denotes the unit tangent vector of γ and $\nabla_{\dot{\gamma}}$ the covariant differentiation along γ with respect to the Riemannian connection ∇ of M. The function κ is called the *curvature* of the curve γ. A smooth curve is said to be a *Frenet curve of order* 2 if it is either a geodesic or a Frenet curve of proper order 2. Needless to say a circle is a Frenet curve of order 2 in a trivial sense. We remark that equation (4.1) is equivalent to the equation in Introduction when κ is constant. For each point $x \in M$, each orthonormal pair (u, v) of unit tangent vectors at x and a positive smooth function κ, there exists locally a unique Frenet curve $\gamma = \gamma(s)$ of proper order 2 on M with initial condition that $\gamma(0) = x$, $\dot{\gamma}(0) = u$ and $\nabla_{\dot{\gamma}}\dot{\gamma}(0) = \kappa v$. For a Frenet curve γ of proper order 2 on a Kähler manifold $(M, \langle\ ,\ \rangle, J)$ with complex structure J the function τ_γ defined by $\tau_\gamma = \langle \dot{\gamma}, JY_s\rangle$ is constant along γ. In fact,

$$\begin{aligned} \nabla_{\dot{\gamma}}\langle \dot{\gamma}, JY_s\rangle &= \langle \nabla_{\dot{\gamma}}\dot{\gamma}, JY_s\rangle + \langle \dot{\gamma}, J\nabla_{\dot{\gamma}}Y_s\rangle \\ &= \kappa(s)\langle Y_s, JY_s\rangle - \kappa(s)\langle \dot{\gamma}, J\dot{\gamma}\rangle = 0. \end{aligned}$$

We call this constant its *complex torsion*. We call γ *Kähler* when $\tau_\gamma = \pm 1$. In other words, we call γ Kähler if $\dot{\gamma}$ and Y span a holomorphic plane (i.e. either $Y = J\dot{\gamma}$ or $Y = -J\dot{\gamma}$). When a circle γ of curvature κ is Kähler, the equation in Introduction reduces to

$$\nabla_{\dot{\gamma}}\dot{\gamma} = \kappa J\dot{\gamma} \quad \text{or} \quad \nabla_{\dot{\gamma}}\dot{\gamma} = -\kappa J\dot{\gamma}. \tag{4.2}$$

It is known that every Kähler circle on a complex projective space lies on a complex projective line and every Kähler circle on a complex hyperbolic space lies on a complex hyperbolic line (see [1] also [2]).

5. Characterization of totally geodesic Kähler immersions

In this section we characterize totally geodesic Kähler immersions by the extrinsic shape of Kähler circles.

Theorem 1. *Let $f : M_n \to \widetilde{M}_N$ be a Kähler isometric immersion of an n-dimensional Kähler manifold M into an N-dimensional Kähler manifold $\widetilde{M}_N$. Then the following conditions are equivalent:*

(1) *f is a totally geodesic immersion.*

(2) *There exists $\kappa > 0$ satisfying that f maps every Kähler circle of curvature κ on M_n to a circle in $\widetilde{M}_N$.*

(3) *There exists $\kappa > 0$ satisfying that f maps every Kähler circle of curvature κ on M_n to a Frenet curve of order 2 in $\widetilde{M}_N$.*

Proof. It suffices to show that the third condition implies the first condition. Let γ be a Kähler circle of curvature κ on M satisfying one of the equations (4.2). By the assumption it is mapped to a Frenet curve $f \circ \gamma$ of proper order 2 in the ambient manifold $\widetilde{M}$, which satisfies the equations

$$\begin{cases} \widetilde{\nabla}_{\dot\gamma}\dot\gamma = \widetilde{\kappa}Y, \\ \widetilde{\nabla}_{\dot\gamma}Y = -\widetilde{\kappa}\,\dot\gamma, \end{cases}$$

where $\widetilde{\nabla}$ denotes the covariant differentiation on $\widetilde{M}$ and $\widetilde{\kappa} = \|\widetilde{\nabla}_{\dot\gamma}\dot\gamma\|$. Here for simplicity we usually denote the curve $f \circ \gamma$ by γ. These guarantee that

$$\widetilde{\kappa}\,\widetilde{\nabla}_{\dot\gamma}\widetilde{\nabla}_{\dot\gamma}\dot\gamma = \widetilde{\kappa}\,\widetilde{\nabla}_{\dot\gamma}(\widetilde{\kappa}Y) = \widetilde{\kappa}'\widetilde{\nabla}_{\dot\gamma}\dot\gamma - \widetilde{\kappa}^3\dot\gamma,$$

where $\widetilde{\kappa}' = \frac{d}{ds}\widetilde{\kappa}$ and s is the arclength parameter of γ. Hence we obtain

$$\widetilde{\kappa}\,(\widetilde{\nabla}_{\dot\gamma}\widetilde{\nabla}_{\dot\gamma}\dot\gamma + \widetilde{\kappa}^2\dot\gamma) = \widetilde{\kappa}'\widetilde{\nabla}_{\dot\gamma}\dot\gamma. \tag{5.1}$$

We calculate the covariant differentiation $\widetilde{\nabla}$ by use of the formulae of Gauss and Weingarten which assure

$$\widetilde{\nabla}_X Z = \nabla_X Z + \sigma(X, Z) \quad \text{and} \quad \widetilde{\nabla}_X\xi = D_X\xi - A_\xi X. \tag{5.2}$$

Here, we define the covariant differentiation $\bar{\nabla}$ of the second fundamental form σ with respect to the connection in (tangent bundle) $\oplus$ (normal bundle) as follows:

$$(\bar{\nabla}_X\sigma)(Y, Z) = D_X(\sigma(Y, Z)) - \sigma(\nabla_X Y, Z) - \sigma(Y, \nabla_X Z). \tag{5.3}$$

It follows from (4.2), (5.2) and (5.3) that

$$\begin{aligned} \widetilde{\nabla}_{\dot\gamma}\widetilde{\nabla}_{\dot\gamma}\dot\gamma &= \nabla_{\dot\gamma}\nabla_{\dot\gamma}\dot\gamma + \sigma(\nabla_{\dot\gamma}\dot\gamma, \dot\gamma) - A_{\sigma(\dot\gamma,\dot\gamma)}\dot\gamma + D_{\dot\gamma}(\sigma(\dot\gamma, \dot\gamma)) \\ &= -\kappa^2\dot\gamma \pm 3\kappa J(\sigma(\dot\gamma, \dot\gamma)) - A_{\sigma(\dot\gamma,\dot\gamma)}\dot\gamma + (\bar{\nabla}_{\dot\gamma}\sigma)(\dot\gamma, \dot\gamma) \end{aligned}$$

and

$$\widetilde{\kappa}^2 = \langle \widetilde{\nabla}_{\dot\gamma}\dot\gamma, \widetilde{\nabla}_{\dot\gamma}\dot\gamma \rangle = \kappa^2 + \|\sigma(\dot\gamma,\dot\gamma)\|^2.$$

We obtain from these equalities and (5.1)

$$\begin{aligned}\widetilde{\kappa}\,\{\|\sigma(\dot\gamma,\dot\gamma)\|^2\dot\gamma \pm 3\kappa J(\sigma(\dot\gamma,\dot\gamma)) - A_{\sigma(\dot\gamma,\dot\gamma)}\dot\gamma + (\bar\nabla_{\dot\gamma}\sigma)(\dot\gamma,\dot\gamma)\}\\ = \widetilde{\kappa}'\{\pm\kappa J\dot\gamma + \sigma(\dot\gamma,\dot\gamma)\}.\end{aligned}$$

Taking the tangential and the normal components for the submanifold M of this equality, we obtain the following:

$$\widetilde{\kappa}(-A_{\sigma(\dot\gamma,\dot\gamma)}\dot\gamma + \|\sigma(\dot\gamma,\dot\gamma)\|^2\dot\gamma) = \pm\widetilde{\kappa}'\kappa J\dot\gamma, \tag{5.4}$$

$$\widetilde{\kappa}\,\{\pm 3\kappa J(\sigma(\dot\gamma,\dot\gamma)) + (\bar\nabla_{\dot\gamma}\sigma)(\dot\gamma,\dot\gamma)\} = \widetilde{\kappa}'\sigma(\dot\gamma,\dot\gamma). \tag{5.5}$$

Take the inner product of both hand-sides of (5.4) with $J\dot\gamma$. We find

$$\begin{aligned}\pm\widetilde{\kappa}'\kappa &= -\widetilde{\kappa}\langle A_{\sigma(\dot\gamma,\dot\gamma)}\dot\gamma, J\dot\gamma\rangle = -\widetilde{\kappa}\langle\sigma(\dot\gamma,\dot\gamma), \sigma(\dot\gamma, J\dot\gamma)\rangle\\ &= -\widetilde{\kappa}\langle\sigma(\dot\gamma,\dot\gamma), J(\sigma(\dot\gamma,\dot\gamma))\rangle = 0.\end{aligned}$$

This implies that the curvature function $\widetilde{\kappa}$ of the curve $f\circ\gamma$ is constant, so that this curve is a circle of positive curvature in the ambient manifold $\widetilde{M}$. This shows the third condition implies the second condition.

For an arbitrary unit tangent vector $v \in TM$ we denote by γ_i $(i = 1, 2)$ Kähler circles of curvature κ with $\dot\gamma_i(0) = v$ and satisfying $\nabla_{\dot\gamma_1}\dot\gamma_1 = \kappa J\dot\gamma_1$, $\nabla_{\dot\gamma_2}\dot\gamma_2 = -\kappa J\dot\gamma_2$, respectively. Following (5.5) we have

$$\begin{aligned}3\kappa J(\sigma(\dot\gamma_1,\dot\gamma_1)) + (\bar\nabla_{\dot\gamma_1}\sigma)(\dot\gamma_1,\dot\gamma_1) = 0,\\ -3\kappa J(\sigma(\dot\gamma_2,\dot\gamma_2)) + (\bar\nabla_{\dot\gamma_2}\sigma)(\dot\gamma_2,\dot\gamma_2) = 0.\end{aligned}$$

Evaluating these at $s = 0$, we obtain

$$3\kappa J(\sigma(v,v)) + (\bar\nabla_v\sigma)(v,v) = 0 = -3\kappa J(\sigma(v,v)) + (\bar\nabla_v\sigma)(v,v),$$

which lead us to $\sigma(v,v) = 0$. As v is an arbitrary unit vector we find that M is totally geodesic in $\widetilde{M}$. □

A curve is said to be a *plane curve* on a Riemannian manifold M, if it is locally contained on some real 2-dimensional totally geodesic submanifold of M. As an immediate consequence of Theorem 1, we obtain the following characterization of totally geodesic Kähler submanifolds in a complete simply connected complex space form $\widetilde{M}_N(c)$.

Corollary. *Let f be a Kähler isometric immersion of a Kähler manifold M_n into a complete simply connected complex space form $\widetilde{M}_N(c)$. Then the following conditions are equivalent:*

(1) *f is a totally geodesic immersion.*

(2) *There exists $\kappa > 0$ satisfying that f maps every Kähler circle of curvature κ on M_n to a circle in $\widetilde{M}_N(c)$.*

(3) *There exists $\kappa > 0$ satisfying that f maps every Kähler circle of curvature κ on M_n to a Frenet curve of order 2 in $\widetilde{M}_N(c)$.*

(4) *There exists $\kappa > 0$ satisfying that f maps every Kähler circle of curvature κ on M_n to a plane curve in $\widetilde{M}_N(c)$.*

Remark 1.

1) In this Corollary every plane curve $f \circ \gamma$ in the fourth condition lies on a complex line $\widetilde{M}_1(c)$ of the ambient space $\widetilde{M}_N(c)$.

2) If we put $k = 0$ in the statements of Theorem 1 and Corollary, these results are no longer true. For example, we consider the second Veronese imbedding $f_2 : \mathbb{C}P^n(c/2) \longrightarrow \mathbb{C}P^{(n^2+3n)/2}(c)$. This non-totally geodesic Kähler isometric imbedding f_2 maps every geodesic on $\mathbb{C}P^n(c/2)$ to a circle of curvature $\sqrt{c}/2$ in a real projective plane $\mathbb{R}P^2(c/4)$.

6. Characterization of Veronese imbeddings

In Theorem 1 it suffices to study *some* circles on Kähler submanifolds. However, in this section we need to observe *all* circles on Kähler submanifolds. We have the following ([15], [5]).

Theorem 2. *Let $f : M_n \to \widetilde{M}_N(c)$ be a Kähler isometric full immersion of an n-dimensional Kähler manifold M_n into an N-dimensional complex space form $\widetilde{M}_N(c)$ of constant holomorphic sectional curvature $c > 0$. Then the following conditions are equivalent:*

(1) *For some positive integer k, the submanifold M_n is locally congruent to $\mathbb{C}P^n(c/k)$, $N = (n+k)!/(n!\,k!) - 1$ and f is locally equivalent to the k-th Veronese imbedding f_k.*

(2) *There exists $\kappa > 0$ satisfying that for each circle γ of curvature κ on the submanifold M_n the curve $f \circ \gamma$ in $\widetilde{M}_N(c)$ has constant first curvature κ_1 along this curve.*

Proof. (1) $\Rightarrow$ (2): For each Veronese imbedding $f_k : \mathbb{C}P^n(c/k) \to \mathbb{C}P^N(c)$ we see that $\|\sigma(u,u)\|^2 = c(k-1)/2k$ for any unit tangent vector u at each point $x \in \mathbb{C}P^n(c/k)$ (see [12]). Then we find that for each circle γ of curvature κ on $\mathbb{C}P^n(c/k)$ the curve $f_k \circ \gamma$ has constant first curvature $\kappa_1 = \sqrt{\kappa^2 + c(k-1)/2k}$ in the ambient manifold $\mathbb{C}P^N(c)$.

(2) ⇒ (1): Let $f : M_n \to \widetilde{M}_N(c)$ be a Kähler isometric full immersion satisfying the second condition. We take an arbitrary orthonormal pair of vectors $u, v \in T_xM$ at an arbitrary point $x \in M$. Let $\gamma_i = \gamma_i(s)$, $s \in I$ $(i = 1, 2)$ be a circle of curvature κ on the submanifold M with initial condition that $\gamma_i(0) = x$, $\dot\gamma_i(0) = u$ and $\nabla_{\dot\gamma_1}\dot\gamma_1(0) = \kappa v$, $\nabla_{\dot\gamma_2}\dot\gamma_2(0) = -\kappa v$. By the second condition the first curvature $\|\widetilde\nabla_{\dot\gamma_i}\dot\gamma_i\|$ of the curve $f \circ \gamma_i$ is constant, so that Gauss formula (5.2) implies $\|\sigma(\dot\gamma_i, \dot\gamma_i)\|$ is constant on I. It follows from (5.3) that

$$0 = \frac{d}{ds}\|\sigma(\dot\gamma_i, \dot\gamma_i)\|^2 = 2\langle D_{\dot\gamma_i}(\sigma(\dot\gamma_i, \dot\gamma_i)), \sigma(\dot\gamma_i, \dot\gamma_i)\rangle$$
$$= 2\langle(\bar\nabla_{\dot\gamma_i}\sigma)(\dot\gamma_i, \dot\gamma_i) + 2\sigma(\nabla_{\dot\gamma_i}\dot\gamma_i, \dot\gamma_i), \sigma(\dot\gamma_i, \dot\gamma_i)\rangle.$$

Evaluating this equation at $s = 0$, we get

$$\langle(\bar\nabla_u\sigma)(u, u), \sigma(u, u)\rangle + 2\kappa\langle\sigma(u, u), \sigma(u, v)\rangle = 0, \tag{6.1}$$
$$\langle(\bar\nabla_u\sigma)(u, u), \sigma(u, u)\rangle - 2\kappa\langle\sigma(u, u), \sigma(u, v)\rangle = 0. \tag{6.2}$$

Thus we can see that $\langle\sigma(u, u), \sigma(u, v)\rangle = 0$ for an arbitrary pair of orthonormal tangent vectors u, v of M, so that the submanifold M is (λ-)isotropic in $\widetilde{M}_N(c)$ at each point $x \in M$ through the isometric immersion f by Lemma A.

Next, we shall show that $\lambda : M \to \mathbb{R}$ is locally constant. It follows from (6.1) and (6.2) that $\langle(\bar\nabla_u\sigma)(u, u), \sigma(u, u)\rangle = 0$ for every unit tangent vector u. Hence we see that λ is constant along each geodesic on the submanifold M, so that λ is locally constant. On the other hand we denote by R (resp. $\widetilde{R}$) the curvature tensor of M (resp. $\widetilde{M}_N(c)$). We recall the Gauss equation

$$\langle\widetilde{R}(X, Y)Z, W\rangle = \langle R(X, Y)Z, W\rangle + \langle\sigma(X, Z), \sigma(Y, W)\rangle - \langle\sigma(X, W), \sigma(Y, Z)\rangle.$$

Since M is a Kähler submanifold in $\widetilde{M}_N(c)$, from this equation and

$$\widetilde{R}(X, Y)Z = \frac{c}{4}(\langle Y, Z\rangle X - \langle X, Z\rangle X + \langle JY, Z\rangle JX - \langle JX, Z\rangle JY + 2\langle X, JY\rangle JZ),$$

we find that the holomorphic sectional curvature $K(X, JX)$ of M determined by a unit vector X is given by

$$K(X, JX) = \langle R(X, JX)JX, X\rangle = c - 2\|\sigma(X, X)\|^2.$$

Thus we can see that our submanifold M is a complex space form. Therefore from the first statement of Theorem B we obtain the first condition. □

Remark 2.

1) We here make mention of other curvatures of a curve $f_k \circ \gamma$ in Theorem 2. It is known that for each circle γ on $\mathbb{C}P^n(c/k)$ the curve $f_k \circ \gamma$ is an orbit of

some one-parameter subgroup of the full isometry group of the ambient space $\mathbb{C}P^N(c)$, so that all curvatures of $f_k \circ \gamma$ are also constant.

2) If we put $\kappa = 0$ in Theorem 2, this result does not hold. For examle, all Kähler submanifolds M's with parallel second fundamental form in $\widetilde{M}_N(c)$ satisfy the condition that every geodesic γ on M, considered as a curve in the ambient manifold $\widetilde{M}_N(c)$, has constant first curvature along γ.

7. Characterization of parallel complex hypersurfaces

As an immediate consequence of Theorem 2 and Theorem B ([4], [9]) we establish the following ([15]).

Theorem 3. *Let M_n be an n-dimensional complex hypersurface of a complex space form $\widetilde{M}_{n+1}(c)$. Then there exists $\kappa > 0$ satisfying that every circle γ of curvature κ on the hypersurface M_n, considered as a curve in $\widetilde{M}_{n+1}(c)$, has constant first curvature κ_1 along this curve if and only if M_n is totally geodesic in $\widetilde{M}_{n+1}(c)$.*

The aim of this section is to generalize Theorem 3. We pay particular attention to *Kähler circles* on a complex hypersurface. The following is a characterization of complex hypersurfaces with parallel second fundamental form in a complex space form.

Theorem 4. *Let M_n be a complex hypersurface of a complex space form $\widetilde{M}_{n+1}(c)$. Then the following conditions are equivalent:*

(1) *There exists $\kappa \geqq 0$ satisfying that every Kähler circle γ of curvature κ on the hypersurface M_n, considered as a curve in $\widetilde{M}_{n+1}(c)$, has constant first curvature κ_1 along this curve.*

(2) *Either M_n is totally geodesic in $\widetilde{M}_{n+1}(c)$, or $c > 0$ and M_n is locally congruent to a complex quadric $Q_n(\mathbb{C})$.*

Proof. (1) $\Rightarrow$ (2): It follows from (6.1), (6.2) that

$$\langle(\bar{\nabla}_u\sigma)(u,u), \sigma(u,u)\rangle \pm 2\kappa\langle\sigma(u,u), \sigma(u,Ju)\rangle = 0,$$

so that

$$\langle(\bar{\nabla}_u\sigma)(u,u), \sigma(u,u)\rangle = 0. \tag{7.1}$$

for every unit tangent vector u of M. By replacing u by Ju, we obtain

$$\langle(\bar{\nabla}_u\sigma)(u,u), J(\sigma(u,u))\rangle = 0. \tag{7.2}$$

Since M is a complex hypersurface, we can see from (7.1) and (7.2) that $(\bar{\nabla}_u \sigma)(u,u) = 0$ for all u. This implies that the second fundamental form σ is parallel. Thus we find that the complex hypersurface M is totally geodesic in $\widetilde{M}_{n+1}(c)$ or locally congruent to a complex quadric $Q_n(\mathbb{C})$ (cf. [10]).

Note that our complex hypersurface M_n has parallel second fundamental form in $\widetilde{M}_{n+1}(c)$. By easy computation we know that every Kähler circle on M_n, consider as a curve in $\widetilde{M}_{n+1}(c)$, has constant first curvature. □

Remark 3.

1) Theorem 4 is a generalization of the result of Ogiue and the first author, which corresponds to the case $\kappa = 0$ ([7]).
2) In case where M is a Hermitian symmetric space, every Kähler circle on M is an orbit of some one-parameter subgroup of the full isometry group of M (see [3]). Moreover, the isometric imbedding of $Q_n(\mathbb{C})$ into $\mathbb{C}P^{n+1}(c)$ is an equivariant mapping. Hence the same reason as in Remark 2(1) shows that each Kähler circle on $Q_n(\mathbb{C})$, as a curve in $\mathbb{C}P^{n+1}(c)$, has constant curvatures.

8. Characterization of the Veronese imbedding of degree 2

Motivated by Remark 1 (2), we are interested in the Veronese imbedding of degree 2. Our aim here is to provide a characterization of this imbedding by the extrinsic shape of *geodesics* on Kähler submanifolds.

Theorem 5. *Let $f : M_n \to \widetilde{M}_N(c)$ be a non-totally geodesic Kähler isometric full immersion of an n-dimensional connected Kähler manifold into an N-dimensional complex space form of constant holomorphic sectional curvature $c > 0$. Then the following conditions are equivalent:*

(1) *M_n is locally congruent to $\mathbb{C}P^n(c/2)$, $N = n(n+3)/2$ and f is locally equivalent to the second Veronese imbedding f_2.*
(2) *The image $f \circ \gamma$ of each geodesic γ on M_n is a circle in $\widetilde{M}_N(c)$.*
(3) *The image $f \circ \gamma$ of each geodesic γ on M_n is a Frenet curve of order 2 in $\widetilde{M}_N(c)$.*

Proof. It suffices to show that the third condition implies the first. Our steps for proof are as follows: We first show that the submanifold M is (λ-)isotropic at its each point in the ambient manifold $\widetilde{M}_N(c)$, next we verify that the function λ is constant on M and finally we calculate the holomorphic sectional curvature of M.

As Lemma A holds in a trivial sense at an arbitrary geodesic point of M, we have only to consider a non-geodesic point $x \in M$ and a unit tangent vector

$u \in T_xM$ with $\sigma(u,u) \neq 0$. We take the geodesic $\gamma = \gamma(s)$ $(s \in I)$ on M with initial condition that $\gamma(0) = x$ and $\dot\gamma(0) = u$. Here, I is a sufficiently small open interval on $\mathbb{R}$ satisfying $\sigma(\dot\gamma(s), \dot\gamma(s)) \neq 0$ for all $s \in I$. We find that the Frenet frame of the curve $f \circ \gamma$ is $\{\dot\gamma, (1/\kappa)\sigma(\dot\gamma,\dot\gamma)\}$, where $\kappa = \|\sigma(\dot\gamma,\dot\gamma)\|$. Since f is a Kähler immersion, we find that $\tau_\gamma = (1/\kappa)\langle J\dot\gamma, \sigma(\dot\gamma,\dot\gamma)\rangle = 0$, where J is the complex structure of $\widetilde{M}_N(c)$. So we can take the totally real totally geodesic surface $M^2(c/4)$ passing x satisfying that the vectors $\dot\gamma(0)$ and $\sigma(\dot\gamma,\dot\gamma)(0)$ span the tangent space $T_xM^2(c/4)$. We here consider the Frenet curve ρ of order 2 on the surface $M^2(c/4)$ passing the point $x = \rho(0)$ with the same curvature $\kappa(s)$ (> 0) and the same initial frame $\{\dot\gamma(0), (1/\kappa(0))\sigma(\dot\gamma,\dot\gamma)(0)\}$. By the uniqueness of solutions for ordinary differential equations we can see that the curve $f \circ \gamma$ locally coincides with ρ, so that it is locally contained in $M^2(c/4)$.

The following discussion on the isotropic property of M is the same as in pp. 40–41 of [13]. As $M^2(c/4)$ is a 2-dimensional totally geodesic submanifold of $\widetilde{M}_N(c)$, the vectors $\dot\gamma(s)$ and $\sigma(\dot\gamma(s), \dot\gamma(s))$ span the tangent space $T_{\gamma(s)}M^2(c/4)$ for each s. This, together with $\widetilde\nabla_{\dot\gamma}(\sigma(\dot\gamma,\dot\gamma)) \in T_{\gamma(s)}M^2(c/4)$, implies

$$\widetilde\nabla_{\dot\gamma}(\sigma(\dot\gamma,\dot\gamma)) = \alpha\dot\gamma + \beta\sigma(\dot\gamma,\dot\gamma) \tag{8.1}$$

for some smooth functions $\alpha = \alpha(s)$ and $\beta = \beta(s)$ on the interval I. Let v be an arbitrary vector at x which is perpendicular to the vector $u = \dot\gamma(0)$. We extend the vector v to a vector field Y on the curve $f \circ \gamma$. Hence (8.1) gives

$$\begin{aligned}\langle \sigma(u,u), \sigma(u,v)\rangle &= \langle \sigma(\dot\gamma,\dot\gamma), \sigma(\dot\gamma, Y)\rangle(0) = \langle \sigma(\dot\gamma,\dot\gamma), \widetilde\nabla_{\dot\gamma}Y\rangle(0) \\ &= -\langle \widetilde\nabla_{\dot\gamma}(\sigma(\dot\gamma,\dot\gamma)), Y\rangle(0) = -\langle \alpha(0)u + \beta(0)\sigma(u,u), v\rangle = 0.\end{aligned}$$

So it follows from Lemma A that the submanifold M is (λ-)isotropic at its each point in $\widetilde{M}_N(c)$. In order to see the function λ is constant on M, we choose an arbitary pair of orthonomal tangent vectors w, z at any fixed point $y \in M$. Assume that $\lambda(y) \neq 0$. We here define a smooth vector field W (resp. Z) on some sufficiently small neighborhood $\mathcal{U}_y$ by using parallel displacement for the vector w (resp. z) along each geodesic with origin y. Note that $\nabla W = \nabla Z = 0$ at the point y and $\langle W, Z\rangle = 0$ on $\mathcal{U}_y$. Hence, at the point y we find

$$\begin{aligned}w(\lambda^2) &= w\langle \sigma(Z,Z), \sigma(Z,Z)\rangle = 2\langle D_w(\sigma(Z,Z), \sigma(Z,Z))\rangle \\ &= 2\langle (\bar\nabla_w\sigma)(z,z), \sigma(z,z)\rangle.\end{aligned}$$

So, using the equation of Codazzi, Lemma A and an equality $\widetilde\nabla_Z(\sigma(Z,Z)) = \alpha_0 Z + \beta_0\sigma(Z,Z)$ which corresponds to (8.1), at the point y we have

$$\begin{aligned} w(\lambda^2) &= 2\langle(\bar{\nabla}_z\sigma)(w,z),\sigma(z,z)\rangle = 2\langle D_z(\sigma(W,Z)),\sigma(z,z)\rangle \\ &= 2\langle\widetilde{\nabla}_z(\sigma(W,Z)),\sigma(z,z)\rangle = -2\langle\sigma(w,z),\widetilde{\nabla}_z(\sigma(Z,Z))\rangle \\ &= -2\langle\sigma(w,z),\alpha_0 z+\beta_0\sigma(z,z)\rangle = 0. \end{aligned}$$

This implies that if $\lambda(y) \neq 0$, then the function λ is constant on $\mathcal{U}_y$. Therefore, from the connectivity of M and the continuity of λ we can see that the function λ is (nonzero) constant on the submanifold M.

On the other hand, the holomorphic sectional curvature $K(X, JX)$ of M determined by a unit tangent vector X is given by

$$K(X,JX) = \langle R(X,JX)JX,X\rangle = c - 2\|\sigma(X,X)\|^2,$$

hence M is a complex space form. Therefore we can see that M_n is locally congruent to $\mathbb{C}P^n(c/2)$, $N = n(n+3)/2$ and f is locally equivalent to the second Veronese imbedding f_2 (for details, see section 3). □

Remark 4.

1) For each Kähler circle γ of positive curvature on $\mathbb{C}P^n(c/2)$ the curve $f_2 \circ \gamma$ through the second Veronese imbedding f_2 has constant three or four positive curvatures along $f_2 \circ \gamma$ in the sense of Frenet formula (see [14]).
2) The characterization problem of the Veronese imbedding $f_k (k \geqq 2)$ by the extrinsic shape of Kähler circles on a Kähler submanifold is still open.

Acknowledgment

The first author is partially supported by Grant-in-Aid for Scientific Research (C) (No. 14540080), Ministry of Education, Science, Sports and Culture. The second author is partially supported by Grant-in-Aid for Scientific Research (C) (No. 14540075), Ministry of Education, Science, Sports and Culture.

References

1. T. Adachi *Kähler magnetic flows on a manifold of constant holomorphic sectional curvature*, Tokyo J. Math. **18**(1995), 473–483.
2. T. Adachi, S. Maeda and S. Udagawa, *Circles in a complex projective space*, Osaka J. Math. **32**(1995), 709–719.
3. ———, *Simpleness and closedness of circles in compact Hermitian symmetric spaces*, Tsukuba J. Math. **24**(2000), 1–13.
4. E. Calabi, *Isometric imbedding of complex manifolds*, Ann. Math. **269**(1982), 481–499.

5. S. Maeda, *A characterization of constant isotropic immersions by circles*, to appear in Archiv der Mathematik.
6. S. Maeda and T. Adachi, *A characterization of the second Veronese embedding into a complex projective space*, Proc. Japan Acad. Sci. **77** Ser. A(2001), 99–102.
7. S. Maeda and K. Ogiue *Geometry of submanifolds in terms of behavior of geodesics*, Tokyo J. Math. **17**(1994), 347–354.
8. K. Mashimo and K. Tojo, *Circles in Riemannian symmetric spaces*, Kodai Math. J. **22**(1999), 1–14.
9. H. Nakagawa and K. Ogiue, *Complex space forms immersed in complex space forms*, Trans. Amer. Math. Soc. **219**(1976), 289–297.
10. H. Nakagawa and R. Takagi, *On locally symmetric Kaehler submanifolds in a complex projective space*, J. Math. Soc. Japan **28**(1976), 638–667.
11. B. O'Neill, *Isotropic and Kaehler immersions*, Canad. J. Math. **17**(1965), 905–915.
12. K. Ogiue, *Differential geometry of Kaehler submanifolds*, Advances in Math. **13**(1974), 73–114.
13. J.S. Pak and K. Sakamoto, *Submanifolds with d-planar geodesic immersed in complex projective spaces*, Tôhoku Math. J. **38**(1986), 297–311.
14. K. Suizu, S. Maeda and T. Adachi, *Characterization of totally geodesic Kähler immersions*, Hokkaido Math. J. **31**(2002), 629–641.
15. ———, *A characterization of Veronese imbeddings into complex projective spaces*, Mathematical Reports of the Academy of Science of the Royal Society of Canada, **24**(2002), 61–66.

ON THE MEILIKHSON THEOREM

MARIN S. MARINOV
Thechnical University of Sofia
Depatrment of Applied Mathematics and Informatics
Sofia 1000, P.O.Box 384, Bulgaria
E-mail: msm@vmei.acad.bg

Let $\mathbb{K}$ be a non-commutative $\mathbb{R}$-algebra of a finite dimension without zero divisors and $f : D \to \mathbb{K}$ be a function on a domain $D \subset \mathbb{K}$. We study a right (left) derivative of the function f and obtain a linear complex system of PDE's with constant coefficients corresponding to the derivative of f. We prove that every continuous differentiable solution of the system is an affine function.

In 1958 Milnor [1] and Kervaire [2] independently from him proved that the dimension n of every division $\mathbb{R}$-algebra is equal to 1, 2, 4 or 8. These are the algebras of the real numbers, of the complex numbers, of the quaternions and of the octonions. In 1948 Meilikhson [3] proved that every right (left) quaternionic differentiable function is an affine function. After that many authors rediscover this result (see for more completely information [4] and [5]). For $n = 8$ (the case of real octonions $\mathbb{O}$) in 1997 Marinov [6] obtains Meilikhson type theorem for octonionic differentiable functions.

Let $\mathbb{K}$ be a non-commutative division $\mathbb{R}$-algebras and $f : G \to \mathbb{K}$ be a function on a domain $G \subset \mathbb{K}$. The function f is called a right (left) differentiable at a point $q^0 \in G$ if the limit

$$f\frac{d}{dq}(q^0) := \lim_{h\to 0, h\neq 0}(\Delta f).h^{-1} \qquad (\frac{d}{dq}f(q^0) := \lim_{h\to 0, h\neq 0} h^{-1}.(\Delta f))$$

exists. Let us remark that , as usual $\Delta f := f(q^0 + h) - f(q^0)$ and the limits are in the Euclidean topology in $\mathbb{R}^n$. If the function $f : G \to \mathbb{K}$ is right (left) differentiable at every point $q \in G$ then the function f is called right (left) differentiable on G. The following problem will be considered:

Problem. To be describe the class of all right (left) differentiable functions $f : G \to \mathbb{K}$.

As a consequence of the results of Milnor and Kervair we have only two cases $\mathbb{K} = \mathbb{H}$ and $\mathbb{K} = \mathbb{O}$.

Remark 1. The differentiability of a $\mathbb{K}$-valued functions have been studied by many authors from different points of view. The book of Gürlebeck and Sprössig [7] and the papers of Habetha [8], [9], Rizza [10] and Snyder [11] are basic for a more complete bibliography. After that in the case $\mathbb{K} = \mathbb{H}$ Malonek [12], Pertici [13], Mitelman and Shapiro [14], Lugojan [15], [16], Gentili, Mariconda and Tarallo [17], Yasukura [21] generalized this notion in several directions.

By the end we shall consider only right differentiability.
For $\mathbb{K} = \mathbb{H}$ we have the following result.

Theorem 1 (Meilikhson-Petrici). *Let $f : G \to \mathbb{H}$ be a function on a domain $G \subset \mathbb{H}$. Then the following conditions are equivalent:*

a) f is an affine right linear function, i. e. there exist $a \in \mathbb{H}$ and $b \in \mathbb{H}$ such that $f(q) = a.q + b$ for every $q \in G$.

b) $g \circ f$ is a left regular (in the sense of Moisil-Fueter) on G for every function g, which is left regular on a neighbourhood of $f(G)$.

c) f is right differentiable on G.

It should be remembered that Meilikhson [3] proved $a) \iff c)$ and Pertici proved that $a) \iff b)$.

Remark 2. Some similar results of the above theorem for some generalizations of a quaternionic derivative are obtained by Lugojan [15], [16] and Marinov [6].

Now let consider the non-associative algebra $\mathbb{O}$ of real octonions with the following multiplication rules for the basic imaginary units $e_1 = i$, $e_2 = j$, $e_3 = k$, $e_4 = e$, e_5, e_6 and e_7

$$e_1^2 = e_2^2 = \cdots = e_7^2 = -1,$$
$$e_m.1 = 1.\, e_m = e_m, \qquad 1 \leq m \leq 7,$$
$$e_m.e_{m+1} = e_{m+3} = -e_{m+1}.e_m \ (\mathrm{mod}\ 7)$$

(here the addition of the indices is taken mod 7).

If we consider (as a vector space) $\mathbb{O} = \mathbb{H} \oplus \mathbb{H}e$ (ortogonal sum with respect to the Euclidean scalar product of $\mathbb{O}$), then every octonion q has the representation $q = q_1 + q_2 e$, where $q_1 = z_1 + z_2 j \in \mathbb{H}$ and $q_2 = z_3 + z_4 j \in \mathbb{H}$ (in view of the fact that $\mathbb{H} = \mathbb{C} \oplus \mathbb{C}j$). Furthemore, if f is an octonion function, then $f = u + v.e$, where u and v are quaternion functions. We note, that $u = f_1 + f_2 j$ and $v = f_3 + f_4 j$, where f_n, $1 \leq n \leq 4$, are complex functions.

Theorem 2. *Let f be a $\mathbb{R}$-Fréchet differentiable function on G. The next statements are equivalent:*

$i)$ *The octonion function f is right differentiable on G.*

$ii)$ *The next conditions hold:*

$$\frac{\partial f}{\partial x_0} e_m = \frac{\partial f}{\partial x_m}, \quad 1 \leq m \leq 7.$$

The proof is a modification of the proof in the quaternionic case (c.f. [2]).

Remark 3. The above conditions for octonionic differentiability are an analogue of the Krylov conditions [2] for a quaternionic differentiability and the classical Cauchy-Riemann conditions for a complex differentiability.

Proposition 1. *Let an octonion function $f = (f_1 + f_2 j) + (f_3 + f_4 j)e$ be right differentiable on G. Then the complex functions $f_m, 1 \leq m \leq 4$, satisfy the system*

$$\frac{\partial f_1}{\partial \bar{z}_1} = \frac{\partial f_2}{\partial z_1} = \frac{\partial f_3}{\partial z_1} = \frac{\partial f_4}{\partial \bar{z}_1} = 0, \qquad \frac{\partial f_1}{\partial z_2} = \frac{\partial f_2}{\partial \bar{z}_2} = \frac{\partial f_3}{\partial z_2} = \frac{\partial f_4}{\partial \bar{z}_2} = 0,$$

$$\frac{\partial f_1}{\partial z_3} = \frac{\partial f_2}{\partial z_3} = \frac{\partial f_3}{\partial \bar{z}_3} = \frac{\partial f_4}{\partial \bar{z}_3} = 0, \qquad \frac{\partial f_1}{\partial \bar{z}_4} = \frac{\partial f_2}{\partial \bar{z}_4} = \frac{\partial f_3}{\partial \bar{z}_4} = \frac{\partial f_4}{\partial \bar{z}_4} = 0,$$

$$\frac{\partial f_2}{\partial \bar{z}_1} = -\frac{\partial f_1}{\partial \bar{z}_2}, \quad \frac{\partial f_1}{\partial z_1} = \frac{\partial f_2}{\partial z_2}, \quad \frac{\partial f_4}{\partial z_1} = \frac{\partial f_3}{\partial \bar{z}_2}, \quad \frac{\partial f_3}{\partial \bar{z}_1} = -\frac{\partial f_4}{\partial z_2},$$

$$\frac{\partial f_1}{\partial z_1} = \frac{\partial f_3}{\partial z_3}, \quad \frac{\partial f_2}{\partial \bar{z}_1} = \frac{\partial f_4}{\partial z_3}, \quad \frac{\partial f_3}{\partial \bar{z}_1} = -\frac{\partial f_1}{\partial \bar{z}_3}, \quad \frac{\partial f_4}{\partial z_1} = -\frac{\partial f_2}{\partial \bar{z}_3},$$

$$\frac{\partial \bar{f}_2}{\partial z_1} = -\frac{\partial f_3}{\partial z_4}, \quad \frac{\partial \bar{f}_1}{\partial \bar{z}_1} = \frac{\partial f_4}{\partial z_4}, \quad \frac{\partial \bar{f}_4}{\partial \bar{z}_1} = -\frac{\partial f_1}{\partial z_4}, \quad \frac{\partial \bar{f}_3}{\partial z_1} = \frac{\partial \bar{f}_2}{\partial z_4}.$$

The proof follows from Theorem 2 after an octonionic calculation.

The next Meilikhson type theorem characterize completely the octonionic differentiability.

Theorem 3. *Let f be a R-Fréchet differentiable octonionic function on a domain $G \subset \mathbb{O}$. Then the following conditions are equivalent:*

$a)$ *f is an affine right linear function, i.e. there exist $a \in \mathbb{O}, b \in \mathbb{O}$ such that $f(q) = a.q + b$ for every $q \in G$.*

$b)$ *f is right differentiable on G.*

Proof. The proof of the implication $a) \Longrightarrow b)$ is obvious.

Now we shall prove $b) \Longrightarrow a)$. Let $f = (f_1 + f_2 j) + (f_3 + f_4 j)e$. First, let G be a polydisk with a center $q^0 = 0$. By an analoug of Hartogs' theorem on separately analyticity (see [18]) we obtain that $f_n \in C^\infty(G)$, $1 \le n \le 4$. Then by the Schwartz's lemma and Proposition 1 we have

$$\frac{\partial^2 f_1}{\partial z_1^2} = \frac{\partial^2 f_1}{\partial z_1 \partial \bar z_2} = \frac{\partial^2 f_1}{\partial z_1 \partial \bar z_3} = \frac{\partial^2 f_1}{\partial z_1 \partial z_4}$$
$$= \frac{\partial^2 f_1}{\partial \bar z_2^2} = \frac{\partial^2 f_1}{\partial \bar z_2 \partial \bar z_3} = \frac{\partial^2 f_1}{\partial \bar z_2 \partial z_4} = \frac{\partial^2 f_1}{\partial \bar z_3^2} = \frac{\partial^2 f_1}{\partial \bar z_3 \partial z_4} = \frac{\partial^2 f_1}{\partial z_4^2} = 0.$$

Therefore the complex function f_1 have representation

$$f_1(z_1, \bar z_2, \bar z_3, z_4) = \alpha_0 + \alpha_1 z_1 + \alpha_2 \bar z_2 + \alpha_3 \bar z_3 + \alpha_4 z_4.$$

Analogously

$$f_2(\bar z_1, z_2, \bar z_3, z_4) = \beta_0 + \beta_1 \bar z_1 + \beta_2 z_2 + \beta_3 \bar z_3 + \beta_4 z_4,$$
$$f_3(\bar z_1, \bar z_2, z_3, z_4) = \gamma_0 + \gamma_1 \bar z_1 + \gamma_2 \bar z_2 + \gamma_3 z_3 + \gamma_4 z_4,$$
$$f_4(z_1, z_2, z_3, z_4) = \delta_0 + \delta_1 z_1 + \delta_2 z_2 + \delta_3 z_3 + \delta_4 z_4.$$

Hence for the above constants we calculate from the equalities in the last three rows of the proposition above

$$\alpha_1 = \beta_2 = \gamma_3 = \bar\delta_4, \quad \beta_1 = -\alpha_2 = -\bar\gamma_4 = \delta_3,$$
$$\gamma_1 = -\alpha_3 = \bar\beta_4 = -\delta_2, \quad \delta_1 = -\bar\alpha_4 = -\beta_3 = \gamma_2.$$

Let

$$a_{11} := \alpha_1, \quad a_{12} := \beta_1, \quad a_{21} := \gamma_1, \quad a_{22} := \delta_1,$$
$$b_{11} := \alpha_0, \quad b_{12} := \beta_0, \quad b_{21} := \gamma_0, \quad b_{22} := \delta_0,$$

and

$$a_1 := a_{11} + a_{12} j, \quad a_2 := a_{21} + a_{22} j,$$
$$b_1 := b_{11} + b_{12} j, \quad b_2 := b_{21} + b_{22} j.$$

Then

$$u = f_1 + f_2 j = a_1(z_1 + z_2 j) - (\bar z_3 - z_4 j)a_2 + b_1,$$
$$v = f_3 + f_4 j = (z_3 + z_4 j)a_1 + a_2(\bar z_1 - z_2 j) + b_2.$$

Therefore

$$\begin{aligned} f(q) &= u + ve \\ &= (a_1 + a_2 e).[(z_1 + z_2 j) + (z_3 + z_4 j)e] + (b_1 + b_2 e) = a.q + b, \end{aligned}$$

where

$$a := a_1 + a_2 e \in \mathbb{O}, \quad b := b_1 + b_2 e \in \mathbb{O}.$$

In the general case, when G is an arbitrary domain, we supply the method of analytic continuation along curves by power series (see [19] for the quaternionic case). □

Remark 4. Having in mind the theory of octonionic regular functions (see [20]) it would be obtained an analoug of Pertici's result from Theorem 1 for regular octonionic functions.

References

1. R. Bott, J. Milnor, On the parallelizability of the spheres, *Bull. Amer. Math. Soc.*, **64**, (1958), 87–89.
2. M. Kervaire, Non-parallelizability of n-sphere for $n = 7$, *Proc. Nat. Acad. Sci. U.S.A.*, **44** (1958), 280–283.
3. A.S. Meilikhson, On the notion of the monogeneous quaternions, *Dokl. AN USSR*, **59**, (1948), 431–434, (in Russian).
4. M.S. Marinov, Meilkhson type theorems I., *Analele Univ. Timisoira, Ser. Mat.-Informstica*, **36** no. 1, (1996), 95–110.
5. H. Snyder, Reference, *Math Reviews*, **49**, 2, N 3162 (1975).
6. M.S. Marinov, Octonionic differentiable functions, *C.R. Acad. bulg. Sci.*, **50**, 5, (1997) 9–12.
7. K. Güdlebeg, W. Sprössig, *Quaternionic Analysis and Elliptic Boundary Value Problems*, Birkhäuser, 1990.
8. K. Habetha, Function theory in algebras, *In: Complex Analysis, Methods, Trends, and Appl.*, Akad.-Verlag, Berlin, (1983), 225–237.
9. ———, Function theory in algebras, *Indianapolis Jour. Math. Phys.*, v. 1-2, (1982–83), 43–64.
10. G.B. Rizza, Monogenic functions on the real algebras and conformal mappings, *Boll. Un. Mat. Ital.*, **4**, 12, Suppl. Fasc. 3, (1975), 437–450.
11. H.H. Snyder, An introduction to theories of regular functions on linear associative algebras, *Lect. Notes Pure Appl. Math.*, **68**, (1982), 75–93.
12. H. Malonek, Hypercomplex differentiability and its applications, *In: Clifford algebras and their Applications in Math. Physics*, (Ed. F. Brackx et al.,) Kluver Acad. Publ., (1993), 141–150.
13. D. Pertici, *Theoria delle funzioni di più variabili quaternioniche*, Tesi di dottorato, Univ. degli Studi di Firence, A.A., (1987–88).
14. I.M. Mitelman, M.V. Shapiro, Differentiation of the Martinelli-Bochner integrals and the notion of hyperderivability, *Math. Nachrichten*, **172**, (1995), 211–238.
15. S. Lugojan, Quaternionic derivability, *An. Univ. din Timisoara*, Ser. Matematice, **29**, 2–3, (1991), 175–190.
16. ———, Quaternionic derivability II, *In: Quaternionic Structures in Mathematics and Physics*, Trieste, 5–9 September 1994, (1996), 217–224.
17. G. Gentili, C. Mariconda, M. Tarallo, Quaternionic regular maps and $\bar{\partial}$-type operators, *Simon Stevin Quaterly J. of Pure and Appl. Math.*, **67**, 3–4, (1993), 333–359.

18. S.G. Krantz, *Function theory of several complex variables*, John Wiley and Sons, (1982).
19. A. Sudbery, Quaternionic analysis, *Math. Proc. Camb. Phil. Soc.*, **85**, (1979), 199–225.
20. K. Nôno, On the octonionic linearization of Laplacian and octonionic function theory, *Bull. Fukuoka Univ. Ed.*, Part III, **37**, (1988), 1–15.
21. O. Yasukura, On Ehresmann's theorem of quaternion holomorphy, *Yokohama Math. Journal*, **49** (2002), 183–169.

DIFFEOMORPHISMS PRESERVING A SKEW-SYMMETRIC CURVATURE OPERATOR IN RIEMANNIAN GEOMETRY

SVILEN M. MARINOV

Department of Mathematics,
University of Chemical Technology and Metallurgy - Sofia,
P.O. Box 30, Sofia 1700, Bulgaria

Dedicated to Professor Grozio Stanilov on his 70th birthday

In this paper we study the problem of preservation of the skew-symmetric curvature operator on Riemannian manifolds of constant sectional curvature, introduced by G. Stanilov.

1. Introduction

Let (M, g) and $(\overline{M}, \overline{g})$ are Riemannian manifolds and M_p, $\overline{M}_q$ are their tangent spaces at $p \in M$ and $q \in \overline{M}$, respectively.

Definition 1. We will call the diffeomorphism $f : M \to \overline{M}$ a Riemannian isometry of (M, g) into $(\overline{M}, \overline{g})$ if $f^*\overline{g} = g$, or equivalently if $\overline{g}(f_*X, f_*Y) = g(X, Y)$ for all differentiable vector fields X, Y on M.

Note. The Riemannian isomerty is equivalent to the notion of isometry in the sense of piecewise differentiable distance function, determined by the Riemannian metric. So the class of Riemannian isometries coincides with the class of distance-preserving isometries.

In [1] is defined the operator

$$S_p(u) := R_p(e_1, e_2, u),$$

where (e_1, e_2) is an orthonormal base for a tangent plane $E^2 \subset M_p$ and $u \in M_p$. In this way well-defined is the operator $S : p \mapsto S_p$ for any $p \in M$.

Theorem 1. *Let S_p is defined with respect to the plane $E^2(x \wedge y) \subset M_p$ and (x, y) is an arbitrary base for $E^2(x \wedge y)$. Then (M, g) is a Riemannian manifold of constant Riemannian curvature, if and only if the operator S_p conserves $E^2(x \wedge y)$ invariant, i.e.*

$$S_p E^2(x \wedge y) = E^2(x \wedge y).$$

Proof. *Necessity*. The necessity of this proof was considered in [2].

Sufficiency. Let us fix arbitrary $p \in M$ and tangent plane $E^2(x \wedge y) \subset M_p$. Then $S_p(x) = -Cy$, $S_p(y) = Cx$, where C is a real constant. For $u = ax + by \in E^2(x \wedge y)$ we have $S_p(u) = C(bx - ay) = C(g_p(y,u)x - g_p(x,u)y)$. Since p to be fixed arbitrary point of M, then (M, g) is a Riemannian manifold of constant Riemannian curvature C. □

2. Main problem

Let (M, g) and $(\overline{M}, \overline{g})$ are both connected Riemannian manifolds with constant Riemannian curvatures K and $\overline{K}$ respectively, and $\dim M = n$, $\dim \overline{M} = m$. Let $f : M \to \overline{M}$ is a diffeomorphism. Then the linear operator $f_{*,p} : M_p \to M_q$, $q = f(p)$, defined for arbitrary $p \in M$, is an isomorphism between the tangent spaces M_p and M_q. Hence $n = m$. The skew-symmetric operator S_p, induced by $E^2 \subset M_p$, maps E^2 onto E^2. On the other hand $f_{*,p}$ maps E^2 onto plane $\overline{E}^2 \subset \overline{M}_q$. In this way we have the existence of the operator $\overline{S}_q : \overline{M}_q \to \overline{M}_q$,

$$\overline{S}_q : u \mapsto \overline{R}_q(\overline{e}_1, \overline{e}_2, \overline{u}),$$

where $(\overline{e}_1, \overline{e}_2)$ is an orthonormal base for $\overline{E}^2$ and $\overline{R}_q$ is the corresponding curvature tensor on $\overline{M}$ at $q \in \overline{M}$.

Definition 2. We will say that the map $f : M \to \overline{M}$ preserves the operator S, if the next conditions are fulfilled:

(i) f is a diffeomorphism;

(ii) $f_{*,p} \circ S_p = \overline{S}_{f(p)} \circ f_{*,p}$ for every $p \in M$ and every $E^2 \subset M_p$.

Main problem. To study the map $f : M \to \overline{M}$ preserving the operator S.

Obviously, f may preserve other curvature operators (c.f. [3]).

3. Theorems

Proposition 1. *If (M, g) and $(\overline{M}, \overline{g})$ are two-dimensional Riemannian manifolds of constant Riemannian curvatures, then every Riemannian isometry of (M, g) onto $(\overline{M}, \overline{g})$ preserves the operator S.*

Proposition 2. *If (M, g) and $(\overline{M}, \overline{g})$ are flat Riemannian manifolds, then every diffeomorphism $f : M \to \overline{M}$ preserves the operator S.*

By the end let $\dim M = \dim \overline{M} \geq 3$ and (M, g) and $(\overline{M}, \overline{g})$ are not flat Riemannian manifolds.

Theorem 2. *Let $f : M \to \overline{M}$ is a Riemannian isometry of (M, g) onto $(\overline{M}, \overline{g})$ and f preserves S. If (M, g) is a Riemannian manifold of constant Riemannian curvature, then $(\overline{M}, \overline{g})$ is also a Riemannian manifold of the same Riemannian curvature.*

Proof. Let $\overline{X}$, $\overline{Y}$, $\overline{Z}$ are arbitrary differentiable vector fields on $\overline{M}$. There exists an unique localization of these fields at arbitrary point $q \in \overline{M}$. We will construct it as follows.

There always exist three tangent vectors $\overline{x}$, $\overline{y}$, $\overline{u}$ on $\overline{M}$ such that $\overline{x} = f_{*,p}x$, $\overline{y} = f_{*,p}y$, $\overline{u} = f_{*,p}u$. The mentioned above localization has the form

$$\overline{X}_q = \overline{x}, \quad \overline{Y}_q = \overline{y}, \quad \overline{Z}_q = \overline{u}.$$

Then

$$\overline{R}_q(\overline{x}, \overline{y}, \overline{u}) = K(\overline{g}_q(\overline{y}, \overline{u})\overline{x} - \overline{g}_q(\overline{x}, \overline{u})\overline{y}), \tag{1}$$

where K is the Riemannian curvature of (M, g). Therefore, $(\overline{M}, \overline{g})$ is a Riemannian manifold of constant Riemannian curvature K. □

Theorem 3. *Let $f : M \to \overline{M}$ is a Riemannian isometry and (M, g), $(\overline{M}, \overline{g})$ are both Riemannian manifolds of equal constant Riemannian curvatures. Then f preserves S.*

Proof. It's enough to fulfil (1) for arbitrary M_p and arbitrary base (x, y) for the tangent plane $E^2(x \wedge y)$. Thus, one can achieve (ii). □

Theorem 4. *Let $f : M \to \overline{M}$ preserves the operator S and (M, g), $(\overline{M}, \overline{g})$ are Riemannian manifolds of equal constant Riemannian curvatures. Then f is a Riemannian isometry.*

Proof. Let $p \in M$, $E^2(x \wedge y) \subset M_p$ and

$$G := g_p(x, x)g_p(y, y) - g_p^2(x, y),$$
$$\overline{G} = \overline{g}_q(\overline{x}, \overline{x})\overline{g}_q(\overline{y}, \overline{y}) - \overline{g}_q^2(\overline{x}, \overline{y}),$$

where $q = f(p)$, $\overline{x} = f_{*,p}x$, $\overline{y} = f_{*,p}y$.

If we choose $u \in M_p$ with the condition $u \in (E^2)^\perp$, then for $\overline{u} = f_{*,p}u$ we have

$$\overline{g}_q(\overline{y}, \overline{u})\overline{x} - \overline{g}_q(\overline{x}, \overline{u})\overline{y} = 0. \tag{2}$$

The vectors $\overline{x}$, $\overline{y}$ form a base $(\overline{x},\overline{y})$ for a tangent plane $\overline{E}^2 \subset \overline{M}_q$. Condition (2) shows that $\overline{u} \in (\overline{E}^2)^{\perp}$. On the other hand $f_{*,p} = \lambda(p)h_p$, where λ is a differentiable function on M, $\lambda \neq 0$ and h_p is an orthogonal operator.

Hence,

$$\lambda^2(p)g_p(x,u) = \sqrt{\frac{\overline{G}}{G}}g_p(x,u).$$

But $\overline{G} = \lambda^2(p)G$. So $\lambda^2(p) = |\lambda(p)|$ and $\lambda(p) = \pm 1$. Since h is a Riemannian isometry, we attain that f is a Riemannian isometry. □

Theorem 5. *Let $f : M \to \overline{M}$ preserves S and f is a Riemannian isometry, when $(\overline{M},\overline{g})$ is a Riemannian manifold of constant Riemannian curvature K. Then (M,g) is also a Riemannian manifold of the same curvature.*

Proof. We consider the Riemannian isometry $f^{-1} : \overline{M} \to M$ of $(\overline{M},\overline{g})$ onto (M,g). Using Theorem 2 we have

$$R_p(x,y,u) = K(g_p(y,u)x - g_p(x,u)y)$$

as an analogue of (1). Then (M,g) is a Riemannian manifold of constant Riemannian curvature K. □

References

1. R. Ivanova, G. Stanilov, A skew-symmetric curvature operator in Riemannian geometry, *Symposia Gaussiana, Conf. A,*, Eds: Behara (Fritsch) Lintz, Walter de Gruyter, 1995, 391–395.
2. R. Ivanova, M. Yawata, On the Riemannian Manifolds of Constant Sectional Curvature, *Tensor*, N.S., **57** (1996), 328–333.
3. G. Stanilov, M. Belger, On some curvature operators in Riemannian Geometry, *Ann. Univ. Sofia St. K. Ohridski, Fac. Math. Informatique*, **88** (1994), 1–16.

CURVATURE TENSORS ON ALMOST CONTACT MANIFOLDS WITH B-METRIC

GALIA NAKOVA

Department of Algebra and Geometry, Faculty of Pedagogics,
University of Veliko Tirnovo "St. St. Cyril and Methodius",
1 Theodosij Tirnovsky Str., 5000 Veliko Tirnovo, Bulgaria
E-mail: gnakova@yahoo.com

A complete decomposition of the space of curvature tensors $\mathcal{R}$ over an almost Hermitian vector space is given in [8]. The space $\mathcal{R}$ over an almost complex vector space with B-metric is decomposed in [7]. The odd-dimensional analogue of the decomposition from [8] is obtained in [3], where 18 factors of $\mathcal{R}$ over an almost contact metric vector space are found. In this paper we give a decomposition of $\mathcal{R}$ over an almost contact vector space with B-metric and we obtain 20 mutually orthogonal and $G \times I$-invariant factors.

1. Introduction

Let $(M^{2n+1}, \varphi, \xi, \eta, g)$ be a $(2n+1)$- dimensional almost contact manifold with B-metric, i.e. (φ, ξ, η) is an almost contact structure [2] and g is a metric [4] on M^{2n+1} such that:

$$\varphi^2 = -id + \eta \otimes \xi; \quad \eta(\xi) = 1; \quad g(\varphi X, \varphi Y) = -g(X,Y) + \eta(X)\eta(Y),$$

where $X, Y \in \mathcal{X}M$.

The associated with g metric $\tilde{g}$ [4] on the manifold is given by

$$\tilde{g}(X,Y) = g(X, \varphi Y) + \eta(X)\eta(Y), \quad X, Y \in \mathcal{X}M.$$

Both metrics g and $\tilde{g}$ are indefinite of signature $(n+1, n)$ [4].

Let $\mathcal{R}$ be the set of all Riemannian-like curvature tensors:

$$\mathcal{R} = \Big\{ R \in \bigotimes^4 V^* : R(X,Y,Z,W) = -R(Y,X,Z,W) = -R(X,Y,W,Z), \\ \underset{(X,Y,Z)}{\sigma} R(X,Y,Z,W) = 0 \Big\}.$$

Further ρ_R, ρ^*_R, ρ''_R, $\tilde{\rho}_R$ are the Ricci-type contractions associated with $R \in \mathcal{R}$

and τ_R, τ_R^*, $\tau_R^{''}$, $\tilde{\tau}_R$ are the corresponding scalar curvatures:

$$\rho_R(X,Y) = \sum_{i=1}^{2n+1} R(e_i, X, Y, e_i), \quad \tau_R = \sum_{i,j=1}^{2n+1} R(e_i, e_j, e_j, e_i);$$

$$\rho_R^*(X,Y) = \sum_{i=1}^{2n+1} R(e_i, X, Y, \varphi e_i), \quad \tau_R^* = \sum_{i,j=1}^{2n+1} R(e_i, e_j, e_j, \varphi e_i);$$

$$\rho_R^{''}(X,Y) = \sum_{i=1}^{2n+1} R(e_i, X, \varphi Y, \varphi e_i), \quad \tau_R^{''} = \sum_{i,j=1}^{2n+1} R(e_i, e_j, \varphi e_j, \varphi e_i);$$

$$\tilde{\rho}_R(X,Y) = R(\xi, X, Y, \xi), \quad \tilde{\tau}_R = \sum_{i=1}^{2n+1} R(\xi, e_i, e_i, \xi),$$

where $\{e_i, e_{2n+1} \equiv \xi\}$, $(i = 1, 2, \ldots, 2n)$ is an orthonormal basis of T_pM, $p \in M$.

Let V be a $(2n+1)$-dimensional vector space with almost contact structure (φ, ξ, η) and B-metric g. The standart representation of the group $G \times I$ ($G = GL(n, C) \cap O(n, n)$, i.e. G consists of the real matrices $\begin{pmatrix} A & B \\ -B & A \end{pmatrix}$ which are in $O(n, n)$) in V induces a natural representation λ of $G \times I$ in the space $\otimes^k V^*$ of all tensors of type $(0, k)$ over V:

$$(\lambda(a)S)(X_1, \ldots, X_k) = S(a^{-1}X_1, \ldots, a^{-1}X_k), a \in G \times I,$$
$$S \in \otimes^k V^*, \; X_1, \ldots, X_k \in V.$$

The metric g induces on $\otimes^k V^*$ an inner product $<,>$, defined by

$$<S_1, S_2> = g^{i_1 j_1} g^{i_2 j_2} \ldots g^{i_k j_k} S_1(e_{i_1}, \ldots, e_{i_k}) S_2(e_{j_1}, \ldots, e_{j_k})$$

for S_1, $S_2 \in \otimes^k V^*$ and $\{e_i, e_{2n+1} \equiv \xi\}$, $(i = 1, 2, \ldots, 2n)$ – a basis of V.

Now, let $hV = \{X \in V : X = hX = -\varphi^2 X\}$. Denoting the restrictions of g and φ on hV with the same letters we obtain the almost complex vector space $\{hV, \varphi, g\}$ of dimension $2n$ with a complex structure φ and B-metric g. Then for an arbitrary $X \in V$ we have $X = x + \eta(X)\xi$, where by x we denote hX.

2. The space $\mathcal{L}$ of the tensors of type $(0, 2)$ over V

The following lemma is known

Lemma 2.1 ([6]). *The decomposition $\mathcal{L} = h\mathcal{L} \oplus v\mathcal{L} \oplus w\mathcal{L}$ is orthogonal and* *invariant under the action of $G \times I$. The subspaces $h\mathcal{L}$, $v\mathcal{L}$, $w\mathcal{L}$ are defined by*

$$h\mathcal{L} = \{L \in \mathcal{L} : L(X,Y) = hL(X,Y) = L(hX,hY) = L(x,y),\ X,Y \in V\},$$
$$v\mathcal{L} = \{L \in \mathcal{L} : L(X,Y) = vL(X,Y) = \eta(X)L(\xi,y) + \eta(Y)L(x,\xi),\ X,Y \in V\},$$
$$w\mathcal{L} = \{L \in \mathcal{L} : L(X,Y) = wL(X,Y) = \eta(X)\eta(Y)L(\xi,\xi),\ X,Y \in V\}.$$

In [7] it is proved that $h\mathcal{L}$ splits into six invariant and mutually orthogonal subspaces $\mathcal{L}_i$, $i = 1, 2, 3, 4, 5, 6$:

$$\mathcal{L}_1 : L(X,Y) = \frac{\operatorname{tr} L}{2n} g(x,y), \quad \operatorname{tr} L = g^{ij} L(e_i, e_j),$$
$$\mathcal{L}_2 : L(X,Y) = -\frac{\operatorname{tr}^* L}{2n} \tilde{g}(x,y), \quad \operatorname{tr}^* L = g^{ij} L(e_i, \varphi e_j),$$
$$\mathcal{L}_3 : L(X,Y) = \frac{1}{4}\{L(x,y) + L(y,x) - L(\varphi x, \varphi y) - L(\varphi y, \varphi x)\} - \frac{\operatorname{tr} L}{2n} g(x,y) + \frac{\operatorname{tr}^* L}{2n} \tilde{g}(x,y),$$
$$\mathcal{L}_4 : L(X,Y) = \frac{1}{4}\{L(x,y) + L(y,x) + L(\varphi x, \varphi y) + L(\varphi y, \varphi x)\},$$
$$\mathcal{L}_5 : L(X,Y) = \frac{1}{4}\{L(x,y) - L(y,x) + L(\varphi x, \varphi y) - L(\varphi y, \varphi x)\},$$
$$\mathcal{L}_6 : L(X,Y) = \frac{1}{4}\{L(x,y) - L(y,x) - L(\varphi x, \varphi y) + L(\varphi y, \varphi x)\}.$$

We denote

$$\mathcal{L}_7 : L(X,Y) = \frac{1}{2}\{vL(X,Y) + vL(Y,X)\},$$
$$\mathcal{L}_8 : L(X,Y) = \frac{1}{2}\{vL(X,Y) - vL(Y,X)\},$$
$$\mathcal{L}_9 : L(X,Y) = wL(X,Y).$$

So we obtain

Lemma 2.2. *The decomposition $\mathcal{L} = \bigoplus_{i=1}^{9} \mathcal{L}_i$ is orthogonal and invariant under the action of $G \times I$. The components of an arbitrary $L \in \mathcal{L}$ are hL, vL, wL.*

3. The space $\mathcal{R}$ of the Riemannian-like curvature tensors over V

Analogously as [1] we prove

Lemma 3.1. *The decomposition* $\mathcal{R} = h\mathcal{R} \oplus v\mathcal{R} \oplus w\mathcal{R}$ *is orthogonal and invariant under the action of* $G \times I$. *The subspaces* $h\mathcal{R}$, $v\mathcal{R}$, $w\mathcal{R}$ *are defined by*

$$\begin{aligned}
h\mathcal{R} = \{R \in \mathcal{R} : R(X,Y,Z,W) &= hR(X,Y,Z,W) = R(hX,hY,hZ,hW) \\
&= R(x,y,z,w)\}, \\
v\mathcal{R} = \{R \in \mathcal{R} : R(X,Y,Z,W) &= vR(X,Y,Z,W) \\
&= -\eta(X)R(z,w,y,\xi) + \eta(Y)R(z,w,x,\xi) \\
&\quad - \eta(Z)R(x,y,w,\xi) + \eta(W)R(x,y,z,\xi)\}, \\
w\mathcal{R} = \{R \in \mathcal{R} : R(X,Y,Z,W) &= wR(X,Y,Z,W) \\
&= -\eta(X)\eta(Z)R(\xi,y,w,\xi) + \eta(X)\eta(W)R(\xi,y,z,\xi) \\
&\quad + \eta(Y)\eta(Z)R(\xi,x,w,\xi) - \eta(Y)\eta(W)R(\xi,x,z,\xi)\}, \\
&\qquad\qquad\qquad\qquad X,Y,Z,W \in V.
\end{aligned}$$

In the geometry of the almost contact manifolds with B-metric the tensors $\psi_i(S)$, $S \in \mathcal{L}$ and π_i, $i = 1, 2, 3, 4, 5$, defined in [6] are essential. It is known [6] that

- $\psi_1(S), \psi_4(S) \in \mathcal{R} \iff S(X,Y) = S(Y,X)$,
- $\psi_2(S), \psi_5(S) \in \mathcal{R} \iff S(X,\varphi Y) = S(Y,\varphi X)$,
- $\psi_3(S) \in \mathcal{R} \iff S(X,Y) = S(Y,X)$ and $S(X,\varphi Y) = S(Y,\varphi X)$.

Let $S \in \mathcal{L}$, $X, Y, Z, W \in V$. We define the tensors

$$\begin{aligned}
\psi_2'(S)(X,Y,Z,W) &= g(Y,\varphi Z)\{S(X,\varphi W) + \eta(W)S(\xi,\varphi X)\} \\
&\quad - g(X,\varphi Z)\{S(Y,\varphi W) + \eta(W)S(\xi,\varphi Y)\} \\
&\quad + g(X,\varphi W)\{S(Y,\varphi Z) + \eta(Z)S(\xi,\varphi Y)\} \\
&\quad - g(Y,\varphi W)\{S(X,\varphi Z) + \eta(Z)S(\xi,\varphi X)\}, \\
\psi_3'(S)(X,Y,Z,W) &= -g(Y,Z)\{S(X,\varphi W) + \eta(W)S(\xi,\varphi X)\} \\
&\quad + g(X,Z)\{S(Y,\varphi W) + \eta(W)S(\xi,\varphi Y)\} \\
&\quad - g(X,W)\{S(Y,\varphi Z) + \eta(Z)S(\xi,\varphi Y)\} \\
&\quad + g(Y,W)\{S(X,\varphi Z) + \eta(Z)S(\xi,\varphi X)\} \\
&\quad - g(X,\varphi W)S(Y,Z) + g(Y,\varphi W)S(X,Z) \\
&\quad - g(Y,\varphi Z)S(X,W) + g(X,\varphi Z)S(Y,W).
\end{aligned}$$

Lemma 3.2. *Let* $S \in \mathcal{L}$. *Then*

i) $h\psi_i'(S) \equiv h\psi_i(S)$, $(i = 2,\ 3)$;

ii) $\psi_2'(S) \in \mathcal{R} \iff hS \in \mathcal{L}_1 \oplus \mathcal{L}_2 \oplus \mathcal{L}_3 \oplus \mathcal{L}_5$;

iii) $\psi_3'(S) \in \mathcal{R} \iff S(X,Y) = S(Y,X)$ *and* $hS \in \mathcal{L}_1 \oplus \mathcal{L}_2 \oplus \mathcal{L}_3$;

iv) $(\psi_1 \pm \psi_2')(S) \in \mathcal{R} \iff S(X,Y) = S(Y,X)$ *and* $hS \in \mathcal{L}_1 \oplus \mathcal{L}_2 \oplus \mathcal{L}_3$.

4. The subspace $h\mathcal{R}$

The decomposition of $h\mathcal{R}$ over V coincides with the decomposition of $\mathcal{R}$ over hV. A complete decomposition of $\mathcal{R}$ over an almost Hermitian vector space in ten orthogonal and $U(n)$-invariant subspaces is obtained in [8]. The space $\mathcal{R}$ over an almost complex vector space with B-metric has been decomposed into eleven mutually orthogonal and $G-$ invariant subspaces ω_i $(i = 1, 2, \dots, 11)$ in [7]. We give another way of finding of ω_i $(i = 1, 2, \dots, 11)$.

We consider the same linear operators L_i $(i = 1, 2, 3)$ on $h\mathcal{R}$ as in [8], defined by

$$
\begin{aligned}
(L_1R)(x,y,z,w) &= \frac{1}{2}\{R(\varphi x,\varphi y,z,w) + R(y,\varphi z,\varphi x,w) + R(\varphi z,x,\varphi y,w)\},\\
(L_2R)(x,y,z,w) &= \frac{1}{2}\{R(x,y,z,w) + R(\varphi x,\varphi y,z,w) + R(\varphi x,y,\varphi z,w)\\
&\qquad + R(\varphi x,y,z,\varphi w)\},\\
(L_3R)(x,y,z,w) &= R(\varphi x,\varphi y,\varphi z,\varphi w), \quad x,y,z,w \in hV.
\end{aligned}
$$

It is easy to check that L_i $(i = 1, 2, 3)$ determines an involutive isometry on $h\mathcal{R}_{i+1}$ and L_i commutes with the action of G, where

$$
\begin{aligned}
h\mathcal{R}_{i+1} &= h\mathcal{R}_i \oplus h\mathcal{R}_i^{\perp}, \quad i = 1,2,3, \quad (h\mathcal{R}_4 = h\mathcal{R}),\\
h\mathcal{R}_i^{\perp} &= \{R \in h\mathcal{R}_{i+1} : R = \frac{1}{2}(R - L_iR) = -L_iR\},\\
h\mathcal{R}_i &= \{R \in h\mathcal{R}_{i+1} : R = \frac{1}{2}(R + L_iR) = L_iR\}.
\end{aligned}
$$

From well known algebraic result it follows

Lemma 4.1. *The decomposition* $h\mathcal{R} = h\mathcal{R}_1 \oplus h\mathcal{R}_1^{\perp} \oplus h\mathcal{R}_2^{\perp} \oplus h\mathcal{R}_3^{\perp}$ *is orthogonal and invariant under the action of* $G \times I$.

The subspaces

$$
\begin{aligned}
(\mathcal{L}_1) &: R(x,y,\varphi z,\varphi w) = -R(x,y,z,w),\\
(\mathcal{L}_2) &: R(x,y,\varphi z,\varphi w) + R(y,z,\varphi x,\varphi w) + R(z,x,\varphi y,\varphi w) = 0,\\
(\mathcal{L}_3) &: R(\varphi x,\varphi y,\varphi z,\varphi w) = R(x,y,z,w)
\end{aligned}
$$

of $\mathcal{R}$ over an almost complex vector space with a complex structure φ and B-metric have been investigated in [5] and $\mathcal{L}_1 \subseteq \mathcal{L}_2 \subseteq \mathcal{L}_3 \subseteq \mathcal{R}$ has been obtained.

In [7] by $\mathcal{L}_1^\perp, \mathcal{L}_2^\perp, \mathcal{L}_3^\perp$ are denoted the orthogonal complement of $\mathcal{L}_1$ in $\mathcal{L}_2$, the orthogonal complement of $\mathcal{L}_2$ in $\mathcal{L}_3$, the orthogonal complement of $\mathcal{L}_3$ in $\mathcal{R}$, respectively.

Let us consider the linear operator $L : \mathcal{L}_2 \mapsto \mathcal{L}_2$, defined by

$$(LR)(x, y, z, w) = R(x, y, \varphi z, \varphi w), \quad x, y, z, w \in hV.$$

The following lemma is true.

Lemma 4.2. *L is an involutive isometry in $\mathcal{L}_2$ and commutes with the action of $G \times I$. We have $\mathcal{L}_2 = h\mathcal{R}_1 \oplus h\mathcal{R}_2^\perp$.*

Proof. It is not difficult to verify that $L^2 = id$ in $\mathcal{L}_2$, $<LR, LR'> = <R, R'>$ for $R, R' \in \mathcal{L}_2$ and $L \circ \lambda(a) = \lambda(a) \circ L$, $a \in G \times I$. Then the decomposition $\mathcal{L}_2 = \mathcal{L}_{21} \oplus \mathcal{L}_{21}^\perp$ is orthogonal and $G \times I$-invariant and its components are

$$\mathcal{L}_{21} = \{R \in \mathcal{L}_2 : R = \frac{1}{2}(R + LR) = LR\},$$
$$\mathcal{L}_{21}^\perp = \{R \in \mathcal{L}_2 : R = \frac{1}{2}(R - LR) = -LR\}.$$

Analogously as [8] we prove $R \in h\mathcal{R}_1 \iff R(x, y, \varphi z, \varphi w) = R(x, y, z, w)$ and $R \in h\mathcal{R}_2^\perp \iff R(x, y, \varphi z, \varphi w) = -R(x, y, z, w)$. Then $\mathcal{L}_{21} \equiv h\mathcal{R}_1$, $\mathcal{L}_{21}^\perp \equiv h\mathcal{R}_2^\perp$ and $\mathcal{L}_2 = h\mathcal{R}_1 \oplus h\mathcal{R}_2^\perp$. □

Taking into account the definitions of $\mathcal{L}_i$ ($i = 1, 3$), $\mathcal{L}_i^\perp$ ($i = 1, 2, 3$), Lemma 4.1 and 4.2 we conclude $\mathcal{L}_1 \equiv h\mathcal{R}_2^\perp$, $\mathcal{L}_3 \equiv h\mathcal{R}_3$, $\mathcal{L}_1^\perp \equiv h\mathcal{R}_1$, $\mathcal{L}_2^\perp \equiv h\mathcal{R}_1^\perp$, $\mathcal{L}_3^\perp \equiv h\mathcal{R}_3^\perp$. Hence $h\mathcal{R} = \mathcal{L}_1 \oplus \mathcal{L}_1^\perp \oplus \mathcal{L}_2^\perp \oplus \mathcal{L}_3^\perp$, i.e. the partial decomposition of $h\mathcal{R}$ from Lemma 4.1 coincides with the partial decomposition of $\mathcal{R}$ over an almost complex vector space with B-metric in [7].

By explicit calculations we obtain the following lemmas.

Lemma 4.3. *Let S be a tensor of type $(0, 2)$. Then*

i) *$h(\psi_1 \pm \psi_2')(S)$ and $h\psi_3'(S)$ are curvature-like tensors iff $hS \in \mathcal{L}_1 \oplus \mathcal{L}_2 \oplus \mathcal{L}_3$;*

ii) *$h(\psi_1 - \psi_2')(S)$ and $h\psi_3'(S)$ belong to $h\mathcal{R}_2^\perp$ for $S \in \mathcal{L} : hS \in \mathcal{L}_1 \oplus \mathcal{L}_2 \oplus \mathcal{L}_3$;*

iii) *$h(\psi_1 + \psi_2')(S) \in h\mathcal{R}_1$ for $S \in \mathcal{L} : hS \in \mathcal{L}_1 \oplus \mathcal{L}_2 \oplus \mathcal{L}_3$;*

iv) *$h(\pi_1 - \pi_2) \in h\mathcal{R}_2^\perp$, $h\pi_3 \in h\mathcal{R}_2^\perp$, $h(\pi_1 + \pi_2) \in h\mathcal{R}_1$.*

Lemma 4.4. *For the Ricci tensors ρ_R, ρ_R^*, ρ_R'', $\tilde{\rho}_R$ of $R \in h\mathcal{R}$ we have*

i) *$R \in h\mathcal{R}_1 \Rightarrow \rho_R = \rho_R'' \in \mathcal{L}_1 \oplus \mathcal{L}_2 \oplus \mathcal{L}_3$, $\rho_R^* \in \mathcal{L}_1 \oplus \mathcal{L}_2 \oplus \mathcal{L}_3$;*

ii) *$R \in h\mathcal{R}_1^\perp \Rightarrow \rho_R = \rho_R'' = \rho_R^* = 0$;*

iii) *$R \in h\mathcal{R}_2^\perp \Rightarrow \rho_R = -\rho_R'' \in \mathcal{L}_1 \oplus \mathcal{L}_2 \oplus \mathcal{L}_3$, $\rho_R^* \in \mathcal{L}_1 \oplus \mathcal{L}_2 \oplus \mathcal{L}_3$;*

iv) $R \in h\mathcal{R}_3^{\perp} \Rightarrow \rho_R \in \mathcal{L}_4$, $\rho''_R \in \mathcal{L}_5$, $\rho^*_R \in \mathcal{L}_4$;

v) $R \in h\mathcal{R} \Rightarrow \tilde{\rho}_R = 0$.

Further we shall describe the components of the decomposition from Lemma 4.1.

Lemma 4.5. *The operator* $p_1 : h\mathcal{R}_2^{\perp} \mapsto h\mathcal{R}_2^{\perp}$,

$$p_1 R = R - \frac{\tau_R}{4n(n-1)} h(\pi_1 - \pi_2) - \frac{\tau^*_R}{4n(n-1)} h\pi_3$$

is a self-conjugate projection, commuting with $G \times I$.

Proof. From Lemma 4.3 it follows that $h(\pi_1 - \pi_2) \in h\mathcal{R}_2^{\perp}$, $h\pi_3 \in h\mathcal{R}_2^{\perp}$. We compute $\tau_{h(\pi_1-\pi_2)} = 4n(n-1)$, $\tau_{h\pi_3} = 0$, $\tau^*_{h(\pi_1-\pi_2)} = 0$, $\tau^*_{h\pi_3} = 4n(n-1)$, $\tau_{p_1R} = \tau^*_{p_1R} = 0$. Then

$$p_1(p_1R) = p_1R - \frac{\tau_{p_1R}}{4n(n-1)} h(\pi_1 - \pi_2) - \frac{\tau^*_{p_1R}}{4n(n-1)} h\pi_3 = p_1R,$$

i.e. p_1 is a projection. After direct computations we check that p_1 has the property $<p_1R_1, R_2> = <R_1, p_1R_2>$, R_1, $R_2 \in h\mathcal{R}_2^{\perp}$ and p_1 commutes with $G \times I$. □

Lemma 4.5 implies the decomposition $h\mathcal{R}_2^{\perp} = \mathrm{Ker}(p_1) \oplus \mathrm{Im}(p_1)$ that is orthogonal and $G \times I$-invariant, where

$$\begin{aligned}
\mathrm{Ker}(p_1) &= \{R \in h\mathcal{R}_2^{\perp} : R = \frac{\tau_R}{4n(n-1)} h(\pi_1 - \pi_2) + \frac{\tau^*_R}{4n(n-1)} h\pi_3\} \\
&= \{R \in h\mathcal{R}_2^{\perp} : \rho_R \in \mathcal{L}_1 \oplus \mathcal{L}_2,\ \rho^*_R \in \mathcal{L}_1 \oplus \mathcal{L}_2,\ \rho''_R \in \mathcal{L}_1 \oplus \mathcal{L}_2\}, \\
\mathrm{Im}(p_1) &= \mathrm{Ker}^{\perp}(p_1) \\
&= \{R' \in h\mathcal{R}_2^{\perp} : R' = p_1R \\
&\qquad = R - \frac{\tau_R}{4n(n-1)} h(\pi_1 - \pi_2) - \frac{\tau^*_R}{4n(n-1)} h\pi_3, R \in h\mathcal{R}_2^{\perp}\} \\
&= \{R \in h\mathcal{R}_2^{\perp} : \rho_R \in \mathcal{L}_3,\ \rho^*_R \in \mathcal{L}_3,\ \rho''_R \in \mathcal{L}_3\} \\
&= \{R \in h\mathcal{R}_2^{\perp} : \tau_R = \tau^*_R = \tau''_R = 0\}.
\end{aligned}$$

Lemma 4.6. *The operator* $p_2 : \mathrm{Ker}(p_1) \mapsto \mathrm{Ker}(p_1)$, $p_2R = R - \dfrac{\tau_R}{4n(n-1)} h(\pi_1 - \pi_2)$ *is a self-conjugate projection, commuting with* $G \times I$.

If we denote

$$\begin{aligned}
\omega_1 &= \mathrm{Ker}(p_2) = \{R \in \mathrm{Ker}(p_1) : R = \frac{\tau_R}{4n(n-1)} h(\pi_1 - \pi_2)\} \\
&= \{R \in \mathrm{Ker}(p_1) : \rho_R \in \mathcal{L}_1,\ \rho^*_R \in \mathcal{L}_2,\ \rho''_R \in \mathcal{L}_1\} \\
&= \{R \in \mathrm{Ker}(p_1) : \tau^*_R = 0\}, \\
\omega_2 &= \mathrm{Im}(p_2) \\
&= \{W' \in \mathrm{Ker}(p_1) : W' = R - \frac{\tau_R}{4n(n-1)} h(\pi_1 - \pi_2),\ R \in \mathrm{Ker}(p_1)\} \\
&= \{W' \in \mathrm{Ker}(p_1) : W' = \frac{\tau^*_R}{4n(n-1)} h\pi_3,\ R \in \mathrm{Ker}(p_1)\} \\
&= \{R \in \mathrm{Ker}(p_1) : \rho_R \in \mathcal{L}_2,\ \rho^*_R \in \mathcal{L}_1,\ \rho''_R \in \mathcal{L}_2\} \\
&= \{R \in \mathrm{Ker}(p_1) : \tau_R = \tau''_R = 0\},
\end{aligned}$$

then Lemma 4.6 implies the decomposition $\mathrm{Ker}(p_1) = \omega_1 \oplus \omega_2$ that is orthogonal and $G \times I$-invariant.

Lemma 4.7. *The operator* $p_3 : \mathrm{Im}(p_1) \mapsto \mathrm{Im}(p_1)$, $p_3 R = R - \frac{1}{2(n-2)} h(\psi_1 - \psi'_2)(\rho_R)$ *is a self-conjugate projection, commuting with* $G \times I$.

If we denote

$$\begin{aligned}
\omega_3 &= \mathrm{Ker}(p_3) = \{R \in \mathrm{Im}(p_1) : R = \frac{1}{2(n-2)} h(\psi_1 - \psi'_2)(\rho_R)\} \\
&= \{R \in \mathrm{Im}(p_1) : \rho_R \in \mathcal{L}_3,\ \rho^*_R \in \mathcal{L}_3,\ \rho''_R \in \mathcal{L}_3\} \\
&= \{R \in Im(p_1) : \tau_R = \tau^*_R = \tau''_R = 0\}, \\
\omega_4 &= \mathrm{Im}(p_3) = \{W'' \in \mathrm{Im}(p_1) : W'' = (p_3 \circ p_1)R,\ R \in h\mathcal{R}_2^{\perp}\} \\
&= \{R \in \mathrm{Im}(p_1) : \rho_R = \rho^*_R = \rho''_R = 0\},
\end{aligned}$$

then Lemma 4.7 implies the decomposition $\mathrm{Im}(p_1) = \omega_3 \oplus \omega_4$ that is orthogonal and $G \times I$-invariant.

From Lemma 4.6 and Lemma 4.7 we obtain immediately

Proposition 4.1. *The decomposition* $h\mathcal{R}_2^{\perp} = \omega_1 \oplus \omega_2 \oplus \omega_3 \oplus \omega_4$ *is orthogonal and* $G \times I$*-invariant.*

Further we consider $h\mathcal{R}_1$.

Lemma 4.8. *The operator* $q_1 : h\mathcal{R}_1 \mapsto h\mathcal{R}_1$, $q_1 R = R - \frac{\tau_R}{4n^2} h(\pi_1 + \pi_2)$ *is a self-conjugate projection, commuting with* $G \times I$.

If we denote

$$\begin{aligned}
\omega_5 &= \mathrm{Ker}(q_1) = \{R \in h\mathcal{R}_1 : R = \frac{\tau_R}{4n^2} h(\pi_1 + \pi_2)\} \\
&= \{R \in h\mathcal{R}_1 : \rho_R = \rho_R^{''} \in \mathcal{L}_1,\, \rho_R^* \in \mathcal{L}_2\} \\
&= \{R \in h\mathcal{R}_1 : \tau_R^* = 0\}, \\
\mathrm{Im}(q_1) &= \{V' \in h\mathcal{R}_1 : V' = \mathrm{Im}(q_1) = R - \frac{\tau_R}{4n^2} h(\pi_1 + \pi_2),\, R \in h\mathcal{R}_1\} \\
&= \{R \in h\mathcal{R}_1 : \rho_R = \rho_R^{''} \in \mathcal{L}_2 \oplus \mathcal{L}_3,\, \rho_R^* \in \mathcal{L}_1 \oplus \mathcal{L}_3\} \\
&= \{R \in h\mathcal{R}_1 : \tau_R = \tau_R^{''} = 0\},
\end{aligned}$$

then Lemma 4.8 implies the decomposition $h\mathcal{R}_1 = \omega_5 \oplus \mathrm{Im}(q_1)$ that is orthogonal and $G \times I$-invariant.

Lemma 4.9. *The operator* $q_2 : \mathrm{Im}(q_1) \mapsto \mathrm{Im}(q_1)$, $q_2 R = R - \frac{1}{2n} h(\psi_1 + \psi_2')(\rho_R)$ *is a self-conjugate projection, commuting with* $G \times I$.

If we denote

$$\begin{aligned}
\omega_6 &= \mathrm{Ker}(q_2) = \{R \in \mathrm{Im}(q_1) : R = \frac{1}{2n} h(\psi_1 + \psi_2')(\rho_R)\} \\
&= \{R \in \mathrm{Im}(q_1) : \rho_R = \rho_R^{''} \in \mathcal{L}_3,\, \rho_R^* \in \mathcal{L}_3\} \\
&= \{R \in \mathrm{Im}(q_1) : \tau_R = \tau_R^{''} = \tau_R^* = 0\}, \\
\omega_7 &= \mathrm{Im}(q_2) = \{V^{''} \in \mathrm{Im}(q_1) : V^{''} = (q_2 \circ q_1) R,\, R \in h\mathcal{R}_1\} \\
&= \{R \in \mathrm{Im}(q_1) : \rho_R = \rho_R^{''} \in \mathcal{L}_2,\, \rho_R^* \in \mathcal{L}_1\} \\
&= \{R \in \mathrm{Im}(q_1) : \tau_R = \tau_R^{''} = 0\},
\end{aligned}$$

then Lemma 4.9 implies the decomposition $\mathrm{Im}(q_1) = \omega_6 \oplus \omega_7$ that is orthogonal and $G \times I$-invariant.

From Lemma 4.8 and Lemma 4.9 we obtain immediately

Proposition 4.2. *The decomposition* $h\mathcal{R}_1 = \omega_5 \oplus \omega_6 \oplus \omega_7$ *is orthogonal and* $G \times I$*-invariant.*

Lemma 4.10. *The operator* $r_1 : h\mathcal{R}_3^{\perp} \mapsto h\mathcal{R}_3^{\perp}$, $r_1 R = R - \frac{1}{2(n-1)} h\{\psi_1(\rho_R) + \psi_2'(\rho_R^{''})\}$ *is a self-conjugate projection, commuting with* $G \times I$.

If we denote

$$\operatorname{Ker}(r_1) = \{R \in h\mathcal{R}_3^{\perp} : R = \frac{1}{2(n-1)} h\{\psi_1(\rho_R) + \psi_2'(\rho_R'')\},$$

$$\begin{aligned}\omega_{11} = \operatorname{Im}(r_1) &= \{S' \in h\mathcal{R}_3^{\perp} : S' = \operatorname{Im}(r_1)\\ &= R - \frac{1}{2(n-1)} h\{\psi_1(\rho_R) + \psi_2'(\rho_R'')\}, R \in h\mathcal{R}_3^{\perp}\}\\ &= \{R \in h\mathcal{R}_3^{\perp} : \rho_R = \rho_R'' = \rho_R^* = 0\},\end{aligned}$$

then Lemma 4.10 implies the decomposition $h\mathcal{R}_3^{\perp} = \operatorname{Ker}(r_1) \oplus \omega_{11}$ that is orthogonal and $G \times I$-invariant.

Lemma 4.11. *The operator* $r_2 : \operatorname{Ker}(r_1) \mapsto \operatorname{Ker}(r_1)$, $r_2 R = R - \frac{1}{2(n-1)} h\psi_1(\rho_R)$ *is a self-conjugate projection, commuting with* $G \times I$.

If we denote

$$\begin{aligned}\omega_9 = \operatorname{Ker}(r_2) &= \{R \in \operatorname{Ker}(r_1) : R = \frac{1}{2(n-1)} h\psi_1(\rho_R)\}\\ &= \{R \in \operatorname{Ker}(r_1) : \rho_R \in \mathcal{L}_4,\ \rho_R^* = \rho_R'' = 0\}\\ &= \{R \in \operatorname{Ker}(r_1) : \tau_R = 0\},\\ \omega_{10} = \operatorname{Im}(r_2) &= \{S'' \in \operatorname{Ker}(r_1) : S'' = R - \frac{1}{2(n-1)} h\psi_1(\rho_R), R \in \operatorname{Ker}(r_1)\}\\ &= \{R \in \operatorname{Ker}(r_1) : \rho_R = 0,\ \rho_R'' \in \mathcal{L}_5,\ \rho_R^* \in \mathcal{L}_4\}\\ &= \{R \in \operatorname{Ker}(r_1) : \tau_R^* = \tau_R'' = 0\},\end{aligned}$$

then Lemma 4.11 implies the decomposition $\operatorname{Ker}(r_1) = \omega_9 \oplus \omega_{10}$ is orthogonal and $G \times I$-invariant. From Lemma 4.10 and Lemma 4.11 we obtain immediately

Proposition 4.3. *The decomposition* $h\mathcal{R}_3^{\perp} = \omega_9 \oplus \omega_{10} \oplus \omega_{11}$ *is orthogonal and* $G \times I$*-invariant.*

Finally, we denote $\omega_8 = h\mathcal{R}_1^{\perp}$. Taking into account Lemma 4.1, Lemma 4.4 – ii), Proposition 4.1 ÷4.3, we have

Theorem 4.1. *The decomposition* $h\mathcal{R} = \omega_1 \oplus \omega_2 \oplus \ldots \oplus \omega_{11}$ *is orthogonal and* $G \times I$*-invariant.*

5. The subspace $v\mathcal{R}$

The decomposition of $v\mathcal{R}$ depends on the decomposition of $\overline{\mathcal{R}} \subset \mathcal{R}$:

$$\overline{\mathcal{R}} = \{R(x, y, z, \xi),\ \ x, y, z \in hV\}.$$

If $R \in \overline{\mathcal{R}}$, then $\rho_R(y,\xi) = \sum_{i=1}^{2n} R(e_i, y, \xi, e_i) = -\sum_{i=1}^{2n} R(e_i, y, e_i, \xi)$.

Lemma 5.1. *The decomposition $\overline{\mathcal{R}} = \overline{\mathcal{R}}_1 \oplus \overline{\mathcal{R}}_1^{\perp} \oplus \overline{\mathcal{R}}_2^{\perp}$ is orthogonal and invariant under the action of $G \times I$. Its components are defined by*

$$\overline{\mathcal{R}}_{i+1} = \overline{\mathcal{R}}_i \oplus \overline{\mathcal{R}}_i^{\perp}, \quad i = 1, 2, 3, \quad (\overline{\mathcal{R}}_3 = \overline{\mathcal{R}}),$$
$$\overline{\mathcal{R}}_i^{\perp} = \{R \in \overline{\mathcal{R}}_{i+1} : R = \frac{1}{2}(R - V_i R) = -V_i R\},$$
$$\overline{\mathcal{R}}_i = \{R \in \overline{\mathcal{R}}_{i+1} : R = \frac{1}{2}(R + V_i R) = V_i R\},$$

where V_i are involutive isometries on $\overline{\mathcal{R}}_{i+1}$, $i = 1, 2, 3$, commuting with $G \times I$ and V_i, $i = 1, 2$ are defined by

$$(V_1 R)(x,y,z,\xi) = \frac{1}{2}\{R(\varphi x, \varphi y, z, \xi) + R(y, \varphi z, \varphi x, \xi) + R(\varphi z, x, \varphi y, \xi)\},$$
$$(V_2 R)(x,y,z,\xi) = \frac{1}{2}\{R(x,y,z,\xi) + R(\varphi x, y, \varphi z, \xi) + R(\varphi x, \varphi y, z, \xi) + R(x, \varphi y, \varphi z, \xi)\}.$$

Using the properties $R(x,y,z,\xi) = -R(y,x,z,\xi)$ and $\underset{(x,y,z)}{\sigma} R(x,y,z,\xi) = 0$ of $R \in \overline{\mathcal{R}}$ we prove

Lemma 5.2. *For $R \in \overline{\mathcal{R}}$ and $x,\ y,\ z \in hV$ the following conditions are equivalent:*

i) $R(\varphi x, \varphi y, z, \xi) = -R(x,y,z,\xi)$;

ii) $R(x, \varphi y, \varphi z, \xi) = -R(x,y,z,\xi)$;

iii) $R(\varphi x, y, \varphi z, \xi) = -R(x,y,z,\xi)$.

Further we give some lemmas which we need.

Lemma 5.3. $R \in \overline{\mathcal{R}}_2^{\perp}$ *iff* $R(x, \varphi y, \varphi z, \xi) = -R(x,y,z,\xi)$.

Lemma 5.4. $R \in \overline{\mathcal{R}}_1$ *iff* $R(\varphi x, \varphi y, z, \xi) = R(x,y,z,\xi)$.

Using Lemma 5.3 and Lemma 5.4 we obtain

Lemma 5.5. *For the curvature-like tensors $(\psi_1 - \psi_2')(S)$, $\psi_3'(S)$, $(\psi_1 + \psi_2')(S)$, $\psi_i(S)$ $(i = 4, 5)$ and π_i $(i = 1,\ 2,\ 3,\ 4,\ 5)$ we have*

i) $(\psi_1 - \psi_2')(S)(x,y,z,\xi)$ *and* $\psi_3'(S)(x,y,z,\xi)$ *belong to* $\overline{\mathcal{R}}_2^{\perp}$;

ii) $(\psi_1 + \psi_2')(S)(x,y,z,\xi) \in \overline{\mathcal{R}}_1$;

iii) $\psi_i(S)(x,y,z,\xi) = 0$ $(i = 4,5)$, $\pi_i(x,y,z,\xi) = 0$ $(i = 1,2,3,4,5)$.

Lemma 5.6. *Let $R \in \overline{\mathcal{R}}$. Then*

i) $R \in \overline{\mathcal{R}}_1 \Rightarrow \rho_R,\ \rho^*_R$ *are symmetric and* $\rho_R(y,\xi) = \rho^*_R(\varphi y, \xi)$;

ii) $R \in \overline{\mathcal{R}}_1^{\perp} \Rightarrow \rho_R = \rho^*_R = 0$;

iii) $R \in \overline{\mathcal{R}}_2^{\perp} \Rightarrow \rho_R,\ \rho^*_R$ *are symmetric and* $\rho_R(y,\xi) = -\rho^*_R(\varphi y, \xi)$.

Now, by using the self-conjugate projections s, t, commuting with $G \times I$, we describe the components of the decomposition from Lemma 5.1.

Lemma 5.7. *The operator* $s : \overline{\mathcal{R}}_1 \mapsto \overline{\mathcal{R}}_1$, $sR = R - \frac{1}{2n}(\psi_1 + \psi_2')(\rho_R)$ *is a self-conjugate projection, commuting with* $G \times I$.

Lemma 5.8. *The operator* $t : \overline{\mathcal{R}}_2^{\perp} \mapsto \overline{\mathcal{R}}_2^{\perp}$, $tR = R - \frac{1}{2(n-1)}(\psi_1 - \psi_2')(\rho_R)$ *is a self-conjugate projection, commuting with* $G \times I$.

From Lemma 5.1, Lemma 5.6 - ii), Lemma 5.7, Lemma 5.8 we have

Proposition 5.1. *The decomposition* $\overline{\mathcal{R}} = \overline{v}_1 \oplus \overline{v}_2 \oplus \overline{v}_3 \oplus \overline{v}_4 \oplus \overline{v}_5$ *is orthogonal and* $G \times I$*-invariant, where*

$$
\begin{aligned}
\overline{v}_1 &= \mathrm{Ker}(s) = \{R \in \overline{\mathcal{R}}_1 : R = \frac{1}{2n}(\psi_1 + \psi_2')(\rho_R)\},\\
\overline{v}_2 &= \mathrm{Im}(s) = \{V' \in \overline{\mathcal{R}}_1 : V' = sR = R - \frac{1}{2n}(\psi_1 + \psi_2')(\rho_R),\ R \in \overline{\mathcal{R}}_1\}\\
&= \{R \in \overline{\mathcal{R}}_1 : \rho_R = \rho^*_R = 0\},\\
\overline{v}_3 &= \overline{\mathcal{R}}_1^{\perp},\\
\overline{v}_4 &= \mathrm{Ker}(t) = \{R \in \overline{\mathcal{R}}_2^{\perp} : R = \frac{1}{2(n-1)}(\psi_1 - \psi_2')(\rho_R)\},\\
\overline{v}_5 &= \mathrm{Im}(t)\\
&= \{V'' \in \overline{\mathcal{R}}_2^{\perp} : V'' = tR = R - \frac{1}{2(n-1)}(\psi_1 - \psi_2')(\rho_R),\ R \in \overline{\mathcal{R}}_2^{\perp}\}\\
&= \{R \in \overline{\mathcal{R}}_2^{\perp} : \rho_R = \rho^*_R = 0\}.
\end{aligned}
$$

Definition 5.1. We denote by $v\mathcal{R}_i = \{R \in v\mathcal{R} : R(x,y,z,\xi) \in \overline{\mathcal{R}}_i\}$, $v\mathcal{R}_i^{\perp} = \{R \in v\mathcal{R} : R(x,y,z,\xi) \in \overline{\mathcal{R}}_i^{\perp}\}$, $i = 1, 2$; $v_j = \{R \in v\mathcal{R} : R(x,y,z,\xi) \in \overline{v}_j\}$, $j = 1, \ldots, 5$).

From the definition of the subspace $v\mathcal{R}$, Definition 5.1, Lemma 5.1 and Proposition 5.1 we obtain

Theorem 5.1. *The decompositions* $v\mathcal{R} = v\mathcal{R}_1 \oplus v\mathcal{R}_1^{\perp} \oplus v\mathcal{R}_2^{\perp}$, $v\mathcal{R} = v_1 \oplus v_2 \oplus v_3 \oplus v_4 \oplus v_5$ *are orthogonal and* $G \times I$*-invariant.*

Lemma 5.9. *For the Ricci tensors ρ_R, ρ^*_R, ρ''_R, $\tilde{\rho}_R$ of $R \in v\mathcal{R}$ we have*

i) $R \in v_i \Rightarrow \rho_R, \rho^*_R$ *belong to* $\mathcal{L}_7$ *and* $\rho''_R(Y,Z) = \rho^*_R(Y,\varphi Z)$, $i = 1,4$;

ii) $R \in v_i \Rightarrow \rho_R = \rho^*_R = \rho''_R = 0$, $i = 2,3,5$;

iii) $R \in v\mathcal{R} \Rightarrow \tilde{\rho}_R = 0$.

6. The subspace $w\mathcal{R}$

The decomposition of $w\mathcal{R}$ depends on the decomposition of $\widetilde{\mathcal{R}} \subset \mathcal{R} : \widetilde{\mathcal{R}} = \{R(\xi,y,z,\xi) = \tilde{\rho}_R(y,z),\ y,\ z \in hV\}$. For $R \in w\mathcal{R}$ $\tilde{\rho}_R(Y,Z) = R(\xi,y,z,\xi)$ and from the properties of $R \in \mathcal{R}$ it follows $R(\xi,y,z,\xi) = R(\xi,z,y,\xi)$. Hence, the decomposition of $\widetilde{\mathcal{R}}$ coincides with the decomposition of the symmetric tensors of $h\mathcal{L}$.

Definition 6.1. We denote by $w_i = \{R \in w\mathcal{R} : \tilde{\rho}_R \in \mathcal{L}_i\}$, $i = 1, 2, 3, 4$.

Using the definition of the subspace $w\mathcal{R}$, Definition 6.1 and Lemma 2.1 we check that w_i, $i = 1, 2, 3, 4$ are mutually orthogonal and $G \times I$-invariant subspaces of $w\mathcal{R}$. Then we have

Theorem 6.1. *The decomposition* $w\mathcal{R} = w_1 \oplus w_2 \oplus w_3 \oplus w_4$ *is orthogonal and* $G \times I$*-invariant.*

Lemma 6.1. *For the Ricci tensors ρ_R, ρ^*_R, ρ''_R, $\tilde{\rho}_R$ of $R \in w\mathcal{R}$ we have*

i) $R \in w_1 \Rightarrow \rho_R \in \mathcal{L}_9$, $\rho^*_R = 0$, $\tilde{\rho}_R \in \mathcal{L}_1$;

ii) $R \in w_2 \Rightarrow \rho_R = 0$, $\rho^*_R \in \mathcal{L}_9$, $\tilde{\rho}_R \in \mathcal{L}_2$;

iii) $R \in w_i \Rightarrow \rho_R = \rho^*_R = 0$, $\tilde{\rho}_R \in \mathcal{L}_i$, $i = 3,\ 4$;

iv) $R \in w\mathcal{R} \Rightarrow \rho''_R = 0$.

Taking into account Lemma 3.1, Theorem 4.1, Theorem 5.1 and Theorem 6.1 we obtain

Theorem 6.2. *Let* $(V^{2n+1},\varphi,\xi,\eta,g)$ *be an almost contact vector space with B-metric. Then the space of all Riemannian-like curvature tensors* $\mathcal{R}$ *over* $(V^{2n+1},\varphi,\xi,\eta,g)$ *is decomposable into* 20 *mutually orthogonal and* $G \times I$*-invariant factors:*

$$\mathcal{R} = \omega_1 \oplus \omega_2 \oplus \omega_3 \oplus \omega_4 \oplus \omega_5 \oplus \omega_6 \oplus \omega_7 \oplus \omega_8 \oplus \omega_9 \oplus \omega_{10} \oplus \omega_{11}$$
$$\oplus\, v_1 \oplus v_2 \oplus v_3 \oplus v_4 \oplus v_5 \oplus w_1 \oplus w_2 \oplus w_3 \oplus w_4.$$

References

1. V. Alexiev, G. Ganchev, *Conformally invariant tensors on an almost contact metric manifold*, Plovdiv Univ. Sci. Works-Math., vol. 23, **2** (1985), 209–219 (in Bulgarian).
2. D. Blair, *Contact manifolds in Riemmannian geometry*, Lecture Notes in Math., **509** (1976), Springer-Verlag, Berlin.
3. G. Ganchev and M. Hristov, *Riemannian curvatures of the four basic classes of real hypersurfaces of a complex space form*, Third international Conference on Geometry, Integrability and Quantization, June 14–23, 2001, Varna, I. Mladenov and G. Naber (Eds), 238–248.
4. G. Ganchev, V. Mihova and K. Gribachev, *Almost contact manifolds with B-metric*, Math. Balkanica. **7** (1993), 262–276.
5. K. Gribachev, G. Djelepov, D. Mekerov, *On some subclasses of generalized B-manifold*, Comptes rendus I' Academie bulgare des Sciences, **38** (1985), 437–440.
6. M. Manev, *On the conformal geometry of almost contact manifolds with B-metric*, Doctor thesis, Plovdiv, (1998) (in Bulgarian).
7. E. Pavlov, H. Hopteriev, *Curvature tensors on generalized B- manifolds*, Universite de Plovdiv "Paissi Hilendarski", Travaux acientifiques, vel. 24, fasc. **1** (1986) - Mathematique, 96–130.
8. F. Tricerri, L. Vanhecke, *Curvature tensors in an almost Hermitian manifolds*, Trans Amer. Math. soc. **277** (1981).

CURVATURE TENSORS IN THE BASIC CLASSES OF REAL ISOTROPIC HYPERSURFACES OF A KAEHLER MANIFOLD WITH B-METRIC

GALIA NAKOVA

Department of Algebra and Geometry, Faculty of Pedagogics,
University of Veliko Tirnovo "St. St. Cyril and Methodius",
1 Theodosij Tirnovsky Str., 5000 Veliko Tirnovo, Bulgaria
E-mail: gnakova@yahoo.com

In this paper we characterize the twenty factors of the space of curvature tensors $\mathcal{R}$ over an almost contact vector space with B-metric by the essential components and equalities for the curvature tensor. We study the basic classes of real isotropic hypersurfaces of a Kaehler manifold with B-metric in terms of their curvature tensors.

1. Introduction

Let $(M^{2n+1}, \varphi, \xi, \eta, g)$ be a $(2n+1)$ - dimensional almost contact manifold with B-metric, i.e. (φ, ξ, η) is an almost contact structure [1] and g is a metric [2] on M^{2n+1} such that:

$$\varphi^2 = -id + \eta \otimes \xi; \quad \eta(\xi) = 1; \quad g(\varphi X, \varphi Y) = -g(X,Y) + \eta(X)\eta(Y),$$

where $X, Y \in \mathcal{X}M$.

The associated with g metric $\tilde{g}$ [2] on the manifold is given by

$$\tilde{g}(X,Y) = g(X, \varphi Y) + \eta(X)\eta(Y), \quad X, Y \in \mathcal{X}M.$$

Both metrics g and $\tilde{g}$ are indefinite of signature $(n+1, n)$ [2].

Let $(V^{2n+1}, \varphi, \xi, \eta, g)$ be a $(2n+1)$-dimensional vector space with almost contact structure (φ, ξ, η) and B-metric g and $hV = \{X \in V : X = hX = -\varphi^2 X\}$. Denoting the restrictions of g and φ on hV with the same letters we obtain the almost complex vector space $\{hV, \varphi, g\}$ of dimension $2n$ with a complex structure φ and B-metric g. Then for an arbitrary $X \in V$ we have $X = x + \eta(X)\xi$, where by x we denote hX.

Let ∇ be the Levi-Civita connection of the metric g. The tensor field F of type (0,3) is defined by [2]

$$F(X,Y,Z) = g((\nabla_X \varphi)Y, Z), \quad X, Y, Z \in \mathcal{X}M.$$

A classification of the almost contact manifolds with B-metric with respect to the tensor F is given in [2], where are defined eleven basic classes $\mathcal{F}_i$, ($i = 1, \ldots, 11$).

Let $\mathcal{R}$ be the set of all Riemannian-like curvature tensors:

$$\mathcal{R} = \{R \in \bigotimes^4 V^* : R(X,Y,Z,W) = -R(Y,X,Z,W) = -R(X,Y,W,Z),$$
$$\underset{(X,Y,Z)}{\sigma} R(X,Y,Z,W) = 0\}.$$

Further ρ_R, ρ^*_R, ρ''_R, $\tilde{\rho}_R$ are the Ricci-type contractions associated with $R \in \mathcal{R}$ and τ_R, τ^*_R, τ''_R, $\tilde{\tau}_R$ are the corresponding scalar curvatures:

$$\rho_R(X,Y) = \sum_{i=1}^{2n+1} R(e_i, X, Y, e_i), \quad \tau_R = \sum_{i,j=1}^{2n+1} R(e_i, e_j, e_j, e_i);$$

$$\rho^*_R(X,Y) = \sum_{i=1}^{2n+1} R(e_i, X, Y, \varphi e_i), \quad \tau^*_R = \sum_{i,j=1}^{2n+1} R(e_i, e_j, e_j, \varphi e_i);$$

$$\rho''_R(X,Y) = \sum_{i=1}^{2n+1} R(e_i, X, \varphi Y, \varphi e_i), \quad \tau''_R = \sum_{i,j=1}^{2n+1} R(e_i, e_j, \varphi e_j, \varphi e_i);$$

$$\tilde{\rho}_R(X,Y) = R(\xi, X, Y, \xi), \quad \tilde{\tau}_R = \sum_{i=1}^{2n+1} R(\xi, e_i, e_i, \xi),$$

where $\{e_i, e_{2n+1} \equiv \xi\}$, ($i = 1, 2, \ldots, 2n$) is an orthonormal basis of T_pM, $p \in M$.

2. Characterization of the twenty factors of $\mathcal{R}$ over an almost contact vector space with B-metric

A decomposition of the space of curvature tensors $\mathcal{R}$ over an almost contact vector space with B-metric into 20 mutually orthogonal and $G \times I$-invariant factors is obtained in [7]. In this section we shall describe everyone of these factors by the essential components (which may not be zero) and complex equalities for R with respect to the standard complex coordinates [6].

Let $(V^{2n+1}, \varphi, \xi, \eta, g)$ be an almost contact vector space with B-metric and $V^C = D^C \oplus \{\xi\}$, where D^C is the complexification of $D = \operatorname{Ker}\eta$ and $\{\xi\} = (\operatorname{Im}\eta)\xi$. We have $D^C = D^{1,0} \oplus D^{0,1}$, where $D^{1,0}(D^{0,1})$ is $i(-i)$-eigenspace of φ and $D^{1,0} = \operatorname{span}\{Z_\alpha = e_\alpha - i\varphi e_\alpha\}$, $D^{0,1} = \operatorname{span}\{Z_{\bar{\alpha}} = e_\alpha + i\varphi e_\alpha\}$, $Z_0 \equiv \xi$, where $\{e_\alpha, \varphi e_\alpha\}$ ($\alpha = 1, \ldots, n$) is an orthonormal basis of V.

Let $\mathcal{L}$ be the vector space of all $(0,2)$-tensors over $(V^{2n+1},\varphi,\xi,\eta,g)$. In [7] it is proved that $\mathcal{L}=\bigoplus_{i=1}^{9}\mathcal{L}_i$, where $\mathcal{L}_i$ are mutually orthogonal and $G\times I$-invariant spaces. Now we give the description of $\mathcal{L}_i$ $(i=1,\ldots,9)$ by the symmetries for the essential components for $L\in\mathcal{L}$:

$$\mathcal{L}_1: L_{\alpha\beta}=L_{\beta\alpha}=\frac{\operatorname{tr}L}{2n}g_{\alpha\beta},$$
$$\mathcal{L}_2: L_{\alpha\beta}=L_{\beta\alpha}=-i\frac{\operatorname{tr}^*L}{2n}g_{\alpha\beta},$$
$$\mathcal{L}_3: L_{\alpha\beta}=L_{\beta\alpha},\quad \operatorname{tr}L=\operatorname{tr}^*L=0,$$
$$\mathcal{L}_4: L_{\bar\alpha\beta}=L_{\beta\bar\alpha},$$
$$\mathcal{L}_5: L_{\bar\alpha\beta}=-L_{\beta\bar\alpha},$$
$$\mathcal{L}_6: L_{\alpha\beta}=-L_{\beta\alpha},$$
$$\mathcal{L}_7: L_{0\alpha}=L_{\alpha0},$$
$$\mathcal{L}_8: L_{0\alpha}=\ \ L_{\alpha0},$$
$$\mathcal{L}_9: L_{00}.$$

In [7] is obtained the partial decomposition of $\mathcal{R}$ over $(V^{2n+1},\varphi,\xi,\eta,g)$ into mutually orthogonal and $G\times I$-invariant factors as follows:

$$\mathcal{R}=h\mathcal{R}_1\oplus h\mathcal{R}_1^{\perp}\oplus h\mathcal{R}_2^{\perp}\oplus h\mathcal{R}_3^{\perp}\oplus v\mathcal{R}_1\oplus v\mathcal{R}_1^{\perp}\oplus v\mathcal{R}_2^{\perp}\oplus w\mathcal{R} \tag{2.1}$$

Proposition 2.1. *The characteristic conditions for the factors from* (2.1) *in terms of the essential components are:*

$$h\mathcal{R}=h\mathcal{R}_3\oplus h\mathcal{R}_3^{\perp}=\{R\in\mathcal{R}; R_{\alpha\beta\gamma\delta}, R_{\bar\alpha\beta\gamma\delta}, R_{\alpha\bar\beta\gamma\bar\delta}, R_{\alpha\beta\bar\gamma\bar\delta} - \textit{essential}\},$$
$$h\mathcal{R}_3^{\perp}=\{R\in h\mathcal{R}; R_{\bar\alpha\beta\gamma\delta} - \textit{essential}\},$$
$$h\mathcal{R}_3=h\mathcal{R}_2\oplus h\mathcal{R}_2^{\perp}=\{R\in h\mathcal{R}; R_{\bar\alpha\beta\gamma\delta}=0\},$$
$$h\mathcal{R}_2^{\perp}=\{R\in h\mathcal{R}_3; R_{\alpha\beta\gamma\delta} - \textit{essential}\},$$
$$h\mathcal{R}_2=h\mathcal{R}_1\oplus h\mathcal{R}_1^{\perp}=\{R\in h\mathcal{R}_3; R_{\alpha\beta\gamma\delta}=0\},$$
$$h\mathcal{R}_1^{\perp}=\{R\in h\mathcal{R}_2; R_{\alpha\beta\bar\gamma\bar\delta}=2R_{\alpha\bar\gamma\beta\bar\delta}\iff R_{\alpha\bar\beta\gamma\bar\delta}=-R_{\alpha\bar\delta\gamma\bar\beta}\},$$
$$h\mathcal{R}_1=\{R\in h\mathcal{R}_2; R_{\alpha\beta\bar\gamma\bar\delta}=0\iff R_{\alpha\bar\beta\gamma\bar\delta}=R_{\alpha\bar\delta\gamma\bar\beta}\};$$
$$v\mathcal{R}=v\mathcal{R}_2\oplus v\mathcal{R}_2^{\perp}=\{R\in\mathcal{R}; R_{\alpha\beta\gamma0}, R_{\bar\alpha\beta\gamma0}, R_{\alpha\beta\bar\gamma0} - \textit{essential}\},$$
$$v\mathcal{R}_2^{\perp}=\{R\in v\mathcal{R}; R_{\alpha\beta\gamma0} - \textit{essential}\},$$
$$v\mathcal{R}_2=v\mathcal{R}_1\oplus v\mathcal{R}_1^{\perp}=\{R\in v\mathcal{R}; R_{\alpha\beta\gamma0}=0\},$$
$$v\mathcal{R}_1^{\perp}=\{R\in v\mathcal{R}_2; R_{\alpha\beta\bar\gamma0} - \textit{essential}\},$$
$$v\mathcal{R}_1=\{R\in v\mathcal{R}_2; R_{\alpha\beta\bar\gamma0}=0\iff R_{\bar\alpha\beta\gamma0}=R_{\bar\alpha\gamma\beta0}\};$$

$$w\mathcal{R} = \{R \in \mathcal{R}; R_{0\alpha\beta 0}, R_{0\bar{\alpha}\beta 0} - \textit{essential}\}.$$

In [7] it is proved that $\mathcal{R}$ over $(V^{2n+1}, \varphi, \xi, \eta, g)$ is decomposable into 20 mutually orthogonal and $G \times I$-invariant factors:

$$\begin{aligned}\mathcal{R} = {} & \omega_1 \oplus \omega_2 \oplus \omega_3 \oplus \omega_4 \oplus \omega_5 \oplus \omega_6 \oplus \omega_7 \oplus \omega_8 \oplus \omega_9 \oplus \omega_{10} \oplus \omega_{11} \\ & \oplus v_1 \oplus v_2 \oplus v_3 \oplus v_4 \oplus v_5 \oplus w_1 \oplus w_2 \oplus w_3 \oplus w_4.\end{aligned} \tag{2.2}$$

Proposition 2.2. *The twenty factors of the space of curvature tensors $\mathcal{R}$ over an almost contact vector space with B-metric $(V^{2n+1}, \varphi, \xi, \eta, g)$ are characterized in terms of the essential components and essential complex equalities for R as follows:*

$$\omega_1 = \mathrm{Ker}\,(p_2) \subset \mathrm{Ker}\,(p_1) : R_{\alpha\beta\gamma\delta} = \frac{\tau_R}{4n(n-1)} 2\{g_{\beta\gamma}g_{\alpha\delta} - g_{\alpha\gamma}g_{\beta\delta}\},$$

$$\rho_R \in \mathcal{L}_1,\ \rho^*_R \in \mathcal{L}_2,\ \rho''_R \in \mathcal{L}_1,\ \tilde{\rho}_R = 0,\ \tau^*_R = 0;$$

$$\omega_2 = \mathrm{Im}\,(p_2) \subset \mathrm{Ker}\,(p_1) : R_{\alpha\beta\gamma\delta} = \frac{\tau^*_R}{4n(n-1)} 2i\{g_{\alpha\gamma}g_{\beta\delta} - g_{\beta\gamma}g_{\alpha\delta}\},$$

$$\rho_R \in \mathcal{L}_2,\ \rho^*_R \in \mathcal{L}_1,\ \rho''_R \in \mathcal{L}_2,\ \tilde{\rho}_R = 0,\ \tau_R = \tau''_R = 0;$$

$$\omega_3 = \mathrm{Ker}\,(p_3) \subset \mathrm{Im}\,(p_1) : R_{\alpha\beta\gamma\delta} = \frac{1}{2(n-2)} 2\{g_{\beta\gamma}(\rho_R)_{\alpha\delta} - g_{\alpha\gamma}(\rho_R)_{\beta\delta}$$

$$+ g_{\alpha\delta}(\rho_R)_{\beta\gamma} - g_{\beta\delta}(\rho_R)_{\alpha\gamma}\}, \rho_R \in \mathcal{L}_3,\ \rho^*_R \in \mathcal{L}_3,\ \rho''_R \in \mathcal{L}_3,\ \tilde{\rho}_R = 0,$$

$$\tau_R = \tau^*_R = \tau''_R = 0;$$

$$\omega_4 = p_3(\mathrm{Im}\,(p_1)) : W''_{\alpha\beta\gamma\delta} = [(p_3 \circ p_1)R]_{\alpha\beta\gamma\delta},$$

$$\rho_{W''} = \rho^*_{W''} = \rho''_{W''} = \tilde{\rho}_{W''} = 0;$$

$$\omega_5 = \mathrm{Ker}\,(q_1) : R_{\alpha\bar{\beta}\gamma\bar{\delta}} = \frac{\tau_R}{4n^2} 2\{-g_{\alpha\gamma}g_{\bar{\beta}\bar{\delta}}\},$$

$$\rho_R \in \mathcal{L}_1,\ \rho^*_R \in \mathcal{L}_2,\ \rho''_R \in \mathcal{L}_1,\ \tilde{\rho}_R = 0,\ \tau^*_R = 0;$$

$$\omega_6 = \mathrm{Ker}\,(q_2) \subset \mathrm{Im}\,(q_1) : R_{\alpha\bar{\beta}\gamma\bar{\delta}} = \frac{1}{2n} 2\{-g_{\alpha\gamma}(\rho_R)_{\bar{\beta}\bar{\delta}} - g_{\bar{\beta}\bar{\delta}}(\rho_R)_{\alpha\gamma}\},$$

$$\rho_R \in \mathcal{L}_3,\ \rho^*_R \in \mathcal{L}_3,\ \rho''_R \in \mathcal{L}_3,\ \tilde{\rho}_R = 0,\ \tau_R = \tau^*_R = \tau''_R = 0;$$

$$\omega_7 = q_2(\mathrm{Im}\,(q_1)) : V''_{\alpha\bar{\beta}\gamma\bar{\delta}} = [(q_2 \circ q_1)R]_{\alpha\bar{\beta}\gamma\bar{\delta}},$$

$$\rho_{V''} \in \mathcal{L}_2, \rho^*_{V''} \in \mathcal{L}_1, \rho''_{V''} \in \mathcal{L}_2, \tilde{\rho}_{V''} = 0, \tau_R = \tau''_R = 0;$$

$$\omega_8 = h\mathcal{R}_1^{\perp} : R_{\alpha\bar{\beta}\gamma\bar{\delta}} = -R_{\alpha\bar{\delta}\gamma\bar{\beta}} - \textit{essential},\ \rho_R = \rho^*_R = \rho''_R = \tilde{\rho}_R = 0;$$

$$\omega_9 = \mathrm{Ker}\,(r_2) \subset \mathrm{Ker}\,(r_1) : R_{\bar{\alpha}\beta\gamma\delta} = \frac{1}{2(n-1)}\{g_{\beta\gamma}(\rho_R)_{\bar{\alpha}\delta} - g_{\beta\delta}(\rho_R)_{\bar{\alpha}\gamma}\},$$

$$\rho_R \in \mathcal{L}_4,\ \rho^*_R = \rho''_R = \tilde{\rho}_R = 0,\ \tau_R = 0;$$

$$\omega_{10} = \mathrm{Im}\,(r_2) \subset \mathrm{Ker}\,(r_1) : S''_{\bar{\alpha}\beta\gamma\delta} = R_{\bar{\alpha}\beta\gamma\delta} - \frac{1}{2(n-1)}\{g_{\beta\gamma}(\rho_R)_{\bar{\alpha}\delta} - g_{\beta\delta}(\rho_R)_{\bar{\alpha}\gamma}\}, \rho_{S''} = 0,\ \rho^*_{S''} \in \mathcal{L}_4,\ \rho''_{S''} \in \mathcal{L}_5,\ \tilde{\rho}_{S''} = 0,$$
$$\tau^*_{S''} = \tau''_{S''} = 0;$$

$$\omega_{11} = \mathrm{Im}\,(r_1) : S'_{\bar{\alpha}\beta\gamma\delta} = R_{\bar{\alpha}\beta\gamma\delta} - \frac{1}{2(n-1)}\{g_{\beta\gamma}(\rho_R - \rho''_R)_{\bar{\alpha}\delta} - g_{\beta\delta}(\rho_R - \rho''_R)_{\bar{\alpha}\gamma}\}, \rho_{S'} = \rho^*_{S'} = \rho''_{S'} = \tilde{\rho}_{S'} = 0;$$

$$v_1(\bar{v}_1 = \mathrm{Ker}\,(s)) : R_{\bar{\alpha}\beta\gamma 0} = \frac{1}{2n}2g_{\beta\gamma}(\rho_R)_{0\bar{\alpha}},\ \rho_R \in \mathcal{L}_7,\ \rho^*_R \in \mathcal{L}_7,\ \tilde{\rho}_R = 0;$$

$$v_2(\bar{v}_2 = \mathrm{Im}\,(s)) : V'_{\bar{\alpha}\beta\gamma 0} = R_{\bar{\alpha}\beta\gamma 0} - \frac{1}{2n}2g_{\beta\gamma}(\rho_R)_{0\bar{\alpha}},$$
$$\rho_{V'} = \rho^*_{V'} = \rho''_{V'} = \tilde{\rho}_{V'} = 0;$$

$$v_3(\bar{v}_3 = \overline{\mathcal{R}_1^{\perp}}) : R_{\alpha\beta\bar{\gamma}0} - \mathit{essential},\ \rho_R = \rho^*_R = \rho''_R = \tilde{\rho}_R = 0;$$

$$v_4(\bar{v}_4 = \mathrm{Ker}\,(t)) : R_{\alpha\beta\gamma 0} = \frac{1}{2(n-1)}2\{g_{\beta\gamma}(\rho_R)_{0\alpha} - g_{\alpha\gamma}(\rho_R)_{0\beta}\},$$
$$\rho_R \in \mathcal{L}_7,\ \rho^*_R \in \mathcal{L}_7,\ \tilde{\rho}_R = 0;$$

$$v_5(\bar{v}_5 = \mathrm{Im}\,(t)) : V''_{\alpha\beta\gamma 0} = R_{\alpha\beta\gamma 0} - \frac{1}{2(n-1)}2\{g_{\beta\gamma}(\rho_R)_{0\alpha} - g_{\alpha\gamma}(\rho_R)_{0\beta}\},$$
$$\rho_{V''} = \rho^*_{V''} = \rho''_{V''} = \tilde{\rho}_{V''} = 0;$$

$$w_1 : R_{0\alpha\beta 0} = \frac{\tilde{\tau}_R}{2n}g_{\alpha\beta},\ \rho_R \in \mathcal{L}_9,\ \rho^*_R = \rho''_R = 0,\ \tilde{\rho}_R \in \mathcal{L}_1;$$

$$w_2 : R_{0\alpha\beta 0} = -i\frac{\tilde{\tau}^*_R}{2n}g_{\alpha\beta},\ \rho_R = \rho''_R = 0,\ \rho^*_R \in \mathcal{L}_9,\ \tilde{\rho}_R \in \mathcal{L}_2;$$

$$w_3 : W_{0\alpha\beta 0} = R_{0\alpha\beta 0} - \frac{\tilde{\tau}_R}{2n}g_{\alpha\beta} + i\frac{\tilde{\tau}^*_R}{2n}g_{\alpha\beta},\ \rho_W = \rho^*_W = \rho''_W = 0,\ \tilde{\rho}_W \in \mathcal{L}_3;$$

$$w_4 : R_{0\bar{\alpha}\beta 0} - \mathit{essential},\ \rho_R = \rho^*_R = \rho''_R = 0,\ \tilde{\rho}_R \in \mathcal{L}_4.$$

3. Riemannian curvature identities in the basic classes of real isotropic hypersurfaces of a Kaehler manifold with B-metric

In [3] real time-like and isotropic hypersurfaces of a complex manifold with B-metric are introduced. The obtained submanifolds are almost contact manifolds with B-metric. Let us recall [3] that the real isotropic hypersurface (or hypersurface of second type) M of an almost complex manifold with B-metric $(M'^{2n+2}, J, g', \tilde{g}')$ is determined by the condition the normal unit N to be isotropic regarding the associated B-metric $\tilde{g}'$ of g', i.e. $\tilde{g}'(N,N) = 0$. Moreover, M is equipped with the almost contact B-metric structure

$$\varphi := J + g'(.,JN)N, \quad \xi := -JN, \quad \eta := -g'(.,JN), \quad g := g'|_M.$$

In [3] it is obtained that every real isotropic hypersurface of a Kaehler manifold with B-metric belongs to the class $\mathcal{F}_4 \oplus \mathcal{F}_5 \oplus \mathcal{F}_6 \oplus \mathcal{F}_8$.

Using the characterization of the basic classes of real isotropic hypersurfaces of a Kaehler manifold with B-metric in terms of their second fundamental form h from [5] and the decomposition of the space $\mathcal{L}$ we have $\mathcal{F}_0 : h = 0$; $\mathcal{F}_i : h = h_i$; $(i = 4, 5, 6, 8)$, where $h_4 \in \mathcal{L}_1$, $h_5 \in \mathcal{L}_2$, $h_6 \in \mathcal{L}_3$, $h_8 \in \mathcal{L}_4$ and

$$\begin{gathered} h_4(X,Y) = \frac{\theta(\xi)}{2n} g(x,y); \quad h_5(X,Y) = -\frac{\theta^*(\xi)}{2n} g(x,\varphi y); \\ h_6(X,Y) = -\frac{1}{2}(L_\xi g)(\varphi x, y); \quad h_8(X,Y) = \frac{1}{2} d\eta(\varphi x, y); \\ (L_\xi g)(X,Y) = (\nabla_X \eta)Y + (\nabla_Y \eta)X, \quad d\eta(X,Y) = (\nabla_X \eta)Y - (\nabla_Y \eta)X, \end{gathered}$$

X, $Y \in \mathcal{X}M$. For the equations of Gauss [4] and Codazzi in the case when the Kaehler manifold with B-metric $(M'^{2n+2}, J, g', \tilde{g}')$ is of constant totally real sectional curvatures ν' and $\tilde{\nu}'$ we have:

$$\begin{aligned} R(X,Y,Z,W) &= \{\nu'[\pi_1 - \pi_2] + \tilde{\nu}'\pi_3\}(X,Y,Z,W) \\ &\quad - h(Y,Z)h(X,W) + h(X,Z)h(Y,W), \\ (\nabla_X h)(Y,Z) - (\nabla_Y h)(X,Z) &= \nu'\{g(Y,\varphi Z)\eta(X) - g(X,\varphi Z)\eta(Y)\} \\ &\quad + \tilde{\nu}'\{g(Y,Z)\eta(X) - g(X,Z)\eta(Y)\}, \end{aligned}$$

where $X, Y, Z, W \in \mathcal{X}M$.

In the sence of [6], we give the following lemma.

Lemma 3.1. *The characteristic conditions for the basic classes of real isotropic hypersurfaces $(M^{2n+1}, \varphi, \xi, \eta, g)$ of a Kaehler manifold with B-metric in terms of the essential components and essential complex equalities for the fundamental tensors, are as follows:*

- $\mathcal{F}_4 : h_{\alpha\beta} = F_{\alpha\beta 0} = F_{\beta\alpha 0} = \dfrac{\theta(\xi)}{2n} g_{\alpha\beta}$;
- $\mathcal{F}_5 : h_{\alpha\beta} = F_{\alpha\beta 0} = F_{\beta\alpha 0} = -i\dfrac{\theta^*(\xi)}{2n} g_{\alpha\beta}$;
- $\mathcal{F}_6 : h_{\alpha\beta} = F_{\alpha\beta 0} = F_{\beta\alpha 0} = -\dfrac{i}{2}(L_\xi g)_{\alpha\beta}$, $\operatorname{tr} h = \operatorname{tr}^* h = 0$;
- $\mathcal{F}_8 : h_{\alpha\bar{\beta}} = F_{\alpha\bar{\beta}0} = F_{\bar{\beta}\alpha 0} = \frac{i}{2}\eta_{\alpha\bar{\beta}} = \frac{i}{2}N_{0\alpha\bar{\beta}} = \frac{i}{2}(Lg)_{0\alpha\bar{\beta}}$.

The equation of Codazzi implies:

Lemma 3.2. *The essential components of the covariant derivative of the second fundamental form h of a real isotropic hypersurface of a Kaehler manifold with B-metric of constant totally real sectional curvatures ν' and $\tilde{\nu}'$ satisfy the following equalities:*

i) $(\nabla_\alpha h)_{BC} - (\nabla_B h)_{\alpha C} = 0, \ B \in \{\beta, \bar{\beta}\}, C \in \{\gamma, \bar{\gamma}, 0\};$

ii) $(\nabla_0 h)_{\alpha B} - (\nabla_\alpha h)_{0B} = 0, \ B \in \{\bar{\beta}, 0\};$

iii) $(\nabla_0 h)_{\alpha\beta} - (\nabla_\alpha h)_{0\beta} = (\tilde{\nu}' + i\nu')g_{\alpha\beta}.$

The Gauss equation and Lemma 3.1 imply

Lemma 3.3. *Let $(M^{2n+1}, \varphi, \xi, \eta, g)$ be a real isotropic hypersurface of a Kaehler manifold with B-metric $(M'^{2n+2}, J, g', \tilde{g}')$ of constant totally real sectional curvatures ν' and $\tilde{\nu}'$. Then the Riemannian curvature tensor R of M satisfies in terms of the essential components the following identities:*

$$
\begin{aligned}
1.\ R_{\alpha\beta\gamma\delta} &= 2(\nu' - i\nu')(g_{\beta\gamma}g_{\alpha\delta} - g_{\alpha\gamma}g_{\beta\delta}) - h_{\beta\gamma}h_{\alpha\delta} + h_{\alpha\gamma}h_{\beta\delta}\\
&= \{\nu'[\pi_1 - \pi_2] + \tilde{\nu}'\pi_3\}_{\alpha\beta\gamma\delta} - h_{\beta\gamma}h_{\alpha\delta} + h_{\alpha\gamma}h_{\beta\delta}\\
2.\ R_{\alpha\beta\bar{\gamma}\bar{\delta}} &= -h_{\beta\bar{\gamma}}h_{\alpha\bar{\delta}} + h_{\alpha\bar{\gamma}}h_{\beta\bar{\delta}}\\
3.\ R_{\bar{\alpha}\beta\gamma\delta} &= -h_{\beta\gamma}h_{\bar{\alpha}\delta} + h_{\bar{\alpha}\gamma}h_{\beta\delta}\\
4.\ R_{\alpha\bar{\beta}\gamma\bar{\delta}} &= -h_{\bar{\beta}\gamma}h_{\alpha\bar{\delta}} + h_{\alpha\gamma}h_{\bar{\beta}\bar{\delta}}\\
5.\ R_{\alpha\beta\gamma 0} &= 0\\
6.\ R_{\bar{\alpha}\beta\gamma 0} &= 0\\
7.\ R_{\alpha\beta\bar{\gamma}0} &= 0\\
8.\ R_{0\alpha\beta 0} &= \nu' g_{\alpha\beta} - i\tilde{\nu}' g_{\alpha\beta} - h_{\alpha\beta}h_{00} + h_{0\beta}h_{\alpha 0}\\
9.\ R_{0\bar{\alpha}\beta 0} &= 0.
\end{aligned}
$$

Definition 3.1. An almost contact manifold with B-metric $(M^{2n+1}, \varphi, \xi, \eta, g)$ with Riemannian curvature tensor R is said to be in the class $h\mathcal{R}_i$, $h\mathcal{R}_i^\perp$, $v\mathcal{R}_j$, $v\mathcal{R}_j^\perp$, $w\mathcal{R}$, ω_k, v_r, w_s if $R \in h\mathcal{R}_i$, $h\mathcal{R}_i^\perp$, $v\mathcal{R}_j$, $v\mathcal{R}_j^\perp$, $w\mathcal{R}$, ω_k,v_r, w_s, respectively, where $i = 1, 2, 3$; $j = 1, 2$;k=1, ..., 11;r=1, ..., 5;s=1, ..., 4.

Using Lemma 3.1 and 3.3, Proposition 2.1 and 2.2, we express the essential components and the decomposition of the Riemannian curvature tensor R:

Theorem 3.1. *Let $(M^{2n+1}, \varphi, \xi, \eta, g)$ be in one of the basic classes $\mathcal{F}_4$, $\mathcal{F}_5$, $\mathcal{F}_6$, $\mathcal{F}_8$ of real isotropic hypersurfaces of a Kaehler manifold with B-metric*

$(M'^{2n+2}, J, g', \tilde{g}')$ of constant totally real sectional curvatures ν' and $\tilde{\nu}'$. Then the Riemannian curvature tensor R satisfies the following identities:

• $\mathcal{F}_4$:

$$\begin{aligned} R_{\alpha\beta\gamma\delta} &= 2\left(\nu' - \frac{\theta^2(\xi)}{8n^2} - i\tilde{\nu}'\right)(g_{\beta\gamma}g_{\alpha\delta} - g_{\alpha\gamma}g_{\beta\delta}), \\ R_{\alpha\bar{\beta}\gamma\bar{\delta}} &= \frac{\theta^2(\xi)}{8n^2}2(g_{\alpha\gamma}g_{\bar{\beta}\bar{\delta}}), \\ R_{0\alpha\beta 0} &= \nu' g_{\alpha\beta} - i\tilde{\nu}' g_{\alpha\beta} \end{aligned}$$

and $M \in \omega_1 \oplus \omega_2 \oplus \omega_5 \oplus w_1 \oplus w_2$;

• $\mathcal{F}_5$:

$$\begin{aligned} R_{\alpha\beta\gamma\delta} &= 2\left(\nu' + \frac{\theta^{*2}(\xi)}{8n^2} - i\tilde{\nu}'\right)(g_{\beta\gamma}g_{\alpha\delta} - g_{\alpha\gamma}g_{\beta\delta}), \\ R_{\alpha\bar{\beta}\gamma\bar{\delta}} &= \frac{\theta^{*2}(\xi)}{8n^2}2(g_{\alpha\gamma}g_{\bar{\beta}\bar{\delta}}), \\ R_{0\alpha\beta 0} &= \nu' g_{\alpha\beta} - i\tilde{\nu}' g_{\alpha\beta} \end{aligned}$$

and $M \in \omega_1 \oplus \omega_2 \oplus \omega_5 \oplus w_1 \oplus w_2$;

• $\mathcal{F}_6$:

$$\begin{aligned} R_{\alpha\beta\gamma\delta} &= 2(\nu' - i\tilde{\nu}')(g_{\beta\gamma}g_{\alpha\delta} - g_{\alpha\gamma}g_{\beta\delta}) + \frac{1}{4}\begin{vmatrix} (L_\xi g)_{\beta\gamma} & (L_\xi g)_{\alpha\gamma} \\ (L_\xi g)_{\beta\delta} & (L_\xi g)_{\alpha\delta} \end{vmatrix}, \\ R_{\alpha\bar{\beta}\gamma\bar{\delta}} &= \frac{1}{4}(L_\xi g)_{\alpha\gamma}(L_\xi g)_{\bar{\beta}\bar{\delta}}, \\ R_{0\alpha\beta 0} &= \nu' g_{\alpha\beta} - i\tilde{\nu}' g_{\alpha\beta} \end{aligned}$$

and $M \in h\mathcal{R}_2^\perp \oplus h\mathcal{R}_1 \oplus w_1 \oplus w_2$;

• $\mathcal{F}_8$:

$$\begin{aligned} R_{\alpha\beta\gamma\delta} &= 2(\nu' - i\tilde{\nu}')(g_{\beta\gamma}g_{\alpha\delta} - g_{\alpha\gamma}g_{\beta\delta}), \\ R_{\alpha\beta\bar{\gamma}\bar{\delta}} &= \frac{1}{4}(\eta_{\beta\bar{\gamma}}\eta_{\alpha\bar{\delta}} - \eta_{\alpha\bar{\gamma}}\eta_{\beta\bar{\delta}}), \\ R_{\alpha\bar{\beta}\gamma\bar{\delta}} &= -\frac{1}{4}\eta_{\bar{\beta}\gamma}\eta_{\alpha\bar{\delta}}, \\ R_{0\alpha\beta 0} &= \nu' g_{\alpha\beta} - i\tilde{\nu}' g_{\alpha\beta} \end{aligned}$$

and $M \in h\mathcal{R}_2 \oplus \omega_1 \oplus \omega_2 \oplus w_1 \oplus w_2$.

References

1. D. Blair, *Contact manifolds in Riemmannian geometry*, Lecture Notes in Math., **509** (1976) Springer-Verlag, Berlin.
2. G. Ganchev, V. Mihova and K. Gribachev, *Almost contact manifolds with B-metric*, Math. Balkanica. **7** (1993), 262–276.
3. M. Manev, *Almost contact B-metric hypersurfaces of Kaehler manifolds with B-metric*, Perspectives of Complex Analysis, Differential Geometry and Mathematical Physics, eds. S. Dimiev and K. Sekigawa, World Scientific, Singapore, 2001, 159–170.
4. ———, *On the curvature properties of almost contact B-metric hypersurfaces of Kaehler manifolds with B-metric*, Plovdiv Univ. Sci. Works – Math., 33, **3** (2001), 61–72.
5. ———, *Classes of real isotropic hypersurfaces of a Kaehler manifold with B-metric*, C. R. Acad. Bulgare Sci., vol. 55, **4** (2002).
6. G. Nakova, *Some remarks on the integrability conditions for almost contact manifolds with B-metric*, Plovdiv Univ. Sci. Works – Math., 33, **3** (2001), 87–94.
7. G. Nakova, *Curvature tensors on almost contact manifolds with B-metric*, Perspectives of Complex Analysis, Differential Geometry and Mathematical Physics, eds. S. Dimiev and K. Sekigawa, World Scientific, Singapore, 2002, (to appear).

A FUNCTION OF DIRECTION IN A WEYL SUBSPACE ASSOCIATED WITH A SET OF ORTHOGONAL VECTOR FIELDS

ABDÜLKADIR ÖZDEĞER

Işık University,
Faculty of Science and Letters, Mathematics Dept.
80670 Maslak, Istanbul, Turkey
E-mail: ozdeger@itu.edu.tr, aozdeger@turk.net

Let W_m be an m-dimensional subspace of an n-dimensional Weyl space W_n. Suppose that $\underset{1}{\mathbf{v}}$, $\underset{2}{\mathbf{v}}$, ..., $\underset{m}{\mathbf{v}}$ are mutually orthogonal smooth vector fields in W_m and that $\mathbf{v}$ is a non-tangential smooth vector field defined on W_m. Consider the $(m+1)$-dimensional net δ defined by $\underset{1}{\mathbf{v}}$, $\underset{2}{\mathbf{v}}$, ..., $\underset{m}{\mathbf{v}}$, $\mathbf{v}$. In this work, we obtain a function of direction associated with the net δ and define a class of curves on W_m in relation to this function of direction.

1. Introduction

An n-dimensional Weyl space, W_n, is a differentiable manifold with a conformal metric tensor g and a torsion-free (symmetric) connection ∇ satisfying the compatibility condition

$$\nabla_\gamma\, g_{\alpha\beta} = 2\,T_\gamma\, g_{\alpha\beta}$$

where T_γ is a covariant vector field called the complementary vector field of W_n. Such a Weyl space will be denoted by $W_n\,(g_{\alpha\beta}, T_\gamma)$.

According to Norden [1], under renormalization of the fundamental tensor of the form

$$\breve{g}_{\alpha\beta} = \lambda^2 g_{\alpha\beta}, \tag{1.1}$$

the complementary vector field T_γ is transformed by the law

$$\breve{T}_\gamma = T_\gamma + \partial_\gamma \ln \lambda, \tag{1.2}$$

where λ is a scalar point function and ∂_γ indicates partial differentiation .

The prolonged (extended) covariant derivative of the satellite A of $g_{\alpha\beta}$,of weight $\{k\}$, is defined by ([2], [3], [4])

$$\dot{\nabla}_\gamma\, A = \nabla_\gamma\, A - k\,T_\gamma\, A \tag{1.3}$$

from which it follows that

$$\dot{\nabla}_\gamma\, g_{\alpha\beta} = 0. \tag{1.4}$$

Let $W_m\,(g_{ij}, T_k)$ [a] be a subspace, with coordinates u^i, of the Weyl space $W_n(g_{\alpha\beta}, T_\gamma)$ with coordinates x^α. Suppose that the metrics of $W_m(g_{ij}, T_k)$ and $W_n(g_{\alpha\beta}, T_\gamma)$ are elliptic and that they are given, respectively, by

$$g_{ij}\, du^i\, du^j \quad \text{and} \quad g_{\alpha\beta}\, dx^\alpha\, dx^\beta$$

which are connected by the relations

$$g_{ij} = g_{\alpha\beta}\, x_i^\alpha\, x_j^\beta, \tag{1.5}$$

where x_i^α denotes the covariant derivative of x^α with respect to u^i.

The prolonged covariant derivatives of the satellite A, relative to $W_m(g_{ij}, T_k)$ and $W_n(g_{\alpha,\beta}, T_\gamma)$, are related by

$$\dot{\nabla}_k\, A = x_k^\gamma\, \dot{\nabla}_\gamma\, A. \tag{1.6}$$

Let the vector fields $\underset{r}{\mathbf{v}}$ ($r = 1, 2, \ldots, m$) belonging to $W_m(g_{ij}, T_k)$ be mutually orthogonal and be normalized by the conditions

$$g_{ij}\, \underset{r}{v^i}\, \underset{r}{v^j} = 1 \quad (r = 1, 2, \ldots, m). \tag{1.7}$$

It can be easily seen that

$$\sum_{r=1}^{m} \underset{r}{v^i}\, \underset{r}{v^j} = g^{ij}, \tag{1.8}$$

g^{ij} being the reciprocal tensor of g_{ij}. If the components of the vector field $\underset{r}{v}$, relative to $W_m(g_{ij}, T_k)$ and $W_n(g_{\alpha\beta}, T_\gamma)$ are respectively denoted by $\underset{r}{v^i}$ and $\underset{r}{v^\alpha}$, then we have

$$\underset{r}{v^\alpha} = \underset{r}{v^i} x_i^\alpha. \tag{1.9}$$

Let $\underset{\sigma}{\mathbf{n}}$ ($\sigma = 1, 2, \ldots, n - m$), be n-m mutually orthogonal vector fields in $W_n(g_{\alpha\beta}, T_\gamma)$ normal to $W_m(g_{ij}, T_k)$. Suppose further that they are normalized by the conditions

$$g_{\alpha\beta}\, \underset{\sigma}{n^\alpha}\, \underset{\sigma}{n^\beta} = 1. \tag{1.10}$$

We then have

$$g_{\alpha\beta}\, \underset{\sigma}{n^\alpha}\, \underset{\mu}{n^\beta} = \delta_{\sigma\mu}, \quad g_{\alpha\beta}\, x_i^\alpha\, \underset{\mu}{n^\beta} = 0 \quad (\delta_{\alpha\beta} : \text{Kronecker delta}). \tag{1.11}$$

---

[a] In the sequel, Latin indices $i, j, k, \ldots$ will run from 1 to m, while the Greek indices α, β, γ will run from 1 to n.

Taking the prolonged covariant derivative of (1.5) with respect to u^k and making use of (1.4), (1.6) and $(1.11)_2$ we obtain

$$g_{\alpha\beta}(\dot{\nabla}_k\, x_i^\alpha)x_j^\beta = 0$$

showing that $\dot{\nabla}_k x_i^\alpha$, regarded as a vector in $W_n(g_{\alpha\beta}, T_\gamma)$, is normal to $W_m(g_{ij}, T_k)$. It may therefore be expressed in the form

$$\dot{\nabla}_k\, x_i^\alpha = \sum_{\sigma=1}^{n-m} \overset{\sigma}{\omega}_{ik}\, \underset{\sigma}{n}^\alpha. \tag{1.12}$$

On the other hand, prolonged covariant differentiation of (1.10) with respect to u^i gives

$$g_{\alpha\beta}(\dot{\nabla}_i\, \underset{\sigma}{n}^\alpha)\underset{\sigma}{n}^\beta = 0 \tag{1.13}$$

so that $\dot{\nabla}_i\, \underset{\sigma}{n}^\alpha$, regarded as a vector in $W_n(g_{\alpha\beta}, T_\gamma)$, is orthogonal to $\underset{\sigma}{n}^\alpha$. It may therefore be written in the form

$$\dot{\nabla}_i\, \underset{\sigma}{n}^\alpha = A_i^j x_j^\alpha + \sum_{\mu=1}^{n-m} \underset{\mu\sigma}{B_i}\underset{\mu}{n}^\alpha, \qquad \underset{\sigma\sigma}{B_i} = 0. \tag{1.14}$$

On taking the prolonged covariant derivative of $(1.11)_2$ with respect to u^j we find that

$$g_{\alpha\beta}(\dot{\nabla}_j\, x_i^\alpha)\underset{\sigma}{n}^\beta + g_{\alpha\beta}x_i^\alpha(\dot{\nabla}_j\, \underset{\sigma}{n}^\beta) = 0.$$

Using (1.12) and (1.14) we find, in virtue of (1.11),

$$\overset{\sigma}{\omega}_{ij} + A_j^k g_{ik} = 0$$

or

$$A_j^k = -\overset{\sigma}{\omega}_{ij}\, g^{ik}.$$

Accordingly, (1.14) becomes

$$\dot{\nabla}_i\, \underset{\sigma}{n}^\alpha = -\overset{\sigma}{\omega}_{ik}\, g^{kj} x_j^\alpha + \sum_{\mu=1}^{n-m} \underset{\mu\sigma}{B_i}\underset{\mu}{n}^\alpha. \tag{1.15}$$

By means of (1.8), (1.9) and (1.15) the absolute derivative of $\underset{\sigma}{n}$ in the direction of $\underset{r}{v}$ is obtained as

$$\underset{r}{v}^\gamma \dot{\nabla}_\gamma\, \underset{\sigma}{n}^\alpha = \underset{r}{v}^i \dot{\nabla}_i\, \underset{\sigma}{n}^\alpha = -\sum_{s=1}^{n} \underset{rs}{\overset{\sigma}{K}}\underset{s}{v}^\alpha + \sum_{\mu=1}^{n-m} (\underset{\mu\sigma}{B_i}\, \underset{r}{v}^i)\underset{\mu}{n}^\alpha, \tag{1.16}$$

where

$$\underset{rs}{\overset{\sigma}{K}} = \overset{\sigma}{\omega}_{ik}\, \underset{r}{v}^i \underset{s}{v}^k. \tag{1.17}$$

2. Generalized Tendency of a Non-tangential Vector Field In a Given Direction In a Weyl Subspace

Let v be a non-tangential smooth vector field defined on $W_m(g_{ij}, T_k)$, normalized by the condition $g_{\alpha\beta}v^\alpha v^\beta = 1$. Then it may be expressed in the form

$$v^\alpha = a^i x_i^\alpha + \sum_{\mu=1}^{n-m} \underset{\sigma}{b}\, \underset{\sigma}{n}^\alpha \tag{2.1}$$

in which a^i are the components (i.e, tangential components of v) of a vector field a in $W_m(g_{ij}, T_k)$ of weight $\{-1\}$ and $\underset{\sigma}{b}$ are smooth functions of weight $\{0\}$ defined on $W_m(g_{ij}, T_k)$. If $\underset{\sigma}{\varphi}$ is the angle between v and $\underset{\sigma}{n}$ we have $\cos \underset{\sigma}{\varphi} = \underset{\sigma}{b}$ and $1 = g_{ij}a^i a^j - \sum_{\sigma} \underset{\sigma}{b}^2$.

Let C be a smooth curve in $W_m(g_{ij}, T_k)$ and suppose that it is an integral curve of the vector field $\underset{r}{v} \in \delta$.

Definition 2.1. The generalized tendency of v in the direction of C is the resolved part, on $\underset{r}{v}$, of the generalized absolute derivative of v in the direction of C, i.e

$$\text{tend}_C\, v = g_{\alpha\beta}\, \underset{r}{v}^\beta (\underset{r}{v}^\gamma \dot{\nabla}_\gamma v^\alpha). \tag{2.2}$$

If $W_m(g_{ij}, T_k)$ is a hypersurface of $W_n(g_{\alpha\beta}, T_\gamma)$, (2.2) will be the negative of the generalized normal curvature of C relative to v which will be denoted by $\underset{n}{\kappa}(v)$. Accordingly, we have

$$\underset{n}{\kappa}(v) = -\text{tend}_C\, v = -g_{\alpha\beta}\, \underset{r}{v}^\beta (\underset{r}{v}^\gamma \dot{\nabla}_\gamma v^\alpha). \tag{2.3}$$

The geodesic vector fields $\underset{r}{\bar{c}}^\alpha$ and $\underset{r}{c}^i$ of the curve C relative to $W_n(g_{\alpha\beta}, T_\gamma)$ and $W_m(g_{ij}, T_k)$ are related by the equation [5]

$$\underset{r}{\bar{c}}^\alpha = \underset{r}{v}^\gamma \dot{\nabla}_\gamma\, \underset{r}{v}^\alpha = \sum_{\sigma=1}^{n-m} \overset{\sigma}{\underset{rr}{K}}\, \underset{\sigma}{n}^\alpha + \underset{r}{c}^i x_i^\alpha, \tag{2.4}$$

$\overset{\sigma}{\underset{rr}{K}}$ being defined by (1.17) is the normal curvature of C.

The functions $\overset{p}{\underset{r}{\eta}}$ defined by

$$\underset{r}{c}^i = \overset{p}{\underset{r}{\eta}}\, \underset{p}{v}^i \tag{2.5}$$

are the geodesic curvatures of the net determined by the vector fields $\underset{1}{\mathbf{v}}, \underset{2}{\mathbf{v}}, \ldots, \underset{m}{\mathbf{v}}$ [4].

Taking the generalized absolute derivative of v, given by (2.1), in the direction of C and using (1.12) and (1.15) we obtain

$$\underset{r}{v}^{\gamma}\dot{\nabla}_{\gamma}v^{\alpha} = \underset{r}{v}^{k}\dot{\nabla}_{k}v^{\alpha} = \underset{r}{v}^{k}\left[\dot{\nabla}_{k}a^{i} - \sum_{\sigma}\underset{\sigma}{b}\,\overset{\sigma}{\omega}_{kl}g^{il}\right]x_{i}^{\alpha} + \sum_{\sigma}\left[(a^{i}\,\overset{\sigma}{\omega}_{ik} + \dot{\nabla}_{k}\underset{\sigma}{b})\underset{\sigma}{n}^{\alpha} + \underset{\sigma}{b}\sum_{\mu}\underset{\sigma\mu}{B_{k}}\,\underset{\mu}{n}^{\alpha}\right]\underset{r}{v}^{k}. \tag{2.6}$$

Putting $a_j = g_{ij}a^i$ in (2.6) and remembering that the prolonged covariant differentiation preserves the metric tensor g, (2.6) can be written in the form

$$\underset{r}{v}^{\gamma}\dot{\nabla}_{\gamma}v^{\alpha} = \underset{r}{v}^{k}\left[(\dot{\nabla}_{k}a_{j})g^{ij} - \sum_{\sigma}\underset{\sigma}{b}\,\overset{\sigma}{\omega}_{kl}g^{il}\right]x_{i}^{\alpha} + \sum_{\sigma}\left[(a^{i}\,\overset{\sigma}{\omega}_{ik} + \dot{\nabla}_{k}\underset{\sigma}{b})\underset{r}{v}^{k} + \underset{r}{v}^{k}\sum_{\mu}\underset{\mu}{b}\,\underset{\mu\sigma}{B_{k}}\right]\underset{\sigma}{n}^{\alpha}. \tag{2.7}$$

Since, according to (1.8)

$$g^{ij} = \sum_{s=1}^{m}\underset{s}{v}^{i}\underset{s}{v}^{j}, \quad g^{il} = \sum_{s=1}^{m}\underset{s}{v}^{i}\underset{s}{v}^{l},$$

(2.7) becomes

$$\underset{r}{v}^{\gamma}\dot{\nabla}_{\gamma}v^{\alpha} = \sum_{s=1}^{m}\left[(\underset{r}{v}^{k}\dot{\nabla}_{k}a_{j})\underset{s}{v}^{j} - \sum_{\sigma=1}^{n-m}\underset{\sigma}{b}\,\overset{\sigma}{\omega}_{kl}\,\underset{r}{v}^{k}\underset{s}{v}^{l}\right]\underset{s}{v}^{\alpha} + \sum_{\sigma=1}^{n-m}\left[(a^{i}\,\overset{\sigma}{\omega}_{ik} + \dot{\nabla}_{k}\underset{\sigma}{b})\underset{r}{v}^{k} + \underset{r}{v}^{k}\sum_{\mu=1}^{n-m}\underset{\mu}{b}\,\underset{\mu\sigma}{B_{k}}\right]\underset{\sigma}{n}^{\alpha}$$

or

$$\underset{r}{v}^{\gamma}\dot{\nabla}_{\gamma}v^{\alpha} = -\sum_{s=1}^{m}\underset{rs}{\overline{\kappa}}\,\underset{s}{v}^{\alpha} + \sum_{\sigma=1}^{n-m}\underset{r}{\overset{\sigma}{\theta}}\,\underset{\sigma}{n}^{\alpha}, \tag{2.8}$$

where

$$\underset{rs}{\overline{\kappa}} = \sum_{\sigma}\underset{\sigma}{b}\,\underset{rs}{\overset{\sigma}{K}} - (\dot{\nabla}_{k}a_{j})\underset{r}{v}^{k}\underset{s}{v}^{j}, \quad \underset{rs}{\overset{\sigma}{K}} = \overset{\sigma}{\omega}_{kl}\,\underset{r}{v}^{k}\underset{s}{v}^{l} \tag{2.9}$$

and

$$\underset{r}{\overset{\sigma}{\theta}} = \underset{r}{v}^{k}\left[\overset{\sigma}{\omega}_{ik}\,a^{i} + \dot{\nabla}_{k}\underset{\sigma}{b} + \sum_{\mu=1}^{n-m}\underset{\mu}{b}\,\underset{\mu\mu\sigma}{B_{k}}\right]. \tag{2.10}$$

Therefore, according to (2.2) and $(2.8)_2$ the generalized tendency of v in the direction of C is

$$\mathrm{tend}_C v = -\underset{rr}{\overline{\kappa}} = \underset{r}{v}^k \underset{r}{v}^j \dot{\nabla}_k a_j - \sum_\sigma \underset{\sigma}{b} \underset{rr}{\overset{\sigma}{K}}, \tag{2.11}$$

showing that $-\underset{rr}{\overline{\kappa}}$, defined by (2.9) with $s = r$, is the generalized tendency of v in the direction of $\underset{r}{C}$.

If summation on r is performed in (2.11), with the help of (1.8), (2.9) and (2.11) we get

$$\begin{aligned} -\sum_{r=1}^m \underset{rr}{\overline{\kappa}} &= -\sum_\sigma \underset{\sigma}{b}\ \overset{\sigma}{\omega}_{kl}\ g^{kl} + (\dot{\nabla}_k a_j) g^{kj} \\ &= -\sum_\sigma \underset{\sigma}{b}\ \overset{\sigma}{\omega}_{kl}\ g^{kl} + \dot{\nabla}_k a^k \\ &= -\sum_\sigma \underset{\sigma}{b}\ \underset{\sigma}{M} + \mathrm{div}_m a, \quad \underset{\sigma}{M} = \overset{\sigma}{\omega}_{kl} g^{kl} \end{aligned}$$

$$-\sum_{r=1}^m \underset{rr}{\overline{\kappa}} = \sum_\sigma \underset{\sigma}{b}\ \mathrm{div}_m \underset{\sigma}{n} + \mathrm{div}_m a, \quad \underset{\sigma}{M} = -\mathrm{div}_m \underset{\sigma}{n} \tag{2.12}$$

where $\underset{\sigma}{M}$ is the resolved part of the mean curvature vector of $W_m(g_{ij}, T_k)$ in the direction of $\underset{\sigma}{n}$.

On the other hand, the left-hand member of (2.12), being the sum of the generalized tendencies of v in m mutually orthogonal directions $\underset{1}{\mathbf{v}}, \underset{2}{\mathbf{v}}, \ldots, \underset{m}{\mathbf{v}}$,is the generalized divergence of v relative to $W_m(g_{ij}, T_k)$ so that we can write

$$\mathrm{div}_m v = \sum_\sigma \underset{\sigma}{b}\,\mathrm{div}_m \underset{\sigma}{n} + \mathrm{div}_m a. \tag{2.13}$$

Special Cases:

Case 2.1. If the non-tangential vector field v is taken as the mean curvature vector M of $W_m(g_{ij}, T_k)$ [5], i.e., if

$$v = M = \sum_\sigma \underset{\sigma}{M} \underset{\sigma}{n},$$

a^i and $\underset{\sigma}{b}$ in (2.1) become zero and $\underset{\sigma}{b} = \underset{\sigma}{M}$. In this case, (2.13) reduces to

$$\mathrm{div}_m M = \sum_\sigma \underset{\sigma}{M}^2 = \sum_\sigma (\mathrm{div}_m \underset{\sigma}{n})^2.$$

Case 2.2. If $W_m(g_{ij}, T_k)$ is a hypersurface of $W_n(g_{\alpha\beta}, T_\gamma)$, then $W_m(g_{ij}, T_k)$ will have only one normal vector field, say n, and the sum of the generalized normal curvatures of the hypersurface for m mutually orthogonal directions in $W_m(g_{ij}, T_k)$ will be an invariant equal to

$$-(b\,\mathrm{div}\,n + \mathrm{div}\,a), \qquad \underset{\sigma}{b} = b.$$

3. A Function of Direction In a Weyl Subspace Associated With The Net δ

The prolonged covariant derivative of $\underset{rr}{\overline{\kappa}}$, given by

$$\underset{rr}{\overline{\kappa}} = -tend_c v = -g_{\alpha\beta}\underset{r}{v}^{\beta}(\underset{r}{v}^{\gamma}\dot{\nabla}_\gamma v^\alpha),$$

in the direction of $\underset{r}{v}$ is

$$\begin{aligned}\underset{r}{v}^i\dot{\nabla}_i\underset{rr}{\overline{\kappa}} &= -g_{\alpha\beta}(\underset{r}{v}^{\delta}\dot{\nabla}_\delta\underset{r}{v}^{\beta})(\underset{r}{v}^{\gamma}\dot{\nabla}_\gamma v^\alpha) - g_{\alpha\beta}\,\underset{r}{v}^{\beta}(\underset{r}{v}^{\delta}\dot{\nabla}_\delta\underset{r}{v}^{\gamma})\dot{\nabla}_\gamma\underset{r}{v}^{\alpha}) \\ &\quad - g_{\alpha\beta}\,\underset{r}{v}^{\beta}\underset{r}{v}^{\gamma}[\underset{r}{v}^{\delta}\dot{\nabla}_\delta(\dot{\nabla}_\gamma\underset{r}{v}^{\alpha})]\,,\end{aligned} \tag{3.1}$$

or, by (2.3) and (2.8)

$$\begin{aligned}\underset{r}{v}^i\dot{\nabla}_i\underset{rr}{\overline{\kappa}} &= -g_{\alpha\beta}(\sum_\sigma\underset{rr}{\overset{\sigma}{K}}\underset{\sigma}{n}^{\beta} + x_i^\beta\underset{r}{c}^i)(-\sum_s\underset{rs}{\bar{\kappa}}\underset{s}{v}^{\alpha} + \sum_\mu\underset{r}{\theta}^{\mu}\underset{\mu}{n}^{\alpha}) \\ &\quad - g_{\alpha\beta}(\sum_\sigma\underset{rr}{\overset{\sigma}{K}}\underset{\sigma}{n}^{\gamma} + x_i^\gamma\underset{r}{c}^i)(\underset{r}{v}^{\beta}\dot{\nabla}_\gamma v^\alpha) \\ &\quad - g_{\alpha\beta}\underset{r}{v}^{\beta}\underset{r}{v}^{\gamma}\underset{r}{v}^{\delta}\dot{\nabla}_\delta(\dot{\nabla}_\gamma v^\alpha).\end{aligned} \tag{3.2}$$

Using (1.9), (1.11), (2.4) and (2.9) in (3.2) we obtain

$$\begin{aligned}\underset{r}{v}^i\dot{\nabla}_i\underset{rr}{\overline{\kappa}} &= -\sum_\sigma\underset{p}{\overset{\sigma}{\theta}}\underset{rr}{\overset{\sigma}{K}} + \underset{rp}{\bar{\kappa}}\,\underset{r}{\overset{p}{\eta}} - g_{\alpha\beta}\underset{r}{v}^{\beta}\sum_\sigma\underset{rr}{\overset{\sigma}{K}}(\underset{\sigma}{n}^{\gamma}\dot{\nabla}_\gamma v^\alpha) \\ &\quad - g_{\alpha\beta}\underset{r}{v}^{\beta}\,\underset{r}{\overset{p}{\eta}}(-\sum_s\underset{ps}{\bar{\kappa}}\,\underset{s}{v}^{\alpha} + \sum_\sigma\underset{p}{\overset{\sigma}{\theta}}\underset{\sigma}{n}^{\alpha}) - g_{\alpha\beta}\underset{r}{v}^{\beta}\underset{r}{v}^{\gamma}\underset{r}{v}^{\delta}\dot{\nabla}_\delta(\dot{\nabla}_\gamma v^\alpha) \\ &= -\sum_\sigma\underset{p}{\overset{\sigma}{\theta}}\underset{rr}{\overset{\sigma}{K}} + \underset{r}{\overset{p}{\eta}}(\underset{pr}{\bar{\kappa}} + \underset{rp}{\bar{\kappa}}) - g_{\alpha\beta}\underset{r}{v}^{\beta}\sum_\sigma\underset{rr}{\overset{\sigma}{K}}(\underset{\sigma}{n}^{\gamma}\dot{\nabla}_\gamma v^\alpha) \\ &\quad - g_{\alpha\beta}\underset{r}{v}^{\beta}\underset{r}{v}^{\gamma}\underset{r}{v}^{\delta}\dot{\nabla}_\delta(\dot{\nabla}_\gamma v^\alpha)\end{aligned}$$

from which it follows that

$$\begin{aligned}&\underset{r}{v}^i\dot{\nabla}_i\underset{rr}{\overline{\kappa}} - \underset{r}{\overset{p}{\eta}}(\underset{pr}{\bar{\kappa}} + \underset{rp}{\bar{\kappa}}) \\ &= -\Big\{\sum_{\sigma,\mu}\Big[\overset{\sigma}{\omega}_{ij}(\overset{\sigma}{\omega}_{lk}\,a^l + \dot{\nabla}_k\underset{\sigma}{b}) + \underset{\mu}{b}\,\underset{\mu\sigma}{B_k}\Big] \\ &\quad + g_{\alpha\beta}\Big[x_k^\beta\sum_\sigma\overset{\sigma}{\omega}_{ij}(\underset{\sigma}{n}^{\gamma}\dot{\nabla}_\gamma v^\alpha) + x_i^\beta\,x_j^\gamma\,x_k^\delta\,\dot{\nabla}_\delta\dot{\nabla}_\gamma v^\alpha\Big]\Big\}\underset{r}{v}^i\,\underset{r}{v}^j\,\underset{r}{v}^k\,.\end{aligned} \tag{3.3}$$

We note that the individual terms on the left hand member of the equation (3.3) depend on the vector fields of the net δ, but the right hand member depends only on the direction of the curve C and consequently the left hand member has the same value for all curves that are tangent to C.

So, we have proved

Theorem 3.1. *The expression*

$$\underset{r}{v}^i \dot{\nabla}_i \underset{rr}{\overline{\kappa}} - \underset{r}{\overset{p}{\eta}}(\underset{pr}{\overline{\kappa}} + \underset{rp}{\overline{\kappa}})$$

is a function of direction in $W_m(g_{ij}, T_k)$ *associated with the net* $\delta = (\underset{1}{\mathbf{v}}, \underset{2}{\mathbf{v}}, \cdots, \underset{m}{\mathbf{v}}, v)$.

A Special Case: Let $W_m(g_{ij}, T_k)$ be a hypersurface of $W_n(g_{\alpha\beta}, T_\gamma)$. Then, $W_m(g_{ij}, T_k)$ has only one normal field and $\underset{rr}{\overline{\kappa}}$ becomes the generalized normal curvature of C relative to the non-tangential vector field v. In this case, we have

$$\underset{\sigma}{n} = n, \quad \overset{\sigma}{\omega}_{ij} = w_{ij}, \quad \underset{\sigma}{b} = b, \quad \underset{\mu\sigma}{B_i} = 0$$

so that the equation (3.3) becomes

$$\begin{aligned} \underset{r}{v}^i \dot{\nabla}_i \underset{rr}{\overline{\kappa}} - \underset{r}{\overset{p}{\eta}}(\underset{pr}{\overline{\kappa}} + \underset{rp}{\overline{\kappa}}) = -\{ & w_{ij}\,(w_{kl}\, a^l + \dot{\nabla}_k b + g_{\alpha\beta}\left[x^\beta_k\, w_{ij}\, n^\gamma\, \dot{\nabla}_\gamma n^\alpha \right. \\ & \left. + x^\beta_i\, x^\gamma_j\, x^\delta_k\, \dot{\nabla}_\delta\, \dot{\nabla}_\gamma n^\alpha\right]\}\underset{r}{v}^i\, \underset{r}{v}^j\, \underset{r}{v}^k. \end{aligned} \tag{3.4}$$

If, in addition, the non-tangential vector field v is identical with the normal vector field n of the hypersurface $W_m(g_{ij}, T_k)$,then according to (2.1) and (2.8), we have

$$a^l = 0, \quad b = 1, \quad \underset{pr}{\overline{\kappa}} = \underset{rp}{\overline{\kappa}} = \underset{pr}{\kappa} = w_{ij}\, \underset{p}{v}^i\, \underset{r}{v}^j$$

and therefore (3.4) reduces to

$$\underset{r}{v}^i \dot{\nabla}_i \underset{rr}{\overline{\kappa}} - 2\, \underset{pr}{\kappa}\, \underset{r}{\overset{p}{\eta}} = -g_{\alpha\beta}\, \{w_{ij}\, x^\beta_k\, (n^\gamma \dot{\nabla}_\gamma\, n^\alpha) + x^\beta_i\, x^\gamma_j\, x^\delta_k\, (\dot{\nabla}_\delta\, \dot{\nabla}_\gamma n^\alpha)\}\underset{r}{v}^i\, \underset{r}{v}^j\, \underset{r}{v}^k \tag{3.5}$$

which is the generalization of the function of direction associated with m congurences of an orthogonal ennuple in a Riemannian hypersurface obtained in [6]. It is clear that $\underset{rr}{\overline{\kappa}}$, in (3.5), is the normal curvature of the hypersurface in the direction of $\underset{r}{v}$.

We finally obtain a function of direction depending only on the curvature properties of the smooth curve C with tangent vector field $\underset{r}{v}$ and not on the mutually orthogonal vector fields $\underset{1}{\mathbf{v}}, \underset{2}{\mathbf{v}}, \ldots, \underset{m}{\mathbf{v}}$ belonging to the subspace $W_m(g_{ij}, T_k)$ of $W_n(g_{\alpha\beta}, T_\gamma)$. To achieve this, we consider $\underset{rr}{\overline{\kappa}}$ given by (2.11), i.e

$$\underset{rr}{\overline{\kappa}} = \sum_\sigma \underset{\sigma}{b} \overset{\sigma}{\underset{rr}{K}} - \underset{r}{v}^k \underset{r}{v}^j \dot{\nabla}_k a_j, \quad \overset{\sigma}{\underset{rr}{K}} = \overset{\sigma}{\omega}_{ij} \underset{r}{v}^i \underset{r}{v}^j. \tag{3.6}$$

For simplicity, we change the notation in the following way:

$$\underset{r}{v} = t, \quad \underset{rr}{\overline{\kappa}} = \bar{\kappa}, \quad \overset{\sigma}{\underset{rr}{K}} = \overset{\sigma}{K} = \overset{\sigma}{\omega}_{ij} t^i t^j.$$

Then (3.6) becomes

$$\bar{\kappa} = \sum_\sigma \underset{\sigma}{b} \overset{\sigma}{K} - t^i t^k \dot{\nabla}_k a_j \tag{3.7}$$

The prolonged covariant derivative of $\bar{\kappa}$ in the direction of C is

$$\begin{aligned} t^l \dot{\nabla}_l \bar{\kappa} &= \sum_\sigma (t^l \dot{\nabla}_l \underset{\sigma}{b}) \overset{\sigma}{\omega}_{ij} t^i t^j \\ &\quad + \underset{\sigma}{b} \left[(\dot{\nabla}_l \overset{\sigma}{\omega}_{ij}) t^l t^i t^j + \overset{\sigma}{\omega}_{ij} (t^j t^l \dot{\nabla}_l t^i + t^i t^l \dot{\nabla}_l t^j) \right] \\ &\quad - \left[t^k (t^l \dot{\nabla}_l t^j) + t^j (t^l \dot{\nabla}_l t^k) \right] \dot{\nabla}_k a_j - t^j t^k t^l \dot{\nabla}_l (\dot{\nabla}_k a_j). \end{aligned} \tag{3.8}$$

By using the Frenet's formula [7]

$$t^l \dot{\nabla}_l t^i = k_g \xi^i$$

and remembering that

$$\overset{\sigma}{\tau}_g = \overset{\sigma}{\omega}_{ij} t^i \xi^j$$

is the generalized geodesic torsion of C relative to the normal vector field $\underset{\sigma}{n}$, the equation (3.8) may be cast into the form

$$\begin{aligned} t^l \dot{\nabla}_l \bar{\kappa} - k_g &\left\{ \sum_\sigma 2 \underset{\sigma}{b} \overset{\sigma}{\tau}_g - (t^k \xi^j + t^j \xi^k) \dot{\nabla}_k a_j \right\} \\ &= \left\{ \sum_\sigma (\underset{\sigma}{b} \dot{\nabla}_k \overset{\sigma}{\omega}_{ij} + \overset{\sigma}{\omega}_{ij} \dot{\nabla}_k \underset{\sigma}{b}) - \dot{\nabla}_l \dot{\nabla}_k a_j \right\} t^i t^j t^k \end{aligned} \tag{3.9}$$

where ξ is the unit principal normal vector field of C normalized by the condition $g_{ij} \xi^i \xi^j = 1$ and k_g is the geodesic curvature of C relative to $W_m(g_{ij}, T_k)$.

Since the right-hand member of (3.9) depends only on the direction of C, so is the left-hand member. Thus, we have

Theorem 3.2. *The expression*

$$t^l \dot{\nabla}_l \bar{\kappa} - k_g \left\{ 2 \sum_\sigma \underset{\sigma}{b} \overset{\sigma}{\tau}_g + (t^k \xi^j + t^j \xi^k) \dot{\nabla}_k a_j \right\}$$

is a function of direction associated with the curve C and the non-tangential vector field v.

In particular, suppose that $W_m(g_{ij}, T_k)$ is a hypersurface of $W_n(g_{\alpha\beta}, T_\gamma)$ and that the non-tangential vector field v is normal to $W_m(g_{ij}, T_k)$. Then, (3.9) becomes

$$t^l\, \dot{\nabla}_l\, \kappa - 2\kappa_g \tau_g = (\dot{\nabla}_l\, \omega_{ij}) t^l\, t^i\, t^j, \tag{3.10}$$

where $\bar{\kappa} = \kappa = \omega_{ij}\, t^i\, t^j$, $\overset{\sigma}{\tau}_g = \tau_g = \omega_{ij}\, t^i\, \xi^j$, $\overset{\sigma}{\omega}_{ij} = \omega_{ij}$ are respectively the normal curvature, the geodesic torsion of C and the coefficients of the second fundamental form of $W_m(g_{ij}, T_k)$.

We call $t^l \dot{\nabla}_l \kappa - 2\kappa_g \tau_g$ the generalized Laguerre's function of direction for the hypersurface $W_m(g_{ij}, T_k)$.

As a generalization of Laguerre lines for a Riemannian hypersurface, defined in [8], to a Weyl hypersurface we give

Definition 3.1. A smooth curve C belonging to the Weyl hypersurface $W_m(g_{ij}, T_k)$ is called a generalized Laguerre line if the generalized Laguerre's function of direction $t^l \dot{\nabla}_l \kappa - 2\kappa_g \tau_g$ vanishes all along C.

According to (3.10), the differential equation of generalized Laguerre lines of the hypersurface is

$$(\dot{\nabla}_l\, \omega_{ij}) t^l\, t^i\, t^j = 0, \tag{3.11}$$

or, since ω_{ij} is a satellite of g of weight $\{1\}$, (3.11) may be written in the form

$$(\nabla_l\, \omega_{ij} - \omega_{ij} T_l) t^l\, t^i\, t^j = 0, \tag{3.12}$$

which is of the first order and of the third degree.

A Weyl hypersurface is called totally umbilical if

$$\omega_{ij} = \frac{M}{m}\, g_{ij}, \tag{3.13}$$

where M is the mean curvature of the hypersurface defined by $M = g^{ij}\omega_{ij}$ and m is its dimension.

For a totally umbilical Weyl hypersurface (3.11) gives

$$\begin{aligned}(\dot{\nabla}_l\, wij) t^l\, t^i\, t^j &= \dot{\nabla}_l\, (\frac{M}{m}\, g_{ij}) t^l\, t^i\, t^j \\ &= \frac{1}{m}\, g_{ij}\, t^i\, t^j\, (t^l\, \dot{\nabla}_l\, M) = \frac{1}{m}\, (t^l\, \dot{\nabla}_l\, M) = 0.\end{aligned} \tag{3.14}$$

We have thus proved

Theorem 3.3. *The mean curvature of a totally umbilical Weyl hypersurface is prolonged covariant constant along the generalized Laguerre lines of the hypersurface.*

On the other hand, if $T = 0$, $W_m(g_{ij}, T_k)$ becomes a locally Riemannian hypersurface and the differential equation (3.12) of the generalized Laguerre lines reduces to

$$(\nabla_l \, \omega_{ij}) t^l t^i t^j = 0. \tag{3.15}$$

Consequently, for a totally umbilical Riemannian hypersurface V_m, (3.15) becomes

$$t^l \nabla_l M = 0$$

stating that M is constant along the Laguerre lines of V_m. Therefore, we have proved

Corollary 3.1. *The mean curvature of a totally umbilical Riemannian hypersurface is constant along the Laguerre lines of the hypersurface.*

References

1. Norden, A.: Affinely Connected Spaces. GRMFL, Moscow, 1976.
2. Norden, A. and Yafarov, S.: Theory of non-geodesic vector fields in two dimensional affinely connected spaces, Izv. Vuzov, Matem., **12** (1974), 29–34
3. Zlatanov, G.: Nets in the $n-$dimensional space of Weyl, C.R.Acad. Bulgare Sci., **41**(10) (1988), 29–32
4. Tsareva, G. and Zlatanov, G.: On the geometry of nets in the n-dimensional space of Weyl, Journal of Geometry, **38** (1990), 182–197.
5. Özdeğer, A.: Chebyshev nets in a subspace of a Weyl space and the generalized Dupin theorem, Webs and Quasigroups (1995), 97–105.
6. ______: A function of direction associated with n congurences of an orthogonal ennuple in a Riemannian hypersurface, Journal of Geometry, 17 (1981),1–6.
7. Hlavaty, V.: Théorie d'immersion d'une W_m dans W_n, Ann. Soc. Polon. Math. V. 21 (1949), 196–206.
8. Özdeğer, A.: Some functions of direction in a subspace of a Riemannian space, Bulletin of the Istanbul Technical University, V. 33, no. 2 (1980), 48–53.

COMPLEX STRUCTURES AND THE QUARK CONFINEMENT

IVANHOE B. PESTOV
Joint Institute for Nuclear Research
Bogoliubov Laboratory of Theoretical Physics
141980, Dubna, Russia
E-mail: pestov@thsun1.jinr.ru

It is argued from geometrical, group-theoretical and physical points of view that in the framework of QCD it is not only necessary but also possible to modify the Dirac equation so that correspondence principle holds valid. The Dirac wave equation for a quark is proposed and some consequences are considered. In particular, it is shown that interquark potential expresses the Coulomb law for the quarks and, in fact, coincides with the known Cornell potential.

1. Introduction

According to the modern standpoint [1], spacetime theory is any one that possesses a mathematical representation whose elements are a smooth four-dimensional manifold M and geometrical objects defined on M. The system of real local coordinates on M is defined as a topological mapping of an open region $U \subset M$ onto the Euclidean four-dimensional space E_4. Thus, the Euclidean four-dimensional space E_4 is a fundamental structure element of the mathematical formalism of contemporary physics. However, it can be shown that E_4 has an underlying structure that is exhibited in the existence of group of transformation that does not coincide with the $SO(4)$ group. In physical space E_3 we consider the group of rotations and dilatations with the generators

$$D = x\frac{\partial}{\partial x} + y\frac{\partial}{\partial y} + z\frac{\partial}{\partial z}, \quad M_1 = z\frac{\partial}{\partial y} - y\frac{\partial}{\partial z}$$
$$M_2 = z\frac{\partial}{\partial x} - x\frac{\partial}{\partial z}, \quad M_3 = -y\frac{\partial}{\partial x} + x\frac{\partial}{\partial y}.$$

We denote this group by $D \otimes SO(3)$. It can be shown that an element of group can be parametrized by the real numbers a, b, c, d; it is suitable to consider them as a quaternion $q = ai + bj + ck + d$. It is easy to verify one-to-one correspondence between the algebras quaternions and the $D \otimes SO(3)$ group. The transformations of the $D \otimes SO(3)$ group in E_3 can be represented as

$$R' = qR\bar{q},$$

where $R = xi+yj+zk$. When we consider $D\otimes SO(3)$ group as linear space it can easily be seen, with the aid of the well-known algebra, that the four-dimensional space of quaternions Q_4 give spinor representation of the group in question. This representation is realized as follows:

$$u' = qu, \qquad (1)$$

where

$$u = u_1 i + u_2 j + u_3 k + u_4.$$

A remarkable property of this transformations is that there is only one point $u = 0$, that is a stable under the transformations given by (1). When we fix any other point, the transformations reduce to the identity transformation. Another important feature of this transformations (1) is that the Euclidean scalar products in Q_4

$$(u, u) = u\bar{u} = u_1^2 + u_2^2 + u_3^2 + u_4^2$$

is invariant under transformations (1), provided that $q\bar{q} = 1$. But this does not mean that Q_4 is Euclidean space because (1) hold. It can be shown, that there is no mapping from Q_4 to four-dimensional Euclidean space. Thus, what we usually call four-dimensional Euclidean by the analogy with a three-dimensional physical space is in fact Q_4. That there is no such mapping follows from the fact that real Dirac matrices γ_i, possessing the properties

$$\gamma_i\gamma_j + \gamma_j\gamma_i = 2\delta_{ij}, \quad i, j = 1, 2, 3, 4$$

do not exist. However, there is mapping from Q_4 to E_3 which can be defined as

$$R = ui\bar{u}.$$

This mapping is known as Hopf mapping. Therefore, the components of a quaternion u are not observable in a direct way, but only through some expressions constructed from these components. In seance, this situation is similar to that with a wave function in quantum mechanics. It is obvious that a three-dimensional sphere in Q_4,

$$u_1^2 + u_2^2 + u_3^2 + u_4^2 = \rho^2$$

inherits the properties of the enveloping space. In view of the unusual properties of four-dimensional space and of, respectively, a three- dimensional sphere S^3, we consider the investigations associated with the last object.

From the geometrical point of view a three-dimensional sphere is a space of constant positive curvature. The Kepler-Coulomb problem in this space has a long history and was first investigated by Schrödinger [2]. On the other hand S^3 is the configuration space for a Top. The quantum-mechanical problem for the free motion of a Top was investigated shortly after the creation of quantum mechanics in its moderns form (see for example [3]). It is obvious that connections between these two directions of investigation are very important. Moreover in the 1930s, it was emphasized by Casimir [4] that, from the physical point of view, the notion of a rigid bogy is as fundamental as the notion of material point. At last, we would like to emphasize that QCD is conceptually a simple theory and that its structure is determined solely by symmetry principles. However, there is no connection between such important phenomena as confinement and quark-lepton symmetry, on one hand, and the first principles of QCD, on the other hand. Despite prolonged and complicated experiments, free quarks have not yet been observed, but it is commonly believed that quarks are true elementary particles like electrons. Experimentalists gradually arrived at the conclusion that the matter is not in the details of the experiments but rather in the fundamental properties of matter, which were sought under various assumptions. For instance, it is hypothesized that quark confinement can be explained by topological methods that have recently found still a wider use in physics. Nevertheless, the most natural and reliable approach to the problem of confinement must be sought in the possibility of modifying the original Dirac equation to take into account the unusual properties of the quarks.

Summarizing all the facts considered above, we conjecture that the configuration space of a quark coinces with the configuration space for a Top. Within this conjecture, we will derive the Dirac wave equation for a quark and consider some of its properties.

2. Formulation of the Problem

The choices of underlying spacetime manifold can be reasoning as followers. Consider a conventional quantum-mechanical operator of the 4-momentum with components

$$P_0 = -i\hbar\frac{\partial}{\partial x^0}, \quad P_1 = -i\hbar\frac{\partial}{\partial x^1}, \quad P_2 = -i\hbar\frac{\partial}{\partial x^2}, \quad P_3 = -i\hbar\frac{\partial}{\partial x^3}, \tag{2}$$

where $x^0 = ct$, $x^1 = x$, $x^2 = y$, $x^3 = z$. Further, the group of translations of spacetime is a finite continuous Lie group with the generators

$$X_0 = \frac{\partial}{\partial x^0}, \quad X_1 = \frac{\partial}{\partial x^1}, \quad X_2 = \frac{\partial}{\partial x^2}, \quad X_3 = \frac{\partial}{\partial x^3}. \tag{3}$$

Besides, the linear operators (3) are generators of a simply transitive group of transformations [5]. In our case it is an Abelian simply transitive group of transformations because

$$[X_a, X_b] = 0, \quad a, b = 0, 1, 2, 3.$$

The connection between(2) and (3) is obvious. If a simply transitive group is a group of transformations of a certain spacetime, this spacetime is called homogeneous, as this group can transform any point of the spacetime to any a priori given point. To make further analysis more transparent, we present the general characteristic of homogeneous spacetime manifolds.

As any vector field with components V^i can be associated with the linear operator $X = V^i \partial / \partial x^i$, the operator X is called the vector field. Then, let the vector fields X_a

$$X_a = V_a^i \frac{\partial}{\partial x^i} \quad (a = 0, 1, 2, 3) \tag{4}$$

are generators of a simply transitive group of transformations of a four-dimensional spacetime manifold M. Indexes from the beginning of the Latin alfavet numerate vectors. In this case

$$[X_a, X_b] = f_{ab}^c X_c, \tag{5}$$

where f_{ab}^c are structure constants of the group in question. The vector fields V_a^i uniquely determine the system of covector fields V_i^a such that

$$V_a^i V_j^a = \delta_j^i, \quad V_i^a V_b^i = \delta_b^a. \tag{6}$$

The simply transitive group induces a natural integrable connection Γ on M with Christoffel symbols

$$\Gamma_{jk}^i = V_a^i \partial_j V_k^a \tag{7}$$

and a natural metrics of the Lorentz signature on M

$$g_{ij} = \eta_{ab} V_i^a V_j^b, \quad g^{ij} = \eta^{ab} V_a^i V_b^j, \tag{8}$$

where $\eta_{ab} = \eta^{ab} = \mathrm{diag}(1, -1, -1, -1)$. From (5) and (7) for the torsion tensor and torsion covector of the connection (7) we obtain

$$T_{jk}^i = \Gamma_{jk}^i - \Gamma_{kj}^i = -f_{bc}^a V_a^i V_j^b V_k^c, \quad T_i = T_{ki}^k = -f_{ab}^a V_i^b.$$

From (5) and (8) we have

$$V_a^i V_{b;i}^j = (f_{ab}^c - \eta_{ad} f_{bc}^d \eta^{ce} - \eta_{bd} f_{ae}^d \eta^{ce}) V_c^j, \tag{9}$$

where semicolon means the covariant derivative with respect to the Levi-Civita connection of the metrics (8).

The manifold M that admits a simply transitive group of transformations and has the metrics (8) will be called the homogeneous spacetime manifold. The Minkowski spacetime is a particular case of homogeneous spacetime manifolds. If a homogeneous spacetime defined by a non-Abelian simply transitive group of transformations has a physical meaning, from (2) and (3) it follows that the operator

$$-i\hbar X_0 = -i\hbar V_0^i \frac{\partial}{\partial x^i}$$

will be analog of the operator P_0 in the Minkowski spacetime.

3. The Dirac wave equation for a quark

Here we will define a homogeneous spacetime manifold and show that it obeys all the required conditions and is of definite interest for the physics of quarks.

In the five-dimensional Minkowski spacetime $M^5_{1,4}$ with Cartesian coordinates x^A (indices denoted by capital letters run through the five values 0, 1, 2, 3, 4) and metrics

$$ds^2 = \eta_{AB} dx^A dx^B = (dx^0)^2 - (dx^1)^2 - (dx^2)^2 - (dx^3)^2 - (dx^4)^2,$$

we will consider the one sheet hyperboloid H^4

$$\eta_{AB} x^A x^B = (x^0)^2 - (x^1)^2 - (x^2)^2 - (x^3)^2 - (x^4)^2 = -a^2, \qquad (10)$$

where a is the radius of H^4, and prove that it is a homogeneous spacetime manifold.

We will use the scalar product $(X, Y) = \eta_{AB} U^A V^B$ for any vector fields $X = U^A \partial_A$ and $Y = V^A \partial_A$ on $M^5_{1,4}$. The vector fields

$$P_A = \delta^C_A \partial_C, \quad M_{AB} = (x_A \delta^C_B - x_B \delta^C_A)\partial_C,$$

where $x_A = \eta_{AB} x^B$, are generators of the Poincare group of the five-dimensional Minkowski spacetime. All vectors fields M_{AB} are orthogonal to the radius-vector $R = x^C \partial_C$, but this is not the case for the vector fields P_A. Representing P_A as the sum of the component aligned with the direction of the radius vector R and the component orthogonal to this direction, we obtain the vector fields

$$M_A = aP_A + \frac{1}{a}(R, P_A)R = (a\delta^C_A + \frac{1}{a} x_A x^C)\partial_C,$$

which are tangent to H^4, because from (10), it follows that $(R, M_A) = 0$ at each point of H^4. The vector fields M_A and M_{AB} are generators of the group of conformal transformations of H^4 because we have

$$[M_A, M_B] = -M_{AB}, \quad [M_A, M_{BC}] = \eta_{AB} M_C - \eta_{AC} M_B. \qquad (11)$$

Let us now introduce the vector fields

$$X_0 = M_0, \quad X_1 = M_{14}+M_{23}, \quad X_2 = M_{24}+M_{31}, \quad X_3 = M_{34}+M_{12}. \quad (12)$$

It is straightforward to see that the vector fields X_0, X_1, X_2, and X_3 are continuous and do not vanish at any point of H^4. Because $(X_a, X_b) = 0$ for $a \neq b$, a, $b = 0, 1, 2, 3$ and

$$(X_0, X_0) = \quad -(X_1, X_1) = \quad -(X_2, X_2) = \quad -(X_3, X_3) = \quad a^2 + x_0^2,$$

the vector fields X_0, X_1, X_2, and X_3 are linearly independent at each point of H^4. From (11), it follows that

$$[X_0, X_i] = 0, \quad [X_i, X_j] = 2e_{ijk}X_k, \quad i, j, k = 1, 2, 3,$$

where e_{ijk} is the completely antisymmetric Levi-Civita symbol specified by the equality $e_{123} = 1$. In this way, we have proven that the one sheet hyperboloid (10) admits a simply transitive group of transformations whose generators are given by (12) and which has only the following nonzero structure constants:

$$f^1_{23} = f^2_{31} = f^3_{12} = 2. \quad (13)$$

Therefore, we will supply H^4 with a metrics of the type (8) and thus transform H^4 into the hyperbolic spacetime $H^4_{1,3}$. From (9) and (13) it follows that the vector field X_0 is absolutely parallel with respect to the Levi-Civita connection on $H^4_{1,3}$ induced by the vector fields (12). For comparison we note that the vector field $X_0 = \partial/\partial x^0$ defined in (2) is also absolutely parallel. Now it is natural to put forward the idea that in the realm of the strong interactions spacetime geometrically can be represented as a one-sheeted hyperboloid (10) in the five-dimensional Minkowski spacetime. A constant a can be interpreted geometrically as the radius of the three-dimensional sphere S^3, which is considered here as a space section $x^0 = 0$. From the physical point of view we treat a as the size of region of quark confinement, since a quark is a pointlike particle in the space of constant positive curvature S^3. For comparison of leptons and quarks we note that free motion of the electron is represented as a straight line in the Euclidean usual space and free motion of the quark is a circumference on the 3d sphere. This correlation between leptons and quarks will be continued with an example of the Coulomb law for these objects. To do this consider the Dirac equation for a quark.

In accordance with the original Dirac equation (see also [6]), we write the Dirac equation in the homogeneous spacetime in the form

$$\gamma^c P_c \psi = \mu \psi, \quad (14)$$

where

$$\gamma^a \gamma^b + \gamma^b \gamma^a = -2\eta^{ab},$$

$$P_c = X_c + \frac{iqa}{\hbar c} A_c - \frac{1}{2} f_c, \quad f_c = f^a_{ac}.$$

Here q is the charge of a particle, and A_c are the components of the vector potential of the electromagnetic field in the basis X_a. In $H^4_{1,3}$, we also have

$$\mu = mca/\hbar.$$

For the time being, we do not specify the values of the structure constants of a simply transitive group of transformations of the spacetime $H^4_{1,3}$. In general, $[X_a, X_b] = f^c_{ab} X_c$; therefore, we have

$$[P_a, P_b] = f^c_{ab} P_c + \frac{iqa}{\hbar c} F_{ab},$$

where

$$F_{ab} = X_a A_b - X_b A_a - f^c_{ab} A_c \tag{15}$$

are the components of the strength tensor of the electromagnetic field in the basis X_a. When the wave equation is established it is not difficult to write the equations of electromagnetic field. The Jacobi identity $[P_a[P_b, P_c]] + [P_b[P_c, P_a]] + [P_c[P_a, P_b]] = 0$ results in the first four Maxwell equations

$$X_a F_{bc} + X_b F_{ca} + X_c F_{ab} + f^d_{ab} F_{cd} + f^d_{bc} F_{ad} + f^d_{ca} F_{bd} = 0 \tag{16}$$

To establish the form of other four Maxwell equations, we set $\tilde{F}^{ab} = \frac{1}{2} e^{abcd} F_{cd}$, where e^{abcd} are components of the antisymmetric Levi-Civita unit tensor in the basis X_a. Then we can write equations (16) in the following equivalent form

$$X_a \tilde{F}^{ab} + f_a \tilde{F}^{ab} + \frac{1}{2} f^b_{ad} \tilde{F}^{ad} = 0 \tag{17}$$

By analogy, from (17) it follows that the remaining Maxwell equations are of the form

$$X_a F^{ab} + f_a F^{ab} + \frac{1}{2} f^b_{ad} F^{ad} = \frac{4\pi a}{c} j^b, \tag{18}$$

where j^b are components of the current vector in the basis X_a.

Now we will write the Maxwell equations in the three-dimensional vector form. As usual, we put

$$j^a = (c\rho, \tilde{\mathrm{j}}), \quad A_a = (\varphi, \quad -\tilde{\mathrm{A}}),$$

$$E_i = F_{0i}, \quad H_i = \frac{1}{2} e_{ijk} F^{jk}, \quad i, j, k = 1, 2, 3.$$

Then from (13) and (15) we obtain

$$\tilde{\mathrm{E}} = -\nabla_0 \tilde{\mathrm{A}} - \nabla\varphi, \quad \tilde{\mathrm{H}} = \mathrm{rot}\,\tilde{\mathrm{A}} = \nabla \times \tilde{\mathrm{A}} - 2\tilde{\mathrm{A}}. \tag{19}$$

where

$$\nabla = (\nabla_1, \nabla_2, \nabla_3), \quad \nabla_0 = X_0, \quad \nabla_i = X_i, \quad i = 1, 2, 3.$$

Considering that $\mathrm{div}\vec{A} = \sum_{i=1}^{3} \nabla_i A_i$, we can recast the Maxwell equations (17) and (18) into the familiar vector form

$$-\nabla_0 \tilde{\mathrm{H}} = \mathrm{rot}\,\tilde{\mathrm{E}}, \quad \mathrm{div}\,\tilde{\mathrm{H}} = 0, \quad \mathrm{rot}\,\tilde{\mathrm{H}} = \nabla_0 \tilde{\mathrm{E}} + \frac{4\pi \mathrm{a}}{\mathrm{c}}\tilde{\mathrm{j}}, \quad \mathrm{div}\,\tilde{\mathrm{E}} = 4\pi \mathrm{a}\rho. \tag{20}$$

Making use of the commutation relations

$$[\nabla_i, \nabla_j] = 2e_{ijk}\nabla_k, \quad i, j, k = 1, 2, 3,$$

we can easily verify the identities

$$\mathrm{div\,rot} = 0, \quad \mathrm{rot\,grad} = 0.$$

In addition, we have

$$\mathrm{div\,grad} = \Delta,$$

where $\triangle$ is the Laplacian on a three-dimensional sphere. Torsion (that is, the non-Abelian character of a simply transitive group of transformations of H^4) manifests itself not only in the definition of the operator rot, (21), but also in the identity

$$(\mathrm{rot} + 1)^2 = -\Delta + 1 + \mathrm{graddiv}.$$

Since the space section of the $H^4_{1,3}$ is a three dimensional sphere, it is interesting to show that the Dirac equation (14) is associated with the Schrodinger equation for a spherical Top. To verify this, we will derive eigenvalues E of the Dirac Hamiltonian in question when there is no electromagnetic field that is, for $F_{ab} = 0$. Squaring equation (14) and using (13), we obtain the following equation for E

$$E^2\psi = m^2c^4\psi - \frac{c^2\hbar^2}{a^2}(\Delta + P)\psi, \tag{21}$$

where

$$P = \Sigma_1\nabla_1 + \Sigma_2\nabla_2 + \Sigma_3\nabla_3$$

and $\Sigma_i = \frac{1}{2}e_{ijk}\gamma^j\gamma^k$. The operator P has properties analogous to those of the operator rot. In particular, we have

$$(P + 1)^2 = -\Delta + 1 \tag{22}$$

Since $\Delta + P = -P(P+1)$, then

$$E^2 = m^2c^4 + p(p+1)\frac{c^2\hbar^2}{a^2},$$

where p is an egenvalue of the operator P.

To determine eigenvalues of the operator P, we consider Hermitian operators acting in the space of solutions to the Dirac equation (14). Generators of the group mutual to the simply transitive group of transformations of the spacetime $H^4_{1,3}$ are given by

$$Y_0 = X_0, \quad Y_1 = M_{14} - M_{23}, \quad Y_2 = M_{24} - M_{31}, \quad Y_3 = M_{34} - M_{12}.$$

This leads to the three Hermitian operators

$$N_i = -\frac{i}{2}Y_i$$

which are analogous to the momentum operators. The remaining three operators,

$$M_i = -\frac{i}{2}(\nabla_i - \Sigma_i) - -\frac{i}{2}(X_i - \Sigma_i) \tag{23}$$

are analogs of the electron angular momentum operators. From (23), it follows that the spin of a particle in question is $\hbar/2$. We have

$$\tilde{\mathrm{M}} \times \tilde{\mathrm{M}} = \mathrm{i}\tilde{\mathrm{M}}, \quad \tilde{\mathrm{N}} \times \tilde{\mathrm{N}} = \mathrm{i}\tilde{\mathrm{N}}$$

and, in addition,

$$2(M^2 + N^2) = (P + \frac{3}{2})^2 - \frac{3}{4}, \quad 2(M^2 - N^2) = P + \frac{3}{2}. \tag{24}$$

Hence, we have the operator equation

$$2(M^2 + N^2) + \frac{3}{4} = 4(M^2 - N^2)^2.$$

Since, $M^2 = l(l+1)$, $N^2 = k(k+1)$, we find from the operator equation that l and k satisfy the equation

$$2[l(l+1) + k(k+1)] + \frac{3}{4} = 4[l(l+1) - k(k+1)]^2.$$

This equation has two solutions, $l = k + \frac{1}{2}$ and $k = l + \frac{1}{2}$, whence it followers that $p = 2[l(l+1) - k(k+1)] - \frac{3}{2} = -2k - 3$. Since S^3 has a metric invariant with respect to the isometrical reflection, $p = 2k + 3$ is an eqenvalue too. For the energy, we then have

$$E^2 = m^2c^4 + n(n+1)\frac{c^2\hbar^2}{a^2} = m^2c^4(1 + n(n+1)\frac{\lambda^2}{a^2}), \tag{25}$$

where $n = 2, 3, \ldots$ and $\lambda = \hbar/mc$. If formula (25) gives the quantum-mechanical value of the energy of the relativistic spherical Top, we can then conclude that, at large a, the moment of inertia $I = ma^2$ is also large, so that the angular velocity is small. Therefore, the nonrelativistic limit can be found from the condition $a \gg \lambda$. In the limit of large a it follows from (25) that

$$E = mc^2 + \frac{L^2}{2I},$$

where $L^2 = n(n+1)\hbar^2$ is the angular momentum of the spherical top and I is its moment of inertia. The last formula is in accordance with the classical formula

$$E = \frac{L^2}{2I}$$

for the energy of a Top. Thus, the formula (25) gives energy of rotation.

Let us now consider the Coulomb law. The Coulomb potential can be derived as a solution of the equations of electrostatics, which are invariant under the group of Euclidean motions, including rotations and translations. In the case being considered, we will seek a Coulomb potential in an analogous manner. From (19) and (20), it follows that for a constant electric field $\operatorname{div}\tilde{\mathrm{E}} = 4\pi \mathrm{a}\rho$, $\tilde{\mathrm{E}} = -\nabla\varphi$, and consequently, φ obeys the equation

$$\Delta\varphi = -4\pi a^2 \rho. \tag{26}$$

As it is well known, the electron Coulomb potential

$$\phi_e(r) = \frac{e}{r}$$

is the fundamental solution to the Laplace equation $\Delta\phi = \operatorname{div}\operatorname{grad}\phi = 0$. In accordance with our conjecture the Coulomb potential for quarks can be derived as follows. Consider the stereographic projection S^3 from point $(0,0,0,-a)$ onto the ball $x^2 + y^2 + z^2 \leq a^2$:

$$x^1 = fx, \quad x^2 = fy, \quad x^3 = fz, \quad x^4 = a(1-f),$$

where $f = 2a^2/(a^2 + r^2)$. Then, it follows that the element of length on the three-dimensional sphere can be represented in the form

$$ds^2 = f^2(dx^2 + dy^2 + dz^2)$$

and hence the Laplace equation on S^3 can be written as follows

$$\Delta\phi = f^{-3}\operatorname{div}(f\operatorname{grad}\phi) = 0.$$

We seek the solution to this equation that is invariant under the transformation of the group $SO(3)$ with generators

$$x^1\frac{\partial}{\partial x^2} - x^2\frac{\partial}{\partial x^1} = x\frac{\partial}{\partial y} - y\frac{\partial}{\partial x}, \quad \text{etc.}$$

This subgroup of the $SO(4)$ group is determined by fixing the point $(0, 0, 0, -a)$. Let us put

$$\psi = f \frac{1}{r} \frac{d\phi}{dr}.$$

Since

$$\Delta\phi = f^{-3}(r\frac{d\psi}{dr} + 3\psi) = r^{-2} f^{-3} \frac{d}{dr}(r^3\psi),$$

then $r^3\psi = c_1 =$ *constant*. Thus, we have

$$\frac{d\phi}{dr} = c_1 \frac{a^2 + r^2}{2a^2r^2} = c_1(\frac{1}{2r^2} + \frac{1}{2a^2})$$

and hence

$$\phi_q = c_1(-\frac{1}{2r} + \frac{r}{2a^2}) + c_2. \tag{27}$$

Generally speaking, expression (27), that we have derived, coincides with the well-known Cornell potential [7]. If we demand that $\phi_e(a) = \phi_q(a)$, then $c_2 = e/a$ and the Coulomb law for quarks has the form

$$\phi_q(r) = q(\frac{1}{2r} - \frac{r}{2a^2}) + \frac{e}{a}, \tag{28}$$

where q is the quark charge.

From our consideration it follows that the idea of rotating matter can be realized in the framework of relativistic quantum mechanics by the Dirac equation (14).

4. Conclusion

As the basic wave equation describing the dynamics of quarks, we have proposed the modified Dirac equation (14), which has been written here in homogeneous coordinates. The conclusion that quarks are described by the wave equation different from the conventional wave equation for electrons is quite natural. In fact, it would be strange if the description of such different particles were based on the same equation.

The physical meaning of the confinement phenomenon is tightly connected with idea of a rotating matter and consists in that quarks possess properties of a quantum-mechanical spherical Top. Among other things, this means that the interquark potential expresses the Coulomb law for quarks and, in fact, coincides with the well-known Cornell potential that was first very successfully used by the Cornell group.

For $a \to \infty$, the theory of electrons can be derived from the theory of quarks, but the electrons are obviously deconfined because in this case, the region of confinement covers the entire Euclidean space. Thus, the symmetry between quarks and leptons has a natural explanation.

Since the kinematics of quarks differs from the kinematics of electrons, decays of hadrons and nuclei such that the energy is conserved, but the momentum is not, are possible.

References

1. K.S. Thorne, D.L. Lee and A.P. Lightman, Phys. Rev. D **7**, (1973), 3563.
2. E. Schrödinger, Proc. R. Irish. Acad. Ser., A **46** (1940), 93.
3. H.A. Kramers and G.P. Itmann, Z. Phys. **53** (1929), 553.
4. H.B. G. Casimir, *Rotation of Rigid Body in Quantum Mechanics*, Leyden, 1931.
5. L.P. Eisenhart, *Continuous Groups of Transformatins*, Princeton University Press, Chapter V, 1933.
6. I.B. Pestov, Aspects of Complex analysis, Differential Geometry, Mathematical Physics and Applications, Eds. S. Dimiev and K. Sekigawa (World Scientific, Singapore, 1999), p. 236.
7. E. Eichten, H. Gottfried, T. Kinoshita, K.D. Lane and T-M. Yan, Phys. Rev. D **21**, (1980), 203.

CHARACTERIZATIONS OF PARALLEL ISOMETRIC IMMERSIONS OF COMPACT RANK ONE SYMMETRIC SPACES INTO SPACE FORMS BY CIRCLES

KAORU SUIZU

Department of Mathematics, Shimane University
Matsue, Shimane, 690-8504, Japan
E-mail: suizu@math.shimane-u.ac.jp

It is known that the parallel immersions of compact rank one symmetric spaces into real space forms are the examples of planar geodesic submanifolds. Namely each geodesic on these symmetric spaces is mapped to a plane curve in the ambient real space form through the parallel immersions Moreover, we know that some particular circles of positive curvature on these submanifolds are also mapped to plane curves. In this paper we consider the converse of these geometric properties of such planar geodesic immersions.

1. Introduction

Let M and $\widetilde{M}$ be Riemannian manifolds and $f : M \to \widetilde{M}$ be an isometric immersion. We recall the definition of planar geodesic immersions. If for each geodesic γ on the submanifold M the curve $f \circ \gamma$ is a plane curve in the ambient manifold $\widetilde{M}$, that is $f \circ \gamma$ is locally contained in a 2-dimensional totally geodesic submanifold of $\widetilde{M}$, then the isometric immersion f is called a *planar geodesic* immersion.

By the result [6] planar geodesic submanifolds of a real space form $\widetilde{M}^m(\tilde{c})(= E^m$, $S^m(\tilde{c})$ or $H^m(\tilde{c}))$ of curvature $\tilde{c}$ are completely classified. If M is a planar geodesic submanifold of $\widetilde{M}^m(\tilde{c})$, then M is totally umbilic in $\widetilde{M}^m(\tilde{c})$ or a compact symmetric space of rank one through immersions with parallel second fundamental form (for details, see Theorem A). We here take care of these embeddings of four projective spaces, namely they are a real projective space $\mathbb{R}P^n$, a complex projective space $\mathbb{C}P^n$, a quaternionic projective space $\mathbb{Q}P^n$ and a Cayley projective plane CaP^2. We shall investigate the extrinsic shape $f \circ \gamma$ for a circle γ on these projective spaces in the ambient real space form $\widetilde{M}^m(\tilde{c})$ under the embedding f. Here a regular curve $\gamma = \gamma(s)$, parametrized by its arclength s, in a Riemannian manifold M is called a *circle* provided that there exist a field $V = V(s)$ of unit vectors along γ and a nonnegative constant k which satisfy

$$\nabla_{\dot{\gamma}}\dot{\gamma} = kV, \quad \nabla_{\dot{\gamma}}V = -k\dot{\gamma}, \tag{1.1}$$

where $\dot{\gamma}$ expresses the velocity vector field $\dot{\gamma}(s)$ of the curve γ at $\gamma(s)$ and $\nabla_{\dot{\gamma}}$ denotes the covariant differentiation along γ with respect to the Riemannian connection ∇ of M. The constant k is called the *curvature* of the circle. A circle of curvature zero is nothing but a geodesic. For an arbitrary point x, an arbitrary orthonormal pair (u, v) of vectors at x and a positive k, by the existence and uniqueness theorem about solutions for ordinary differential equations, there exists locally a unique circle $\gamma = \gamma(s)$ with initial condition that $\gamma(0) = x$, $\dot{\gamma}(0) = u$ and $V(0) = v$. If M is complete, there exists globally such a circle.

It is known that for each circle γ of positive curvature on $\mathbb{R}P^n$ the curve $f \circ \gamma$ is a helix of proper order 3 or 4 in $\widetilde{M}^m(\tilde{c})$, so that it is never a circle in the ambient space (see Proposition 4). On the contrary, we know that some circles of $\mathbb{C}P^n$, $\mathbb{Q}P^n$ or CaP^2 are mapped to circles in the ambient space $\widetilde{M}^m(\tilde{c})$ under parallel embeddings (see Propositions 1, 2 and 3).

Our purpose of this paper is to give some characterizations of parallel embeddings of a complex projective space, a quaternionic projective space and a Cayley projective plane by observing the extrinsic shape of particular circles (Theorems 1, 2 and 3).

2. Preliminaries

We first review the Frenet formula for a smooth Frenet curve in a Riemannian manifold M. A smooth curve $\gamma = \gamma(s)$ parametrized by its arclength s is called a *Frenet curve of proper order* d if there exist orthonormal frame fields $\{V_1 = \dot{\gamma}, \ldots, V_d\}$ along γ and positive functions $\kappa_1, \ldots, \kappa_{d-1}$ satisfying the following system of ordinary equations

$$\nabla_{\dot{\gamma}} V_j = -\kappa_{j-1} V_{j-1} + \kappa_j V_{j+1}, \quad j = 1, \ldots, d, \tag{2.1}$$

where $V_0 \equiv V_{d+1} \equiv 0$. We call Equation (2.1) the Frenet formula for the Frenet curve γ. The functions κ_j ($j = 1, \ldots, d-1$) and the orthonormal frame $\{V_1, \ldots, V_d\}$ are called the curvatures and the Frenet frame of γ, respectively.

A Frenet curve is called a *Frenet curve of order* d if it is a Frenet curve of proper order $r(\leq d)$. For a Frenet curve of order d which is of proper order $r(\leq d)$, we use the convention in (2.1) that $\kappa_j \equiv 0$ ($r \leq j \leq d-1$), and $V_j \equiv 0$ ($r+1 \leq j \leq d$). We call a smooth Frenet curve a *helix* when all its curvatures are constant. A helix of order 1 is nothing but a geodesic and a helix of order 2 is a circle.

Next we introduce the notion of curves of order 2, which plays a key role in this paper. Let M be a Riemannian manifold with Riemannian metric $\langle\ ,\ \rangle$. A smooth curve γ on M parametrized by its arclength s is called a *curve of order* 2 if it satisfies the following differential equation

$$\|\nabla_{\dot\gamma}\dot\gamma\|^2\{\nabla_{\dot\gamma}\nabla_{\dot\gamma}\dot\gamma + \|\nabla_{\dot\gamma}\dot\gamma\|^2\dot\gamma\} = \langle\nabla_{\dot\gamma}\dot\gamma, \nabla_{\dot\gamma}\nabla_{\dot\gamma}\dot\gamma\rangle\nabla_{\dot\gamma}\dot\gamma. \tag{2.2}$$

A geodesic with unit speed, which should be called a curve of order 1, satisfies (2.2) hence is a curve of order 2 in our terminology and a circle satisfies also (2.2) hence is a curve of order 2. Every smooth curve on a Euclidean plane $\mathbb{R}^2$ parametrized by its arclength is also a curve of order 2. When we consider a curve γ of order 2 which is not a geodesic, we admit the case that the curve γ has an inflection point $\gamma(s_0)$, that is, $(\nabla_{\dot\gamma}\dot\gamma)(s_0) = 0$.

We next recall the notion of isotropic immersions. An isometric immersion $f : M \to \widetilde{M}$ is called to be *isotropic* at $x \in M$ if $\|\sigma(X,X)\|/\|X\|^2 (= \lambda(x))$ does not depend on the choice of $X(\neq 0) \in T_xM$. If the immersion is isotropic at every point, then the immersion is said to be isotropic. When the function $\lambda = \lambda(x)$ is constant on M, we call M a constant (λ-)isotropic submanifold. Note that a totally umbilic immersion is isotropic, but not vice versa. It is well-known that for f be an isometric immersion of M into $(\widetilde{M}, \langle\ ,\ \rangle)$, f is isotropic at $x \in M$ if and only if the second fundamental form σ of f satisfies $\langle\sigma(X,X), \sigma(X,Y)\rangle = 0$ for an arbitrary orthogonal pair $X, Y \in T_xM$ (see [5]).

Finally we review fundamental equations in submanifold theory. Let M be an n-dimensional Riemannian submanifold of $\widetilde{M}^m$ with metric $\langle\ ,\ \rangle$. We denote by ∇ and $\widetilde\nabla$ the covariant differentiations of M and $\widetilde{M}$, respectively. Then the second fundamental form σ of the immersion is defined by

$$\sigma(X,Y) = \widetilde\nabla_X Y - \nabla_X Y, \tag{2.3}$$

where X and Y are vector fields tangent to M. For a vector field ξ normal to M, we write

$$\widetilde\nabla_X \xi = -A_\xi X + \nabla^\perp_X \xi, \tag{2.4}$$

where $-A_\xi X$ (resp. $\nabla^\perp_X \xi$) denotes the tangential (resp. the normal) component of $\widetilde\nabla_X \xi$. We define the covariant differentiation $\bar\nabla$ of the second fundamental form σ with respect to the connection in (tangent bundle)+(normal bundle) as follows:

$$(\bar\nabla_X \sigma)(Y,Z) = \nabla^\perp_X(\sigma(Y,Z)) - \sigma(\nabla_X Y, Z) - \sigma(Y, \nabla_X Z). \tag{2.5}$$

We call that an isometric immersion $f : M \to \widetilde{M}$ is *parallel* if the second fundamental form of f is parallel, i.e., $(\bar\nabla_X \sigma)(Y,Z) = 0$ for all tangent vector fields X, Y and Z on M.

3. Extrinsic shape of circles on some projective spaces

The following theorem classifies all planar geodesic submanifolds in a real space form [6].

Theorem A. *Let M^n be a Riemannian submanifold of a real space form $\widetilde{M}^m(\tilde{c})$ through an isometric immersion f. Suppose that f is a planar geodesic immersion. Then M^n is totally umbilic in $\widetilde{M}^m(\tilde{c})$ or M^n is locally congruent to a compact symmetric space of rank one which is immersed into some totally umbilic submanifold of $\widetilde{M}^m(\tilde{c})$ through the parallel minimal immersion. This parallel minimal immersion is locally equivalent either to the first standard minimal embedding of one of the compact symmetric spaces of rank one or to the second standard minimal immersion of a sphere.*

It follows from above Theorem that every geodesic on such a planar geodesic submanifold is mapped to a circle on $\widetilde{M}^m(\tilde{c})$. Consequently it should be emphasized that we cannot distinguish these planar geodesic immersions by the *extrinsic shape of geodesics* on submanifolds.

We here review the definition of some particular circles in a Kähler manifold and a quaternionic Kähler manifold.

Let γ be a circle in a Kähler manifold M. It follows from (1.1) that $\langle J\dot{\gamma}, V\rangle$ is constant along γ, where J is the complex structure of M. Therefore it makes sense to define a *Kähler circle* as a circle γ satisfying the condition that $\dot{\gamma}$ and V span a holomorphic plane, that is, $V = J\dot{\gamma}$ or $V = -J\dot{\gamma}$. Note that if γ is a Kähler circle, then (1.1) reduces to

$$\nabla_{\dot{\gamma}}\dot{\gamma} = kJ\dot{\gamma} \quad \text{or} \quad \nabla_{\dot{\gamma}}\dot{\gamma} = -kJ\dot{\gamma}.$$

The extrinsic shape of Kähler circles through the parallel embedding of a complex projective space is known as follows ([3]):

Proposition 1. *Let $f = f_2 \circ f_1 : \mathbb{C}P^n\big(2nc/(n+1)\big) \xrightarrow{f_1} S^{n(n+2)-1}(c) \xrightarrow{f_2} \widetilde{M}^m(\tilde{c})$ be an isometric parallel embedding of $\mathbb{C}P^n\big(2nc/(n+1)\big)$ into a real space form $\widetilde{M}^m(\tilde{c})$ $(c \geq \tilde{c})$. Here f_1 is the first standard minimal embedding of $\mathbb{C}P^n\big(2nc/(n+1)\big)$ into $S^{n(n+2)-1}(c)$ and f_2 is a totally umbilic embedding of $S^{n(n+2)-1}(c)$ into $\widetilde{M}^m(\tilde{c})$. Then f maps every Kähler circle γ of $\mathbb{C}P^n\big(2nc/(n+1)\big)$ to a circle in $\widetilde{M}^m(\tilde{c})$, so that in particular the curve $f \circ \gamma$ is a curve of order 2 in the ambient space $\widetilde{M}^m(\tilde{c})$.*

Let M be a quaternionic Kähler manifold with local basis $\{I, J, K\}$ of quaternionic structure and let γ be a circle in M. Then I, J and K satisfy

$$\begin{cases} \nabla_{\dot\gamma} I = \quad qJ \ -rK \\ \nabla_{\dot\gamma} J = -qI \qquad\quad +pK \\ \nabla_{\dot\gamma} K = \ \ rI \ -pJ, \end{cases} \tag{3.1}$$

for some functions p, q, r along γ. We see from (1.1) and (3.1) that $\langle V, I\dot\gamma\rangle^2 + \langle V, J\dot\gamma\rangle^2 + \langle V, K\dot\gamma\rangle^2$ is constant along γ ([1]). Therefore it makes sense to consider a circle γ satisfying the condition that V is a linear combination of $I\dot\gamma$, $J\dot\gamma$ and $K\dot\gamma$ at each point of γ. In fact, V is a linear combination of $I\dot\gamma$, $J\dot\gamma$ and $K\dot\gamma$ if and only if $\langle V, I\dot\gamma\rangle^2 + \langle V, J\dot\gamma\rangle^2 + \langle V, K\dot\gamma\rangle^2 = 1$. A *quaternionic circle* is, by definition, a circle with such a property.

The extrinsic shape of quaternionic circles through the parallel embedding of a quaternionic projective space is known as follows ([3]):

Proposition 2. *Let* $f = f_2 \circ f_1 : \mathbb{Q}P^n\big(2nc/(n+1)\big) \xrightarrow{f_1} S^{n(2n+3)-1}(c) \xrightarrow{f_2} \widetilde{M}^m(\tilde c)$ *be an isometric parallel embedding of* $\mathbb{Q}P^n\big(2nc/(n+1)\big)$ *into a real space form* $\widetilde{M}^m(\tilde c)$ $(c \geq \tilde c)$. *Here* f_1 *is the first standard minimal embedding of* $\mathbb{Q}P^n\big(2nc/(n+1)\big)$ *into* $S^{n(2n+3)-1}(c)$ *and* f_2 *is a totally umbilic embedding of* $S^{n(2n+3)-1}(c)$ *into* $\widetilde{M}^m(\tilde c)$. *Then* f *maps every quaternionic circle* γ *of* $\mathbb{Q}P^n\big(2nc/(n+1)\big)$ *to a circle in* $\widetilde{M}^m(\tilde c)$, *so that in particular the curve* $f \circ \gamma$ *is a curve of order* 2 *in the ambient space* $\widetilde{M}^m(\tilde c)$.

We next consider some particular circle in Cayley projective plane. For a Frenet curve γ of proper order 2 in a locally symmetric space M, which satisfies $\nabla_{\dot\gamma}\dot\gamma = \kappa V$, $\nabla_{\dot\gamma} V = -\kappa\dot\gamma$ with a positive function κ, the sectional curvature $K(\dot\gamma, V)$ of the plane spanned by $\dot\gamma$ and V is constant along γ. When M is a complex space form $M_n(c)$, it is an equivalent invariant to the complex torsion τ which is defined by $\tau = \langle\dot\gamma, JV\rangle$, by the following relation: $K(\dot\gamma, V) = c(1 + 3\tau^2)/4$. For a Frenet curve of proper order 2 in a Cayley projective plane $CaP^2(c)$ of maximal sectional curvature c, we call it *Cayley* if it satisfies $K(\dot\gamma, V) \equiv c$, and *totally real* if it satisfies $K(\dot\gamma, V) \equiv c/4$.

It is known that the parallel embedding f of $CaP^2(c)$ into a real space form $\widetilde{M}^m(\tilde c)$ is decomposed as:

$$f : CaP^2(c) \xrightarrow{f_1} S^{25}(3c/4) \xrightarrow{f_2} \widetilde{M}^m(\tilde c), \tag{3.2}$$

where $3c \geq 4\tilde c$, and f_1 is the first standard minimal embedding of $CaP^2(c)$ into $S^{25}(3c/4)$ and f_2 is a totally umbilic embedding of $S^{25}(3c/4)$ into $\widetilde{M}^m(\tilde c)$.

The extrinsic shape of circles on $CaP^2(c)$ in the ambient manifold $\widetilde{M}^m(\tilde c)$ through the parallel embedding f was studied in [2].

Proposition 3. *The extrinsic shape* $f \circ \gamma$ *of each circle* γ *of curvature* $k(> 0)$ *on* $CaP^2(c)$ *in* $\widetilde{M}^m(\tilde{c})$ *through the isometric embedding* f *defined by* (3.2) *is as follows*:

(1) *When* γ *is Cayley, the curve* $f \circ \gamma$ *is a plane curve in* $\widetilde{M}^m(\tilde{c})$, *in particular it is a circle of curvature* $\sqrt{k^2 + c - \tilde{c}}$ *in this space.*

(2) *When* γ *is totally real, the curve* $f \circ \gamma$ *is a helix of proper order* 3 *or proper order* 4 *in* $\widetilde{M}^m(\tilde{c})$.

(3) *When* γ *is neither Cayley nor totally real, the curve* $f \circ \gamma$ *is a helix of proper order* 5 *or proper order* 6 *in* $\widetilde{M}^m(\tilde{c})$.

4. Characterizations of parallel embeddings

We consider the converse of Proposition 1 to obtain a characterization of the parallel embedding of a complex projective space ([9]).

Theorem 1. *Let* M *be a non-flat Kähler manifold of real dimension* $2n$ (≥ 4) *which is immersed into a real space form* $\widetilde{M}^m(\tilde{c})$. *If there exists* $k > 0$ *and all Kähler circles of curvature* k *on* M *are mapped to curves of order* 2 *in* $\widetilde{M}^m(\tilde{c})$, *then* M *is locally congruent to a complex projective space embedded into some sphere in* $\widetilde{M}^m(\tilde{c})$ *through the first standard minimal embedding.*

Proof. We denote by $f : M \to \widetilde{M}^m(\tilde{c})$ the isometric immersion which satisfies our assumption. Let x be any point of M. In the following, take and fix a unit vector $X \in T_xM$.

Let $\gamma = \gamma(s)$ be a Kähler circle of curvature k on M satisfying the equation $\nabla_{\dot{\gamma}}\dot{\gamma} = kJ\dot{\gamma}$ with initial condition that $\gamma(0) = x$, $\dot{\gamma}(0) = X$ and $(\nabla_{\dot{\gamma}}\dot{\gamma})(0) = kJX$. By the assumption it is mapped to the curve $f \circ \gamma$ of order 2 in the ambient manifold $\widetilde{M}$, the curve $f \circ \gamma$ satisfies the differential equation

$$||\widetilde{\nabla}_{\dot{\gamma}}\dot{\gamma}||^2\{\widetilde{\nabla}_{\dot{\gamma}}\widetilde{\nabla}_{\dot{\gamma}}\dot{\gamma} + ||\widetilde{\nabla}_{\dot{\gamma}}\dot{\gamma}||^2\dot{\gamma}\} = \langle\widetilde{\nabla}_{\dot{\gamma}}\dot{\gamma}, \widetilde{\nabla}_{\dot{\gamma}}\widetilde{\nabla}_{\dot{\gamma}}\dot{\gamma}\rangle\widetilde{\nabla}_{\dot{\gamma}}\dot{\gamma}. \tag{4.1}$$

Here for simplicity we usually denote the curve $f \circ \gamma$ by γ. It follows from (2.3), (2.4) and (2.5) that

$$\begin{aligned}\widetilde{\nabla}_{\dot{\gamma}}\widetilde{\nabla}_{\dot{\gamma}}\dot{\gamma} &= \widetilde{\nabla}_{\dot{\gamma}}\big(\nabla_{\dot{\gamma}}\dot{\gamma} + \sigma(\dot{\gamma}, \dot{\gamma})\big)\\ &= \nabla_{\dot{\gamma}}\nabla_{\dot{\gamma}}\dot{\gamma} + \sigma(\nabla_{\dot{\gamma}}\dot{\gamma}, \dot{\gamma}) - A_{\sigma(\dot{\gamma},\dot{\gamma})}\dot{\gamma} + \nabla^{\perp}_{\dot{\gamma}}(\sigma(\dot{\gamma}, \dot{\gamma}))\\ &= -k^2\dot{\gamma} + 3k(\sigma(J\dot{\gamma}, \dot{\gamma})) - A_{\sigma(\dot{\gamma},\dot{\gamma})}\dot{\gamma} + (\bar{\nabla}_{\dot{\gamma}}\sigma)(\dot{\gamma}, \dot{\gamma}).\end{aligned} \tag{4.2}$$

We here see that $||\widetilde{\nabla}_{\dot{\gamma}}\dot{\gamma}||^2 = k^2 + ||\sigma(\dot{\gamma}, \dot{\gamma})||^2 > 0$, hence by putting $\widetilde{\kappa} = ||\widetilde{\nabla}_{\dot{\gamma}}\dot{\gamma}||$ we get

$$\widetilde{\kappa}\widetilde{\kappa}' = \frac{1}{2}\frac{d}{ds}(\widetilde{\kappa})^2 = \frac{1}{2}\frac{d}{ds}\langle\widetilde{\nabla}_{\dot{\gamma}}\dot{\gamma}, \widetilde{\nabla}_{\dot{\gamma}}\dot{\gamma}\rangle = \langle\widetilde{\nabla}_{\dot{\gamma}}\dot{\gamma}, \widetilde{\nabla}_{\dot{\gamma}}\widetilde{\nabla}_{\dot{\gamma}}\dot{\gamma}\rangle,$$

where $\widetilde{\kappa}' = \frac{d}{ds}\widetilde{\kappa}$. Thus, it follows from (4.1) and (4.2) that

$$\begin{aligned}\widetilde{\kappa}\{3k\sigma(J\dot{\gamma},\dot{\gamma}) - A_{\sigma(\dot{\gamma},\dot{\gamma})}\dot{\gamma} + (\bar{\nabla}_{\dot{\gamma}}\sigma)(\dot{\gamma},\dot{\gamma}) + \|\sigma(\dot{\gamma},\dot{\gamma})\|^2\dot{\gamma}\} \\ = \widetilde{\kappa}'\{kJ\dot{\gamma} + \sigma(\dot{\gamma},\dot{\gamma})\}.\end{aligned} \tag{4.3}$$

Considering the tangential and the normal components for the submanifold M in Equation (4.3), we obtain the following :

$$\widetilde{\kappa}(-A_{\sigma(\dot{\gamma},\dot{\gamma})}\dot{\gamma} + \|\sigma(\dot{\gamma},\dot{\gamma})\|^2\dot{\gamma}) = \widetilde{\kappa}'kJ\dot{\gamma}, \tag{4.4}$$

$$\widetilde{\kappa}\{3k\sigma(J\dot{\gamma},\dot{\gamma}) + (\bar{\nabla}_{\dot{\gamma}}\sigma)(\dot{\gamma},\dot{\gamma})\} = \widetilde{\kappa}'\sigma(\dot{\gamma},\dot{\gamma}). \tag{4.5}$$

Taking the inner product of both hand-side of (4.4) with $J\dot{\gamma}$, so we have

$$-\widetilde{\kappa}\langle\sigma(\dot{\gamma},\dot{\gamma}),\sigma(\dot{\gamma},J\dot{\gamma})\rangle = \widetilde{\kappa}'k. \tag{4.6}$$

Note that k is constant. As we have

$$\begin{aligned}\widetilde{\kappa}\widetilde{\kappa}' = \frac{1}{2}\frac{d}{ds}(\widetilde{\kappa})^2 &= \langle\nabla^{\perp}_{\dot{\gamma}}(\sigma(\dot{\gamma},\dot{\gamma})),\sigma(\dot{\gamma},\dot{\gamma})\rangle \\ &= \langle 2k\sigma(\dot{\gamma},J\dot{\gamma}) + (\bar{\nabla}_{\dot{\gamma}}\sigma)(\dot{\gamma},\dot{\gamma}),\sigma(\dot{\gamma},\dot{\gamma})\rangle,\end{aligned}$$

we obtain from (4.6)

$$-\widetilde{\kappa}^2\langle\sigma(\dot{\gamma},\dot{\gamma}),\sigma(\dot{\gamma},J\dot{\gamma})\rangle = k\langle 2k\sigma(\dot{\gamma},J\dot{\gamma}) + (\bar{\nabla}_{\dot{\gamma}}\sigma)(\dot{\gamma},\dot{\gamma}),\sigma(\dot{\gamma},\dot{\gamma})\rangle.$$

Therefore we have

$$(3k^2 + \|\sigma(\dot{\gamma},\dot{\gamma})\|^2)\langle\sigma(\dot{\gamma},\dot{\gamma}),\sigma(\dot{\gamma},J\dot{\gamma})\rangle + k\langle\sigma(\dot{\gamma},\dot{\gamma}),(\bar{\nabla}_{\dot{\gamma}}\sigma)(\dot{\gamma},\dot{\gamma})\rangle = 0.$$

Evaluating the above equation at $s = 0$ we find

$$\begin{aligned}(3k^2 + \|\sigma(X,X)\|^2)\langle\sigma(X,X),\sigma(X,JX)\rangle \\ + k\langle\sigma(X,X),(\bar{\nabla}_X\sigma)(X,X)\rangle = 0.\end{aligned} \tag{4.7}$$

We here apply the above discussion to the Kähler circle $\nabla_{\dot{\delta}}\dot{\delta} = -kJ\dot{\delta}$ with initial condition that $\delta(0) = x$, $\dot{\delta}(0) = X$ and $(\nabla_{\dot{\delta}}\dot{\delta})(0) = -kJX$, we find the following corresponding to (4.7):

$$\begin{aligned}-(3k^2 + \|\sigma(X,X)\|^2)\langle\sigma(X,X),\sigma(X,JX)\rangle \\ + k\langle\sigma(X,X),(\bar{\nabla}_X\sigma)(X,X)\rangle = 0.\end{aligned} \tag{4.8}$$

As $k > 0$, from (4.7) and (4.8) we see $\langle\sigma(X,X),\sigma(X,JX)\rangle = 0$. Therefore, evaluating (4.6) at $s = 0$, we get $\widetilde{\kappa}'(0) = 0$. It follows from (4.4) at $s = 0$ that $A_{\sigma(X,X)}X = \|\sigma(X,X)\|^2X$ for each unit vector $X \in T_xM$ at any point $x \in M$. So, M is an isotropic submanifold of $\widetilde{M}(\tilde{c})$.

Moreover, from $\widetilde{\kappa}'(0) = 0$ we know by (4.5) that

$$(\bar{\nabla}_X\sigma)(X,X) = -3k\cdot\sigma(X,JX).$$

On the other hand, considering the Kähler circle δ again, we get

$$(\bar{\nabla}_X \sigma)(X, X) = 3k \cdot \sigma(X, JX).$$

Hence we have

$$(\bar{\nabla}_X \sigma)(X, X) = 0, \tag{4.9}$$

$$\sigma(X, JX) = 0. \tag{4.10}$$

Since $X \in TM$ is an arbitrary unit vector, by the equation of Codazzi $(\bar{\nabla}_X)(Y, Z) = (\bar{\nabla}_Y)(X, Z)$, Equation (4.9) means $\bar{\nabla}\sigma = 0$. Replacing X by $X + JX$ in (4.10), we get

$$\sigma(JX, JX) = \sigma(X, X) \tag{4.11}$$

for all $X \in TM$. Let R denote the curvature tensor of M, then it follows from (4.10), (4.11) and the equation of Gauss that

$$\begin{aligned}\langle R(X, JX)JX, X\rangle &= \tilde{c} + \langle \sigma(X, X), \sigma(JX, JX)\rangle - \|\sigma(X, JX)\|^2 \\ &= \tilde{c} + \|\sigma(X, X)\|^2\end{aligned}$$

for each unit vector X. Since M is isotropic, this means that M is a complex space form. Therefore our assertion follows from the results of [4] and [10]. □

For a quaternionic Kähler manifold we obtain the following result similar to Theorem 1 ([9]).

Theorem 2. *M be a non-flat quaternionic Kähler manifold of real dimension $4n$ (≥ 8) which is immersed into a real space form $\widetilde{M}^m(\tilde{c})$. If there exists $k > 0$ and all quaternionic circles of curvature k on M are mapped to curves of order 2 in $\widetilde{M}^m(\tilde{c})$, then M is locally congruent to a quaternionic projective space embedded into some sphere in $\widetilde{M}^m(\tilde{c})$ through the first standard minimal embedding.*

Proof. We denote by $f : M \to \widetilde{M}^m(\tilde{c})$ the isometric immersion which satisfies our assumption. Let x be any point of M. In the following, take and fix a unit vector $X \in T_x M$.

Let $\gamma = \gamma(s)$ be a quaternionic circle of curvature k satisfying the equations:

$$\nabla_{\dot{\gamma}}\dot{\gamma} = kV, \quad \nabla_{\dot{\gamma}}V = -k\dot{\gamma}, \quad V(s) = \lambda I\dot{\gamma} + \mu J\dot{\gamma} + \nu K\dot{\gamma},$$

where λ, μ and ν are functions along γ satisfying $\lambda^2 + \mu^2 + \nu^2 = 1$ with initial condition that $\gamma(0) = x$, $\dot{\gamma}(0) = X$ and $V(0) = Y = \lambda(0)IX + \mu(0)JX +$

$\nu(0)KX$ on M. Then by an argument similar to the proof of Theorem 1, at the point x we obtain

$$\begin{aligned}(3k^2 + \|\sigma(X,X)\|^2)\langle\sigma(X,X),\sigma(X,Y)\rangle \\ + k\langle\sigma(X,X),(\bar{\nabla}_X\sigma)(X,X)\rangle = 0,\end{aligned} \tag{4.12}$$

which corresponds to Equation (4.7).

Next, we shall study another quaternionic circle δ of the same curvature k with initial condition that $\delta(0) = x$, $\dot{\delta}(0) = X$ and $V(0) = -Y$. Then we get the following identity corresponding to (4.12):

$$\begin{aligned}-(3k^2 + \|\sigma(X,X)\|^2)\langle\sigma(X,X),\sigma(X,Y)\rangle \\ + k\langle\sigma(X,X),(\bar{\nabla}_X\sigma)(X,X)\rangle = 0.\end{aligned} \tag{4.13}$$

As $k > 0$, from (4.12) and (4.13) we get

$$\langle\sigma(X,X),\sigma(X,Y)\rangle = 0. \tag{4.14}$$

On the other hand, in (4.4) setting $\lambda I\dot{\gamma} + \mu J\dot{\gamma} + \nu K\dot{\gamma}$ in place of $J\dot{\gamma}$ and $s = 0$, we obtain

$$\widetilde{\kappa}(0)(-A_{\sigma(X,X)}X + \|\sigma(X,X)\|^2X) = \widetilde{\kappa}'(0)kY. \tag{4.15}$$

Taking the inner product of this identity and the vector $Y = \lambda(0)IX + \mu(0)JX + \nu(0)KX$, we see that

$$-\widetilde{\kappa}(0)\langle\sigma(X,X),\sigma(X,Y)\rangle = \widetilde{\kappa}'(0)k.$$

This, combined with (4.14), shows $\widetilde{\kappa}'(0) = 0$ and from (4.15) we see that M is isotropic submanifold of $\widetilde{M}(\tilde{c})$. Hence, in (4.5) setting $\lambda I\dot{\gamma} + \mu J\dot{\gamma} + \nu K\dot{\gamma}$ in place of $J\dot{\gamma}$ and $s = 0$, we have

$$(\bar{\nabla}_X\sigma)(X,X) = 0 \quad \text{and} \quad \sigma(X, \lambda(0)IX + \mu(0)JX + \nu(0)KX) = 0.$$

Since $X \in TM$ is an arbitrary and $\lambda(0)$, $\mu(0)$ and $\nu(0)$ are real numbers satisfying $\lambda^2 + \mu^2 + \nu^2 = 1$, these imply that M is parallel and

$$\sigma(X,IX) = \sigma(X,JX) = \sigma(X,KX) = 0 \tag{4.16}$$

for all $X \in TM$. Replacing X by $X + IX$ and so on in Equation (4.16), we obtain

$$\sigma(IX,IX) = \sigma(JX,JX) = \sigma(KX,KX) = \sigma(X,X), \tag{4.17}$$

and replacing X by $X + IX$ and so on, we have

$$\sigma(IX,JX) = \sigma(JX,KX) = \sigma(KX,IX) = 0. \tag{4.18}$$

Let R denote the curvature tensor of M, then it follows from (4.16), (4.17), (4.18) and the equation of Gauss that

$$\begin{aligned}&\langle R(X, \lambda IX + \mu JX + \nu KX)(\lambda IX + \mu JX + \nu KX), X\rangle \\ &= \tilde{c} + \langle \sigma(X,X), \sigma(\lambda IX + \mu JX + \nu KX, \lambda IX + \mu JX + \nu KX)\rangle \\ &\quad - \|\sigma(X, \lambda IX + \mu JX + \nu KX)\|^2 \\ &= \tilde{c} + \|\sigma(X,X)\|^2\end{aligned}$$

for arbitrary unit vector X and arbitrary number λ, μ and ν satisfying $\lambda^2 + \mu^2 + \nu^2 = 1$. Since M is isotropic, this shows that the submanifold M is a quaternionic space form. Thus our assertion follows from the results of [4] and [10]. □

Proposition 3 tells us that if a circle γ is not Cayley, the curve $f \circ \gamma$ is not a curve of order 2 in the ambient manifold $\widetilde{M}^m(\tilde{c})$. Motivated by this fact, we shall characterize the parallel embeddings of a Cayley projective plane into real space forms by Cayley circles ([8]).

Theorem 3. *Let M be an open set of $CaP^2(c)$ and $f : M \longrightarrow \widetilde{M}^m(\tilde{c})$ be an isometric immersion into a real space form $\widetilde{M}^m(\tilde{c})$. If there exists $k > 0$ and all Cayley circles of curvature k on M are mapped to curves of order 2 in $\widetilde{M}^m(\tilde{c})$, then M is a parallel submanifold of $\widetilde{M}^m(\tilde{c})$, so that our isometric immersion of M into $\widetilde{M}^m(\tilde{c})$ is given by* (3.2).

Proof. We denote by $f : M \to \widetilde{M}^m(\tilde{c})$ the isometric immersion which satisfies our assumption. Let x be any point of M. In the following, take and fix a unit vector $X \in T_xM$.

Let $\gamma = \gamma(s)$ be a Cayley circle of curvature k satisfying the equations:

$$\nabla_{\dot\gamma}\dot\gamma = kV, \quad \nabla_{\dot\gamma}V = -k\dot\gamma$$

with initial condition that $\gamma(0) = x$, $\dot\gamma(0) = X$ and $V(0) = Y \in T_xM$. Then by an argument similar to the proof of Theorem 1, at the point x we obtain

$$\widetilde{\kappa}(-A_{\sigma(X,X)}X + \|\sigma(X,X)\|^2 X) = \widetilde{\kappa}' kY, \tag{4.19}$$

$$\widetilde{\kappa}\{3k\sigma(Y,X) + (\bar\nabla_X\sigma)(X,X)\} = \widetilde{\kappa}'\sigma(X,X) \tag{4.20}$$

which corresponds to Equations (4.4) and (4.5) respectively. Thus, trying the similar way to get the equation (4.7), we obtain

$$\begin{aligned}(3k^2 + \|\sigma(X,X)\|^2)\langle\sigma(X,X), \sigma(X,Y)\rangle& \\ + k\langle\sigma(X,X), (\bar\nabla_X\sigma)(X,X)\rangle &= 0\end{aligned} \tag{4.21}$$

at the point x.

Next, we shall study another Cayley circle δ of the same curvature k with initial condition that $\delta(0) = x$, $\dot{\delta}(0) = X$ and $V(0) = -Y$. Then we get the following identity corresponding to (4.21):

$$-(3k^2 + \|\sigma(X,X)\|^2)\langle\sigma(X,X),\sigma(X,Y)\rangle + k\langle\sigma(X,X),(\bar{\nabla}_X\sigma)(X,X)\rangle = 0. \tag{4.22}$$

As $k > 0$, from (4.21) and (4.22) we get

$$\langle\sigma(X,X),\sigma(X,Y)\rangle = 0. \tag{4.23}$$

On the other hand, taking the inner product of (4.19) and the vector Y, we see that

$$-\widetilde{\kappa}(0)\langle\sigma(X,X),\sigma(X,Y)\rangle = \widetilde{\kappa}'(0)k.$$

This, combined with (4.23), shows $\widetilde{\kappa}'(0) = 0$. So, from (4.20) we get

$$(\bar{\nabla}_X\sigma)(X,X) = -3k\cdot\sigma(X,X)$$

and considering the Cayley circle δ again, we get

$$(\bar{\nabla}_X\sigma)(X,X) = 3k\cdot\sigma(X,X).$$

Hence $(\bar{\nabla}_X\sigma)(X,X) = 0$. Since $X \in TM$ is an arbitrary unit vector, we obtain $\bar{\nabla}\sigma = 0$. therefore our assertion follows from the result of [10]. □

5. Extrinsic shape of circles on a real projective space

Until now we have no characterizations of the parallel minimal embedding f of an n-dimensional real projective space $\mathbb{R}P^n\big(nc/2(n+1)\big)$ of curvature $nc/2(n+1)$ into a sphere $S^{n(n+3)/2-1}(c)$ of curvature c by the extrinsic shape of circles on the submanifold. It follows Theorem A that for each geodesic γ on $\mathbb{R}P^n$ the curve $f\circ\gamma$ is a plane curve which is nothing but a circle of the same curvature in the ambient space (see [6]). So when γ be a circle on a real projective space $\mathbb{R}P^n$ we shall study the extrinsic shape $f\circ\gamma$ through the parallel embedding f in the ambient real space form. Note that every circle on $\mathbb{R}P^n$ is locally contained in some totally geodesic $\mathbb{R}P^2$ of $\mathbb{R}P^n$. So it is enough to study the case of $n = 2$.

The following clarifies the extrinsic shape of circles of positive curvature on $\mathbb{R}P^n$ in $\widetilde{M}(\tilde{c})$ ([7]).

Proposition 4. *Let* $f = f_2\circ f_1 : \mathbb{R}P^2(c/3) \xrightarrow{f_1} S^4(c) \xrightarrow{f_2} \widetilde{M}^m(\tilde{c})$ *be an isometric parallel embedding of* $\mathbb{R}P^2(c/3)$ *into a real space form* $\widetilde{M}^m(\tilde{c})$ $(c \geq \tilde{c})$. *Here* f_1 *is the first standard minimal embedding of* $\mathbb{R}P^2(c/3)$ *into* $S^4(c)$ *and* f_2 *is a totally umbilic embedding of* $S^4(c)$ *into* $\widetilde{M}^m(\tilde{c})$. *Then*

(I) When $c = \tilde{c}$,

(i) f *maps each circle of curvature* $\sqrt{c}/\sqrt{6}$ *to a helix of proper order* 3 *of curvatures* $\sqrt{c}/\sqrt{2}$, $\sqrt{c}$.

(ii) f *maps each circle of curvature* $k \neq \sqrt{c}/\sqrt{6}$ *to a helix of proper order* 4 *of curvatures*

$$\sqrt{3k^2 + c}/\sqrt{3}, \quad 3k\sqrt{c}/\sqrt{3k^2 + c}, \quad |6k^2 - c|/\sqrt{3(3k^2 + c)}.$$

(II) When $c > \tilde{c}$, f *maps each circle of curvature* k *to a helix of proper order* 4 *of curvatures*

$$\sqrt{3k^2 + 4c - 3\tilde{c}}/\sqrt{3}, \quad 3k\sqrt{c}/\sqrt{3k^2 + 4c - 3\tilde{c}},$$
$$\sqrt{4(3k^2 + c)^2 - 3\tilde{c}(12k^2 + c)}/\sqrt{3(3k^2 + 4c - 3\tilde{c})}.$$

Proof. Let $f : M^n(c_1) \to \widetilde{M}^m(\tilde{c_2})$ be a λ-isotropic immersion. Then the equation of Gauss for the second fundamental form σ of f is given by

$$\begin{aligned}\langle\sigma(X,Y),\sigma(Z,W)\rangle &- \langle\sigma(X,W),\sigma(Y,Z)\rangle \\ &= (c_1 - c_2)(\langle X,Y\rangle\langle Z,W\rangle - \langle X,W\rangle\langle Y,Z\rangle)\end{aligned} \tag{5.1}$$

for any vector fields X, Y, Z, W on the submanifold $M^n(c_1)$. On the other hand, exchanging X and Y in (5.1), we get

$$\begin{aligned}\langle\sigma(X,Y),\sigma(Z,W)\rangle &- \langle\sigma(X,Z),\sigma(Y,W)\rangle \\ &= (c_1 - c_2)(\langle X,Y\rangle\langle Z,W\rangle - \langle X,Z\rangle\langle Y,W\rangle).\end{aligned} \tag{5.2}$$

Since f is λ-isotropic, for all vector fields X on M, we have $\langle\sigma(X,X),\sigma(X,X)\rangle = \lambda^2\langle X,X\rangle\langle X,X\rangle$, which is equivalent to

$$\begin{aligned}\langle\sigma(X,Y),\sigma(Z,W)\rangle &+ \langle\sigma(X,Z),\sigma(Y,W)\rangle + \langle\sigma(X,W),\sigma(Y,Z)\rangle \\ &= \lambda^2(\langle X,Y\rangle\langle Z,W\rangle + \langle X,Z\rangle\langle Y,W\rangle + \langle X,W\rangle\langle Y,Z\rangle).\end{aligned} \tag{5.3}$$

Summing up (5.1), (5.2) and (5.3), we obtain

$$\begin{aligned}\langle\sigma(X,Y),\sigma(Z,W)\rangle = &\frac{c_1 - c_2}{3}(2\langle X,Y\rangle\langle Z,W\rangle - \langle X,Z\rangle\langle Y,W\rangle - \langle X,W\rangle\langle Y,Z\rangle) \\ &+ \frac{\lambda^2}{3}(\langle X,Y\rangle\langle Z,W\rangle + \langle X,Z\rangle\langle Y,W\rangle + \langle X,W\rangle\langle Y,Z\rangle).\end{aligned}$$

Since our isometric embedding f given by the assumption of Proposition 4 is a $\sqrt{4c - 3\tilde{c}}/\sqrt{3}$ - isotropic (parallel) embedding, it satisfies that

$$\begin{aligned}&\langle\sigma(X,Y),\sigma(Z,W)\rangle \\ &= \frac{2c - 3\tilde{c}}{3}\langle X,Y\rangle\langle Z,W\rangle + \frac{c}{3}(\langle X,W\rangle\langle Y,Z\rangle + \langle X,Z\rangle\langle Y,W\rangle).\end{aligned} \tag{5.4}$$

Let γ be a circle of curvature k on $\mathbb{R}P^2(c/3)$ satisfying (1.1). We see from (2.3) and (5.4) that $\widetilde{\nabla}_{\dot\gamma}\dot\gamma = \kappa_1 V_2$, where

$$\kappa_1 = \frac{\sqrt{3k^2+4c-3\tilde c}}{\sqrt{3}}, \quad V_2 = \frac{\sqrt{3}}{\sqrt{3k^2+4c-3\tilde c}}(kV + \sigma(\dot\gamma,\dot\gamma)).$$

Differentiating V_2, we obtain

$$\widetilde{\nabla}_{\dot\gamma} V_2 = -\kappa_1\dot\gamma + \frac{3\sqrt{3}k}{\sqrt{3k^2+4c-3\tilde c}}\sigma(\dot\gamma, V).$$

Therefore, from (5.4) if we put

$$\kappa_2 = \frac{3k\sqrt{c}}{\sqrt{3k^2+4c-3\tilde c}}, \quad V_3 = \frac{\sqrt{3}}{\sqrt{c}}\sigma(\dot\gamma, V),$$

then we have $\widetilde{\nabla}_{\dot\gamma}V_2 = -\kappa_1\dot\gamma + \kappa_2 V_3$. Similarly, differentiating V_3, we obtain $\widetilde{\nabla}_{\dot\gamma}V_3 = -\kappa_2 V_2 + \kappa_3 V_4$, where

$$\kappa_3 = \frac{\sqrt{4(3k^2+c)^2 - 3\tilde c(12k^2+c)}}{\sqrt{3(3k^2+4c-3\tilde c)}}$$

and

$$V_4 = \frac{(6k^2-4c+3\tilde c)cV - 3k(3k^2+c-3\tilde c)\sigma(\dot\gamma,\dot\gamma)}{\sqrt{c(3k^2+4c-3\tilde c)\{4(3k^2+c)^2-3\tilde c(12k^2+c)\}}} + \frac{3k\sqrt{3k^2+4c-3\tilde c}\,\sigma(V,V)}{\sqrt{c\{4(3k^2+c)^2-3\tilde c(12k^2+c)\}}}.$$

Note that if $c = \tilde c$, then

$$4(3k^2+c)^2 - 3\tilde c(12k^2+c) = (6k^2-c)^2.$$

Finally, differentiating V_4, we find $\widetilde{\nabla}_{\dot\gamma}V_4 = -\kappa_3 V_3$. So we get the conclusion. □

References

1. T. Adachi, *Circles on quaternionic space forms*, J. Math. Soc. Japan **48**(1996), 205–227.
2. T. Adachi and S. Maeda, *Circular helices in a standard sphere*, New Zealand J. Math. **28**(1999), 1–6.
3. T Adachi, S. Maeda and K. Ogiue, *Extrinsic shape of circles and standard imbeddings of projective spaces*, Manuscripta Math. **93**(1997), 267–272.
4. D. Ferus, *Immersions with parallel second fundamental form*, Math. Z. **140**(1974), 87–92.
5. B. O'Neill *Isotropic and Kaehler immersions*, Canadian J. Math. (1965), 905–915.
6. K. Sakamoto, *Planar geodesic immersions*, Tôhoku Math. J. **29**(1977), 25–56.

7. K. Suizu, *Characterizations of parallel imbeddings of projective spaces into space forms by circles*, Math. Japonica **55**(2002), 43–52.
8. K. Suizu, S. Maeda and T. Adachi, *Characterization of totally geodesic Kähler immersions*, Hokkaido Math. J. **31**(2002), 629–641.
9. ———, *Study of submanifolds by curves of order 2*, Mem. Fac. Sci. Eng. Shimane Univ. Ser. B Math. Sci. **35**(2002), 1–21.
10. M. Takeuchi, *Parallel submanifolds of space forms*, Manifolds and Lie groups, in honor of Y. Matsusima, Birkhäuser, Boston (1981), 429–447.

POINCARÉ FORMULAE OF SUBMANIFOLDS IN COMPLEX SPACE FORMS

HIROYUKI TASAKI*

Institute of Mathematics, University of Tsukuba
Tsukuba, Ibaraki 305-8571 Japan
E-mail: tasaki@math.tsukuba.ac.jp

We consider the integral of the volume of intersection $M \cap gN$ of submanifolds M and gN in a complex space form G/K for g in G and describe this integral by the multiple Kähler angles of M and N. This integral formula is called a Poincaré formula. For this purpose we define the multiple Kähler angle and show some fundamental properties of it.

1. Introduction

For submanifolds M and N in a Riemannian homogeneous space G/K which satisfy $\dim M + \dim N \geq \dim(G/K)$, we consider a function on G:

$$g \mapsto \mathrm{vol}(M \cap gN),$$

which is measurable with respect to the invariant measure on G. An equation between the integral of this function and a certain geometric invariant of M and N is called a Poincaré formula. The first one was discovered at the end of the nineteenth century in the case where G/K was the plane and where M and N were plane curves. In this case $\mathrm{vol}(M \cap gN)$ is the intersection number of M and gN and the integral of this intersection number is the product of the lengths of M and N multiplied by a universal constant. This formula has been extended to the cases where G/K is a sphere, a real space form and where M and N are complex submanifolds in a complex space form G/K. Poincaré formulae in real space forms have several applications in variational problems of submanifolds. In order to apply Poincaré formulae to variational problems we need its explicit description. Unfortunately there are few Poincaré formulae which have explicit descriptions. So we consider a problem to describe explicitly Poincaré formulae

*partially supported by Grants-in-Aid for Scientific Research, The Ministry of Education, Science, Sports and Culture, Japan.

of submanifolds in Riemannian homogeneous spaces. In this note we shall show some explicit descriptions of Poincaré formulae of submanifolds in complex space forms.

2. Formulation of Howard

First we review the formulation of Poincaré formulae of submanifolds in Riemannian homogeneous spaces obtained by Howard [1].

Let E be a real vector space of finite dimension with inner product $\langle\ ,\ \rangle$. The inner product on E naturally induces an inner product on the exterior algebra $\wedge^* E$ on E. If $e_1, \dots, e_n$ is an orthonormal basis of E,

$$\left\{ e_{i_1} \wedge \cdots \wedge e_{i_k} \;\middle|\; 1 \leq i_1 < \cdots < i_k \leq k \right\}$$

is an orthonormal basis of $\wedge^k E$ with respect to the induced inner product on $\wedge^* E$.

For points x and y in a Riemannian homogeneous space G/K we take subspaces $V \subset T_x(G/K)$ and $W \subset T_y(G/K)$. We take an orthonormal basis v_1, $\dots$, v_p of V and an orthonormal basis $w_1, \dots, w_q$ of W, and elements g_x, g_y in G which satisfy $g_x o = x$, $g_y o = y$. We define $\sigma_K(V, W)$ by

$$\sigma_K(V, W) = \int_K \|dg_x^{-1}(v_1 \wedge \cdots \wedge v_p) \wedge dk^{-1} dg_y^{-1}(w_1 \wedge \cdots \wedge w_q)\| d\mu(k).$$

The norm of the above integrand is the norm determined from the induced inner product on $\wedge^* T_o(G/K)$. Because of the norm, the integrand is independent of the choice of the orthonormal bases $v_1, \dots, v_p$ of V and $w_1, \dots, w_q$ of W. Moreover $\sigma_K(V, W)$ is independent of the choice of g_x, g_y, since it is integrated on K. Hence $\sigma_K(V, W)$ depends only on V and W. Using this we can show the Poincaré formula of submanifolds in a Riemannian homogeneous space obtained by Howard [1].

Theorem 2.1. *We assume that G is unimodular, that is, the modular function $|\det \mathrm{Ad}_G(g)|$ is equal to 1 for any $g \in G$. For submanifolds M and N which satisfy $\dim M + \dim N \geq \dim(G/K)$, we have*

$$\int_G \mathrm{vol}(M \cap gN) d\mu(g) = \int_{M \times N} \sigma_K(T_x^\perp M, T_y^\perp N) d\mu(x, y).$$

Two Riemannian homogeneous spaces which have equivalent linear isotropy representations have same Poincaré formulae of submanifolds, because σ_K is dependent only on the linear isotropy representation of K on $T_o(G/K)$. Howard calls it a *transfer principle*. Typical examples are symmetric spaces of compact type and their noncompact duals. They have same Poincaré formulae of submanifolds.

σ_K is invariant under the action of G by its definition. In order to investigate σ_K it is sufficient to consider the values of σ_K on the subspaces in the tangent space $T_o(G/K)$ at the origin o of G/K. We can regard σ_K as a function of pairs of two subspaces in $T_o(G/K)$. Then σ_K is invariant under the linear isotropy action of K. Using this Howard showed some cases where σ_K is constant. For example, in the case where G/K is a real space form, that is, a simply connected complete Riemannian manifold of constant curvature, K acts transitively on the Grassmann manifold consisting of subspaces in $T_o(G/K)$, so σ_K is constant. Hence the right hand side of the Poincaré formula in this case is the product of the volumes of two submanifolds multiplied by a universal constant.

Theorem 2.2. *Let G/K be a real space form of dimension n. For a submanifold M of dimension p and a submanifold N of dimension q in G/K which satisfy $p+q \geq n$, we have*

$$\int_G \mathrm{vol}(M \cap gN)d\mu(g) = \frac{\mathrm{vol}(S^{p+q-n})\mathrm{vol}(SO(n+1))}{\mathrm{vol}(S^p)\mathrm{vol}(S^q)}\mathrm{vol}(M)\mathrm{vol}(N).$$

In the case where G/K is a complex space form K acts transitively on the Grassmann manifold consisting of complex subspaces in $T_o(G/K)$, so σ_K is constant for complex subspaces. Hence the right hand side of the Poincaré formula for complex submanifolds in this case is the product of the volumes of two submanifolds multiplied by a universal constant.

Theorem 2.3. *Let G/K be a complex space form of dimension n. For a complex submanifold M of dimension p and a complex submanifold N of dimension q in G/K which satisfy $p+q \geq n$, we have*

$$\int_G \mathrm{vol}(M \cap gN)d\mu(g) = \frac{\mathrm{vol}(\mathbf{C}P^{p+q-n})\mathrm{vol}(U(n+1))}{\mathrm{vol}(\mathbf{C}P^p)\mathrm{vol}(\mathbf{C}P^q)}\mathrm{vol}(M)\mathrm{vol}(N).$$

In the case where M and N represent their homology classes $\mathrm{vol}(M \cap gN)$ is a homological invariant of M and N, but in general $\mathrm{vol}(M \cap gN)$ changes dependently on g. This Poincaré formula was first obtained by Santaló [8] and reformulated by Howard [1]. This has been extended to more general Hermitian homogeneous spaces ([3] and [6]), which will be explained in Section 5.

When the action of K on the Grassmann manifold consisting of subspaces in $T_o(G/K)$ is not transitive, we need to investigate the action on the Grassmann manifold consisting of subspaces in $T_o(G/K)$ and describe σ_K explicitly. So we divide our problem to the following two problems:

(1) To define invariants on the Grassmann manifold $G_k(T_o(G/K))$ under the action of K.

(2) To describe σ_K by the invariants obtained in (1).

In the case of the complex space form the author solved the problem (1) by defining the multiple Kähler angle, which will be defined in Section 4, but the problem (2) is in progress. We shall explain about some results concerning these problems in the case of the complex space form.

3. Kähler angle

Theorem 2.3 shows the Poincaré formulae only for complex submanifolds. We consider Poincaré formulae for general real submanifolds in the complex space forms in this section and the next. There exists an invariant of submanifolds of dimension 2, which is called Kähler angle. We can formulate Poincaré formula of submanifolds of dimension 2 and codimension 2 in the complex space form using Kähler angle. So we define the notion of Kähler angle and explain its fundamental properties.

We denote by ω the standard Kähler form on $\mathbf{C^n}$. For a real vector subspace V of dimension 2 in $\mathbf{C^n}$ we take an orthonormal basis v_1, v_2 of V and define the Kähler angle $\theta(V)$ of V by $\theta(V) = \cos^{-1}|\omega(v_1, v_2)|$, which is independent of the choice of v_1, v_2. The range of the Kähler angle is $[0, \pi/2]$. Since the Kähler form ω is invariant under the action of the unitary group $U(n)$, the Kähler angle $\theta(V)$ is also invariant under the action of $U(n)$. The Kähler angle of V is 0, if and only if V is a complex subspace. The Kähler angle of V is $\pi/2$, if and only if V and $\sqrt{-1}V$ are orthogonal. Thus the Kähler angle $\theta(V)$ measures how close to a complex subspace V is.

When we consider Poincaré formulae in the complex space forms, according to the transfer principle it is sufficient to consider those in the complex projective spaces. The linear isotropy representation of the complex projective space $\mathbf{C}P^n = U(n+1)/U(1) \times U(n)$ is equivalent to the representation of $U(1) \times U(n)$ on $\mathbf{C^n}$ defined by

$$(z, A)v = zvA^* \qquad ((z, A) \in U(1) \times U(n),\ v \in \mathbf{C^n}).$$

Here we regard elements of $\mathbf{C^n}$ as row vectors of complex numbers. We denote by $\mathbf{e_1}, \ldots, \mathbf{e_n}$ the standard unitary basis of $\mathbf{C^n}$.

For θ which satisfies $0 \leq \theta \leq \pi/2$ we denote by G^n_θ the set consisting of real 2-dimensional vector subspaces of Kähler angle θ. Since the action of $U(n)$ preserves the Kähler angle of a real 2-dimensional vector subspace, $U(n)$ acts on G^n_θ. Kang and the author [4] obtained the following property of the Kähler angle.

Lemma 3.1. *The action of $U(n)$ on G^n_θ is transitive. We put*

$$V_\theta = \mathrm{span}_{\mathbf{R}}\{\mathbf{e_1}, \cos\theta\sqrt{-1}\mathbf{e_1} + \sin\theta\mathbf{e_2}\}.$$

Then $G_\theta^n = U(n) \cdot V_\theta$.

This lemma means that the Kähler angle of real vector subspace of dimension 2 in $\mathbf{C^n}$ is a complete invariant under the action of $U(n)$, that is, for real vector subspaces V and W of dimension 2 there exists $g \in U(n)$ such that $W = g \cdot V$ if and only if $\theta(V) = \theta(W)$. Using the Kähler angle the author [10] has obtained the following Poincaré formula.

Theorem 3.1. *For any real 2-dimensional submanifold M and any real $(2n-2)$-dimensional submanifold N in the complex space form G/K of complex dimension n, we have*

$$\begin{aligned}
&\int_G \#(M \cap gN)d\mu(g) \\
&= \frac{\mathrm{vol}(U(n+1))}{\mathrm{vol}(\mathbf{C}P^1)\mathrm{vol}(\mathbf{C}P^{n-1})} \\
&\cdot \int_{M\times N} \left(\frac{1}{4}(1+\cos^2\theta_x)(1+\cos^2\tau_y) + \frac{n}{4(n-1)}\sin^2\theta_x \sin^2\tau_y\right) d\mu(x,y).
\end{aligned}$$

Here θ_x is the Kähler angle of T_xM and τ_y is the Kähler angle of $T_y^\perp N$.

In the case where N is a complex submanifold Kang-T. [4] proved this theorem and in the case where $n = 2$ Kang-T. [5] proved this. We explicitly calculated the integral of σ_K on the isotropy group K in the proofs of these results including Theorem 3.1. Using the canonical form of real vector subspace obtained in Lemma 3.1, we can make the integrand of σ_K simpler. By the coarea formula of integrations and root systems we can reduce an integral on a manifold to that on another manifold of lower dimension. At the last step of the integration we calculate a certain integral on the sphere of dimension 2. A result of Archimedes which claims that we can define an area preserving map from a sphere of dimension 2 to the cylinder of the same radius and the same height makes the integral simpler. We draw horizontal rays from the axis of rotation to the outside. Such rays define a map from the sphere to the cylinder except the north pole and south pole. This map is area preserving.

4. Multiple Kähler angle

In order to formulate Poincaré formulae of general real submanifolds in complex space forms, we define the multiple Kähler angle of a real vector subspace in a complex vector space, which is a generalization of the Kähler angle. The multiple Kähler angle is a complete invariant of real vector subspaces.

We denote by ω the standard Kähler form of $\mathbf{C^n}$ similarly to the last section. The author [9] has generalized the notion of the Kähler angle as follows:

Definition 4.1. Let $1 < k \leq n$. For a real vector subspace V of dimension k in $\mathbf{C^n}$ we consider a canonical form of the restriction of ω to V as an alternating 2-form, that is, we take an orthonormal basis $\alpha^1, \ldots, \alpha^k$ of the dual space of V which satisfies

$$\omega|_V = \sum_{i=1}^{[k/2]} \cos\theta_i \alpha^{2i-1} \wedge \alpha^{2i}, \qquad 0 \leq \theta_1 \leq \cdots \leq \theta_{[k/2]} \leq \pi/2.$$

Then we put $\theta(V) = (\theta_1, \ldots, \theta_{[k/2]})$ and we call it the *multiple Kähler angle* of V. In the case where $n < k \leq 2n-1$, for a real vector subspace V of dimension k in $\mathbf{C^n}$ we define the multiple Kähler angle of V by $\theta(V) = \theta(V^\perp)$.

Salavessa [7] has also considered the same notion in the case where $n = k$ and n is even. We remark some fundamental properties of the multiple Kähler angle which directly follow from the definition.

Remark 4.1. Let $k \leq n$. For a real vector subspace V of dimension k in $\mathbf{C^n}$ the followings hold.

(1) The action of the unitary group $U(n)$ preserves the multiple Kähler angle, that is, for any $g \in U(n)$ we have $\theta(gV) = \theta(V)$.

(2) If $k = 2$, the multiple Kähler angle is nothing but the Kähler angle.

(3) $\theta(V) = (0, \ldots, 0)$ holds if and only if there exists a complex vector subspace of complex dimension $[k/2]$ in V. When k is even, $\theta(V) = (0, \ldots, 0)$ holds if and only if V itself is a complex vector subspace.

(4) $\theta(V) = (\pi/2, \ldots, \pi/2)$ holds if and only if V and $\sqrt{-1}V$ are orthogonal.

We denote by $G_k^{\mathbf{R}}(\mathbf{C^n})$ the real Grassmann manifold consisting of real vector subspaces of dimension k in $\mathbf{C^n}$. The author [9] has showed the following fundamental property of the multiple Kähler angle.

Proposition 4.1. *Let $k \leq n$. For $\theta = (\theta_1, \ldots, \theta_{[k/2]})$ which satisfies $0 \leq \theta_1 \leq \cdots \leq \theta_{[k/2]} \leq \pi/2$ we put*

$$G_{k;\theta}^n = \left\{ V \in G_k^{\mathbf{R}}(\mathbf{C^n}) \;\middle|\; \theta(\mathbf{V}) = (\theta_1, \ldots, \theta_{[\mathbf{k/2}]}) \right\}.$$

Since the natural action of $U(n)$ on $G_k^{\mathbf{R}}(\mathbf{C^n})$ preserves the multiple Kähler angle, this action induces that of $U(n)$ on $G_{k;\theta}^n$. The assertion of this proposition is that *the action of $U(n)$ on $G^n_{k;\theta}$ is transitive. Moreover we put*

$$V^k_\theta = \sum_{i=1}^{[k/2]} span_{\mathbf{R}}\{\mathbf{e_{2i-1}}, \cos\theta_{\mathbf{i}}\sqrt{-\mathbf{1}}\mathbf{e_{2i-1}} + \sin\theta_{\mathbf{i}}\mathbf{e_{2i}}\} \quad (+\mathbf{R}\mathbf{e_k}),$$

where the last term is added only when k is odd. Then $G^n_{k;\theta} = U(n) \cdot V^k_\theta$ holds.

When $n < k \leq 2n - 1$, we put

$$V^k_\theta = (V^{2n-k}_\theta)^\perp.$$

By Proposition 4.1 $U(n)$ acts transitively on

$$G^n_{k;\theta} = \{\, V \in G^{\mathbf{R}}_k(\mathbf{C^n}) \mid \theta(\mathbf{V}) = (\theta_1, \ldots, \theta_{[\mathbf{k}/2]}) \,\}.$$

Moreover $G^n_{k;\theta} = U(n) \cdot V^k_\theta$ holds.

The definition of the multiple Kähler angle depends only on the Hermitian structure of $\mathbf{C^n}$, so we can consider the multiple Kähler angle of any real submanifold in an almost Hermitian manifold. Using the multiple Kähler angle we formulate Poincaré formulae of any real submanifolds in the complex space forms [9].

Theorem 4.1. *For natural numbers p and q which satisfy $p \leq 2n \leq p + q$ and $q \leq 2n \leq p + q$ we define $\sigma^n_{p,q}$ by*

$$\sigma^n_{p,q}(\theta^{(p)}, \theta^{(q)}) = \int_{U(1)\times U(n)} \sigma(V^{2n-p}_{\theta^{(p)}}, k^{-1} \cdot V^{2n-q}_{\theta^{(q)}}) d\mu(k)$$
$$(\theta^{(p)} \in \mathbf{R}^{[\min\{\mathbf{p}, \mathbf{2n-p}\}/\mathbf{2}]},\ \theta^{(\mathbf{q})} \in \mathbf{R}^{[\min\{\mathbf{q}, \mathbf{2n-q}\}/\mathbf{2}]}).$$

Then for any real submanifold M of dimension p and any real submanifold N of dimension q we have

$$\int_{U(n+1)} \mathrm{vol}(M \cap gN) d\mu(g) = \int_{M\times N} \sigma^n_{p,q}(\theta(T_x M), \theta(T_y N)) d\mu(x, y).$$

If two submanifolds M and N in Theorem 4.1 satisfy $\dim M + \dim N = 2n$, then we can describe the Poincaré formulae [11] in more explicit way as follows:

Theorem 4.2. *For a natural number p which satisfies $1 \leq p \leq n$ there exists a polynomial*

$$P^n_p(x_1, \ldots, x_{[p/2]}, y_1, \ldots, y_{[p/2]})$$

which satisfies the following conditions

(1) The degree of the polynomial $P^n_p(x_i, y_j)$ with respect to each x_i and y_j is at most 1.

(2) For any permutation α *of* $\{1, \ldots, [p/2]\}$ *the following identities hold.*

$$P_p^n(x_i, y_j) = P_p^n(x_{\alpha(i)}, y_j) = P_p^n(x_i, y_{\alpha(j)}) = P_p^n(y_j, x_i).$$

(3) For any real submanifold M *of dimension* p *and any real submanifold* N *of dimension* $2n - p$ *in the complex space form* G/K *of complex dimension* n *we have*

$$\int_G \#(M \cap gN) d\mu(g) = \int_{M \times N} P_p^n(\cos^2 \theta_i(T_x M), \cos^2 \theta_j(T_y N)) d\mu(x, y).$$

By the result at page 21 of Howard [1]

$$P_1^n = \frac{2\mathrm{vol}(U(1))\mathrm{vol}(U(n))\mathrm{vol}(S^{2n})}{\mathrm{vol}(S^1)\mathrm{vol}(S^{2n-1})}$$

holds, so P_1^n is constant. This follows from the fact that $U(n)$ acts transitively on the set of unit vectors in $\mathbf{C^n}$.

Theorem 3.1 shows the following identity.

$$P_2^n(x, y) = \frac{\mathrm{vol}(U(n+1))}{\mathrm{vol}(\mathbf{C}P^1)\mathrm{vol}(\mathbf{C}P^{n-1})} \frac{1}{4(n-1)} \cdot [(2n-1) - (x+y) + (2n-1)xy].$$

Using Theorem 4.2 we can get the following identity stated in T. [11].

Theorem 4.3.

$$P_3^3(x, y) = \frac{\mathrm{vol}(U(4))}{\mathrm{vol}(\mathbf{R}P^3)^2} \left(4 - \frac{4}{3}(x+y) + \frac{20}{9}xy\right).$$

Proof. Theorem 4.2 implies that $P_3^3(x, y)$ is of the form:

$$P_3^3(x, y) = a_0 + a_1(x+y) + a_2 xy.$$

In order to determine the coefficients a_0, a_1 and a_2 we calculate the explicit values of $P_3^3(0,0)$, $P_3^3(1,0)$ and $P_3^3(1,1)$.

Howard [1] has showed that $\#(\mathbf{R}P^n \cap g\mathbf{R}P^n) = n + 1$ for almost all $g \in U(n+1)$ in the complex projective space $\mathbf{C}P^n$ at pages 26, 27. Since the multiple Kähler angle of $\mathbf{R}P^3$ in $\mathbf{C}P^3$ is equal to $\pi/2$ and $\cos(\pi/2) = 0$, we have

$$\int_{U(4)} \#(\mathbf{R}P^3 \cap g\mathbf{R}P^3) d\mu(g) = \int_{\mathbf{R}P^3 \times \mathbf{R}P^3} P_3^3(0,0) d\mu(x, y),$$

which implies $4\mathrm{vol}(U(4)) = P_3^3(0,0)\mathrm{vol}(\mathbf{R}P^3)^2$ and

$$P_3^3(0,0) = \frac{\mathrm{vol}(U(4))}{\mathrm{vol}(\mathbf{R}P^3)^2} \cdot 4.$$

Let M be a real hypersurface in $\mathbf{C}P^2$ and regard $\mathbf{C}P^2$ as a submanifold in $\mathbf{C}P^3$ in a natural way. Then we can regard M as a submanifold in $\mathbf{C}P^3$ and its multiple Kähler angle is constant and equal to 0. The invariance of the measure of $U(4)$ implies

$$\begin{aligned}
&\int_{U(4)} \#(M \cap g\mathbf{R}P^3)d\mu(g) \\
&= \frac{1}{\mathrm{vol}(U(3))}\int_{U(4)}\left(\int_{U(3)} \#(M \cap hg\mathbf{R}P^3)d\mu(h)\right)d\mu(g).
\end{aligned}$$

As Howard [1] pointed out on page 20, $g\mathbf{R}P^3 \cap \mathbf{C}P^2$ is isometric to $\mathbf{R}P^1$ for almost all g in $U(4)$. Since M is a real hypersurface in $\mathbf{C}P^2$, the result about P_1^n mentioned above implies

$$\begin{aligned}
&\int_{U(3)} \#(M \cap hg\mathbf{R}P^3)d\mu(h) \\
&= \int_{U(3)} \#(M \cap hg\mathbf{R}P^3 \cap \mathbf{C}P^2)d\mu(h) \\
&= \int_{U(3)} \#(M \cap h(g\mathbf{R}P^3 \cap \mathbf{C}P^2))d\mu(h) \\
&= \frac{2\mathrm{vol}(U(1))\mathrm{vol}(U(2))\mathrm{vol}(S^4)}{\mathrm{vol}(S^1)\mathrm{vol}(S^3)}\mathrm{vol}(M)\mathrm{vol}(\mathbf{R}P^1),
\end{aligned}$$

which is independent of g. So we get

$$\begin{aligned}
&P_3^3(1,0)\mathrm{vol}(M)\mathrm{vol}(\mathbf{R}P^3) \\
&= \int_{M\times\mathbf{R}P^3} P_3^3\left(\cos^2 0, \cos^2\frac{\pi}{2}\right)d\mu(x,y) \\
&= \int_{U(4)} \#(M \cap g\mathbf{R}P^3)d\mu(g) \\
&= \frac{\mathrm{vol}(U(4))}{\mathrm{vol}(U(3))}\frac{2\mathrm{vol}(U(1))\mathrm{vol}(U(2))\mathrm{vol}(S^4)}{\mathrm{vol}(S^1)\mathrm{vol}(S^3)}\mathrm{vol}(M)\mathrm{vol}(\mathbf{R}P^1)
\end{aligned}$$

and

$$\begin{aligned}
P_3^3(1,0) &= \frac{\mathrm{vol}(U(4))}{\mathrm{vol}(\mathbf{R}P^3)} \frac{2\mathrm{vol}(U(1))\mathrm{vol}(U(2))\mathrm{vol}(S^4)}{\mathrm{vol}(U(3))\mathrm{vol}(S^1)\mathrm{vol}(S^3)} \mathrm{vol}(\mathbf{R}P^1) \\
&= \frac{\mathrm{vol}(U(4))}{\mathrm{vol}(\mathbf{R}P^3)^2} \frac{\mathrm{vol}(U(1))\mathrm{vol}(U(2))\mathrm{vol}(S^4)\mathrm{vol}(\mathbf{R}P^1)}{\mathrm{vol}(U(3))\mathrm{vol}(S^1)} \\
&= \frac{\mathrm{vol}(U(4))}{\mathrm{vol}(\mathbf{R}P^3)^2} \frac{\mathrm{vol}(S^4)\mathrm{vol}(\mathbf{R}P^1)}{\mathrm{vol}(\mathbf{C}P^2)\mathrm{vol}(S^1)} \\
&= \frac{\mathrm{vol}(U(4))}{\mathrm{vol}(\mathbf{R}P^3)^2} \frac{2\pi\mathrm{vol}(S^4)\mathrm{vol}(\mathbf{R}P^1)}{\mathrm{vol}(S^5)\mathrm{vol}(S^1)} \\
&= \frac{\mathrm{vol}(U(4))}{\mathrm{vol}(\mathbf{R}P^3)^2} \frac{\pi\mathrm{vol}(S^4)}{\mathrm{vol}(S^5)} \\
&= \frac{\mathrm{vol}(U(4))}{\mathrm{vol}(\mathbf{R}P^3)^2} \cdot \frac{8}{3}.
\end{aligned}$$

Similarly we can obtain

$$\begin{aligned}
&\int_{U(4)} \#(M \cap gM) d\mu(g) \\
&= \frac{1}{\mathrm{vol}(U(3))} \int_{U(4)} \left(\int_{U(3)} \#(M \cap hgM) d\mu(h) \right) d\mu(g)
\end{aligned}$$

and

$$\begin{aligned}
&\int_{U(3)} \#(M \cap hgM) d\mu(h) \\
&= \int_{U(3)} \#(M \cap hgM \cap \mathbf{C}P^2) d\mu(h) \\
&= \int_{U(3)} \#(M \cap h(gM \cap \mathbf{C}P^2)) d\mu(h) \\
&= \frac{2\mathrm{vol}(U(1))\mathrm{vol}(U(2))\mathrm{vol}(S^4)}{\mathrm{vol}(S^1)\mathrm{vol}(S^3)} \mathrm{vol}(M)\mathrm{vol}(gM \cap \mathbf{C}P^2).
\end{aligned}$$

Since

$$\begin{aligned}
&\int_{U(4)} \mathrm{vol}(gM \cap \mathbf{C}P^2) d\mu(g) \\
&= \frac{1}{\mathrm{vol}(U(3))} \int_{U(4)} \left(\int_{U(3)} \mathrm{vol}(M \cap hg\mathbf{C}P^2) d\mu(h) \right) d\mu(g),
\end{aligned}$$

we have to calculate the following integral.

$$\begin{aligned}
&\int_{U(3)} \mathrm{vol}(M \cap hg\mathbf{C}P^2)d\mu(h) \\
&= \int_{U(3)} \mathrm{vol}(M \cap hg\mathbf{C}P^2 \cap \mathbf{C}P^2)d\mu(h) \\
&= \int_{U(3)} \mathrm{vol}(M \cap h(g\mathbf{C}P^2 \cap \mathbf{C}P^2))d\mu(h) \\
&= \frac{\mathrm{vol}(U(1))\mathrm{vol}(U(2))\mathrm{vol}(S^1)\mathrm{vol}(S^4)}{\mathrm{vol}(S^2)\mathrm{vol}(S^3)}\mathrm{vol}(M)\mathrm{vol}(\mathbf{C}P^1),
\end{aligned}$$

because $g\mathbf{C}P^2 \cap \mathbf{C}P^2$ is isometric to $\mathbf{C}P^1$ for almost all g in $U(4)$. Using these results we get

$$\begin{aligned}
&P_3^3(1,1)\mathrm{vol}(M)^2 \\
&= \int_{M\times M} P_3^3(\cos^2 0, \cos^2 0)d\mu(x,y) \\
&= \int_{U(4)} \#(M \cap gM)d\mu(g) \\
&= \frac{2\mathrm{vol}(U(1))\mathrm{vol}(U(2))\mathrm{vol}(S^4)}{\mathrm{vol}(U(3))\mathrm{vol}(S^1)\mathrm{vol}(S^3)} \\
&\quad \times \frac{\mathrm{vol}(U(4))\mathrm{vol}(U(1))\mathrm{vol}(U(2))\mathrm{vol}(S^1)\mathrm{vol}(S^4)}{\mathrm{vol}(U(3))\mathrm{vol}(S^2)\mathrm{vol}(S^3)}\mathrm{vol}(M)^2\mathrm{vol}(\mathbf{C}P^1) \\
&= \frac{2\mathrm{vol}(U(1))\mathrm{vol}(U(2))\mathrm{vol}(S^4)}{\mathrm{vol}(U(3))\mathrm{vol}(S^3)} \\
&\quad \times \frac{\mathrm{vol}(U(4))\mathrm{vol}(U(1))\mathrm{vol}(U(2))\mathrm{vol}(S^4)}{\mathrm{vol}(U(3))\mathrm{vol}(S^2)\mathrm{vol}(S^3)}\mathrm{vol}(M)^2\mathrm{vol}(\mathbf{C}P^1)
\end{aligned}$$

and

$$\begin{aligned}
&P_3^3(1,1) \\
&= \frac{2\mathrm{vol}(U(1))\mathrm{vol}(U(2))\mathrm{vol}(S^4)}{\mathrm{vol}(U(3))\mathrm{vol}(S^3)} \\
&\quad \times \frac{\mathrm{vol}(U(4))\mathrm{vol}(U(1))\mathrm{vol}(U(2))\mathrm{vol}(S^4)}{\mathrm{vol}(U(3))\mathrm{vol}(S^2)\mathrm{vol}(S^3)}\mathrm{vol}(\mathbf{C}P^1) \\
&= \mathrm{vol}(U(4))\frac{2\mathrm{vol}(S^4)}{\mathrm{vol}(\mathbf{C}P^2)\mathrm{vol}(S^3)}\frac{\mathrm{vol}(S^4)}{\mathrm{vol}(\mathbf{C}P^2)\mathrm{vol}(S^2)\mathrm{vol}(S^3)}\mathrm{vol}(\mathbf{C}P^1) \\
&= \frac{\mathrm{vol}(U(4))}{\mathrm{vol}(\mathbf{R}P^3)^2}\frac{\mathrm{vol}(S^4)}{\mathrm{vol}(\mathbf{C}P^2)}\frac{\mathrm{vol}(S^4)}{2\mathrm{vol}(\mathbf{C}P^2)\mathrm{vol}(S^2)}\mathrm{vol}(\mathbf{C}P^1)
\end{aligned}$$

$$
\begin{aligned}
&= \frac{\mathrm{vol}(U(4))}{\mathrm{vol}(\mathbf{R}P^3)^2}\left(\frac{\mathrm{vol}(S^4)}{\mathrm{vol}(\mathbf{C}P^2)}\right)^2 \frac{\mathrm{vol}(\mathbf{C}P^1)}{2\mathrm{vol}(S^2)} \\
&= \frac{\mathrm{vol}(U(4))}{\mathrm{vol}(\mathbf{R}P^3)^2}\left(\frac{2\pi\mathrm{vol}(S^4)}{\mathrm{vol}(S^5)}\right)^2 \frac{\mathrm{vol}(S^3)}{4\pi\mathrm{vol}(S^2)} \\
&= \frac{\mathrm{vol}(U(4))}{\mathrm{vol}(\mathbf{R}P^3)^2}\left(\frac{16}{3}\right)^2 \frac{1}{8} \\
&= \frac{\mathrm{vol}(U(4))}{\mathrm{vol}(\mathbf{R}P^3)^2}\cdot\frac{32}{9}.
\end{aligned}
$$

In the description

$$P_3^3(x,y) = a_0 + a_1(x+y) + a_2xy$$

we substitute $(x,y) = (0,0), (1,0), (1,1)$:

$$
\begin{aligned}
a_0 &= \frac{\mathrm{vol}(U(4))}{\mathrm{vol}(\mathbf{R}P^3)^2}\cdot 4 \\
a_0 + a_1 &= \frac{\mathrm{vol}(U(4))}{\mathrm{vol}(\mathbf{R}P^3)^2}\cdot\frac{8}{3} \\
a_0 + 2a_2 + a_3 &= \frac{\mathrm{vol}(U(4))}{\mathrm{vol}(\mathbf{R}P^3)^2}\cdot\frac{32}{9}
\end{aligned}
$$

and obtain

$$P_3^3(x,y) = \frac{\mathrm{vol}(U(4))}{\mathrm{vol}(\mathbf{R}P^3)^2}\left(4 - \frac{4}{3}(x+y) + \frac{20}{9}xy\right).$$

This completes the proof of the theorem. □

Kang [2] has shown a partial description of P_4^4 as follows:

Theorem 4.4.

$$P_4^4(x_1,x_2,1,1) = \frac{\mathrm{vol}(U(5))}{\mathrm{vol}(\mathbf{C}P^2)^2}\cdot\frac{1}{8}(3 + x_1 + x_2 + 3x_1x_2).$$

5. Other spaces

Generalizations of Theorem 2.3 were recently obtained as follows [3].

Theorem 5.1. *Let G/K a homogeneous almost Hermitian manifold of complex dimension n. Assume that G is a unimodular Lie group and that K acts irreducibly on the exterior algebra $\wedge^p T_o^{(1,0)}(G/K)$ of degree p. Then for any almost* *complex submanifold M of complex dimension p in G/K and for any almost complex submanifold N in G/K with* $\dim M + \dim N = \dim(G/K)$, *we have*

$$\int_G \#(M \cap gN) d\mu_G(g) = \frac{\mathrm{vol}(K)}{\binom{n}{p}} \mathrm{vol}(M)\mathrm{vol}(N).$$

In the case where G/K is an irreducible compact Hermitian symmetric space, each p in the following list satisfies the assumption of Theorem 5.1.

	compact type	
A III	$SU(l)/S(U(m) \times U(l-m))$	any p (if $m = 1$) $p = 1$ (if $m \neq 1$)
D III	$SO(2l)/U(l)$	$p = 1, 2$
BD I	$SO(2l)/SO(2) \times SO(2l-2)$ $SO(2l+1)/SO(2) \times SO(2l-1)$	$p \neq l-1$ any p
C I	$Sp(l)/U(l)$	$p = 1, 2$
E III	$(\mathfrak{e}_{6(-78)}, \mathfrak{so}(10) + \mathbf{R})$	$p = 1, 2, 3$
E VII	$(\mathfrak{e}_{7(-133)}, \mathfrak{e}_6 + \mathbf{R})$	$p = 1, 2, 3, 4$

In the case of the complex space form K acts irreducibly on $\wedge^p T_o^{(1,0)}(G/K)$ of every degree p. So the following theorem obtained by Sakai [6] is a generalization of Theorem 2.3.

Theorem 5.2. *Let G/K be an irreducible Hermitian symmetric space of complex dimension n. Let M and N be complex submanifolds of complex codimensions p and q in G/K which satisfy $p + q \leq n$. Assume that K acts irreducibly on $\wedge^p T_o^{(1,0)}(G/K)$ or $\wedge^q T_o^{(1,0)}(G/K)$. Then we have*

$$\int_G \mathrm{vol}(M \cap gN) d\mu(g) = \frac{(n-p)!(n-q)!\mathrm{vol}(K)}{n!(n-p-q)!} \mathrm{vol}(M)\mathrm{vol}(N).$$

References

1. R. Howard, The kinematic formula in Riemannian homogeneous spaces, Mem. Amer. Math. Soc., No. 509, vol. 106 (1993).
2. H. Kang, Integral geometry in the four dimensional complex projective space, preprint, 2001.
3. H. Kang, T. Sakai, M. Takahashi and H. Tasaki, Poincaré formulas of complex submanifolds, preprint, 2002.
4. H. Kang and H. Tasaki, Integral geometry of real surfaces in complex projective spaces, Tsukuba J. Math., vol. 25, no. 1 (2001), 155–164.
5. H. Kang and H. Tasaki, Integral geometry of real surfaces in the complex projective plane, Geom. Dedicata, vol. 90 (2002), 99–106.

6. T. Sakai, Poincaré formula in irreducible Hermitian symmetric spaces, preprint, 2002.
7. Isabel M.C. Salavessa, On the Kaehler angles of submanifolds, preprint (math. DG/0204349).
8. L.A. Santaló, Integral geometry in Hermitian spaces, Amer. J. Math., vol. 74 (1952), 423–434.
9. H. Tasaki, Generalization of Kähler angle and integral geometry in complex projective spaces, in Steps in Differential Geometry, Proceedings of Colloquium on Differential Geometry, Debrecen, 2000 (edited by L. Kozma, P.T. Nagy and L. Tamássy), Published by the Institute of Mathematics and Informatics, University of Debrecen, 2001, pp.349–361. Available electronically at: `http://www.emis.de/proceedings/CDGD2000/`
10. ———, Integral geometry of submanifolds of real dimension two and codimension two in complex projective spaces, Contemporary Math., vol. 308 (2002), 315–327.
11. ———, Generalization of Kähler angle and integral geometry in complex projective spaces II, to appear in Math. Nachr.

CURVATURE OPERATORS IN THE RELATIVITY*

VESELIN VIDEV

Trakia University
Department of Mathematics and Physics
6000 Stara Zagora, Bulgaria
E-mail address:videv@uni-sz.bg

YULIAN TSANKOV

Sofia University "St. Kliment Ohridski",
Department of Geometry
Faculty of Mathematics and Informatics
Bul. "D.Baucher" 5
1164 Sofia, Bulgaria
E-mail address:ucankov@fmi.uni-sofia.bg

In the present note using the pointwise conditions on the eigenvalues of the Jacobi operator and a skew-symmetric Stanilov operator we characterize the Petrov gravitation fields which are solutions of the Einstein equation in Relativity.

1. Recall of definitions

In the Riemannian geometry the very important role play a linear operators which is defined by the curvature tensor, say R, of the Riemannian manifold M with metric g, say (M, g). The curvature tensor R is induced by the uniquely Levi-Civita connection on M, and as we know R has the following algebraically properties [7]:

$$R(x,y,z,u) = -R(y,x,z,u), \tag{1}$$

$$R(x,y,z,u) = -R(x,y,u,z), \tag{2}$$

$$R(x,y,z,u) + R(y,z,x,u) + R(z,x,y,u) = 0, \tag{3}$$

* Partially supported under contract MM/804 NSF

which is the reason to called R an algebraic curvature tensor. Third of these properties is called first Bianchi identity, we have either the second Bianci identity [7]:

$$\sigma_{x,y,z}(\nabla_x R)(y, z, u) = 0$$

where σ is a cyclic sum over x, y, z. In all properties (1)-(3) of the Riemannian curvature tensor R above stated an arbitrary tangent vectors x, y, z, u, which belongs to the tangent space M_p, in a point $p \in M$. Using R we can define a different curvature operators in the tangent space M_p, which can be of an arbitrary nature, but in our consideration we prefer to use a linear operators on the tangent space M_p . It is clear that M_p (at any point $p \in M$) is a standard linear vector space, which has the same dimension as the manifold M, and which is equipted with an inner product g, induced by the corresponding metric g of the manifold M. These operators are simple examples of the linear endomorphisms in the tangent space M_p, at some point $p \in M$. In the present note we will use the following curvature operators:

Jacobi operator

$$R_X(u) = R(u, X, X), \tag{4}$$

defined for an unit tangent vector $X \in M_p$;

Stanilov operator

$$S_{E^2}(u) = R(u, X, Y), \tag{5}$$

defined for any orthonormal basis X, Y in the tangent two-dimensional subspace $E^2 \subset M_p$.

The so-defined curvature operators are the reason of the definition to *pointwise Osserman* and *pointwise IP Riemannian manifolds*, which are manifolds such that all eigenvalues of respectively *Jacobi operator* and *Stanilov operator*, to be a pointwise constants function, at any point of the manifold. The class of the Riemannian pointwise Osserman manifolds was defined from Osserman [5], and the class of the IP Riemannian manifolds was defined from Gilkey, Leahy and Sadofsky [2].

The theory of these operators starts from the so-called Osserman conjecture, that is: *if a Riemannian manifold M has a pointwise eigenvalues of all Jacobi operators R_X, defined in the tangent spaces M_p, at every points $p \in M$, then M is a two-point homogeneous space [5] ?* Robart Osserman first state this hypothesis suppose the globally constants for the eigenvalues of the Jacobi operator. The corresponding pointwise hypothesis was defined from Gilkey, Swann, and Vanchecke, which first showed that there is a four-dimensional pointwise Osserman manifolds which are not globally Osserman (the definition of the globally

Osserman manifolds is the same as the definition of the pointwise Osserman manifolds [4]. Further the most of the investigations on the manifolds which has a pointwise constant eigenvalues of the Jacobi operator, was madden from Peter Gilkey, which generalize in the purely algebraically context a pointwise Osserman spaces, using only an algebraically properties of the curvature tensor R, and defining the Jacobi operator as a linear endomorphism in an arbitrary real (standard) vector space [1].

In this context, Gilkey state the problem of the pointwise constancy of the characteristicall polinomyal of the Jacoby operator, when the manifold (M, g) is a Lorentzian manifold (as well as (M, g) is a pseudo-Euclidian manifold) respectively the tangent vector space is a Minkovski space or a pseudo-Euclidian space. First result in the Lorentzian setting is that the classes of the pointwise Osserman Lorentzian and globally Osserman Lorentzian manifolds coincides and they are Lorentzian manifolds of constant sectional curvature [3]. This is a reason to surch some generalization of the Osserman conjecture in the Lorentzian setting.

Further we will consider the case when M is a four-dimensional Lorentzian manifold using the characteristically coefficients of the Jacobi operator and Stanilov operator. The classes of the Riemannian and Lorentzian pointwise Osserman manifolds are irreducible, but state the conditions on the pointwise constancy of any of the characteristically coefficients of any Jacobi operator in Lorentzian setting we obtain some reducible spaces. In the Lorentzian case they are manifolds open from Petrov as a solution of the Einstein equation in the Relativity [6]. The manifolds open from Petrov are Einstein Lorentzian spaces, which is an important condition for our results done.

2. Characterization of Petrov manifolds using characteristically coefficients of Jacobi and Stanilov curvature operators and Thorpe canonical forms

Let (M, g) be a 4-dimensional Einstein Lorentzian manifold with a signature $(+,+,+,-)$. From the geometrically point of view (mathematical) problem of the Relativity is to find such types of 4-dimensional Lorentzian manifolds which are solutions of the Einstein equation of the gravitation fields in the space-times theory, namely

$$\rho_{ij} - \frac{1}{2}\tau g_{ij} = -\lambda T_{ij} \tag{6}$$

where ρ is Ricci tensor on the manifold, τ is a scalar curvature, and T is the stress-energy tensor. For the free-mass spaces we have that T is vanishing and then either $\rho = \kappa g$, which is the geometrically definition for the Einstein (gravitation) spaces.

In the relativity theory A.Z. Petrov define three types of Einstein fields denoted they by T_1, T_2, T_3 and which are solutions of the field equation (6).

Einstein manifold) is of type T_i (where $i = 1, 2, 3$), if the non-zero components of the curvature tensor R, given by one of the following formulas [6]

For type T1

$$\begin{aligned} R_{1212} &= -R_{3434} = \alpha_1, \\ R_{1313} &= -R_{2424} = \alpha_2, \\ R_{2323} &= -R_{1414} = \alpha_3. \end{aligned} \tag{7}$$

For type T2

$$\begin{aligned} R_{1212} &= -R_{3434} = \alpha_1, \\ R_{1313} &= -R_{2424} = \alpha_2 + 1, \\ R_{2323} &= -R_{1414} = \alpha_2 - 1, \\ R_{3114} &= -R_{3224} = 1. \end{aligned} \tag{8}$$

For type T3

$$\begin{aligned} &R_{1212} = -R_{3434} = R_{1313} = -R_{2424} = R_{2323} = -R_{1414} = \alpha, \\ &R_{3114} = -R_{3224} = 1, \\ &R_{2443} = -R_{2113} = 1, \end{aligned} \tag{9}$$

with respect to one special basis e_1, e_2, e_3, e_4 ($e_4 \in {}^{-}S_pM$), in the tangent space M_p, at any points $p \in M$. This basis in the classical differential geometry was called Thorpe-basis. Here ${}^{-}S_pM$ denote the set of all unit timelike tangent vectors in M_p, at a point $p \in M$. Let Jacobi operator R_X has (in this four-dimensional case) the following characteristically equation

$$c(c^3 + J_1c^2 + J_2c + J_3) = 0,$$

where X is an unit space-like or timelike, neither isotropic tangent vector in M_p, at a point $p \in M$.

We proved the following [9]

Theorem 1. *Let (M, g) is a four-dimensional Lorentzian manifold of constant sectional curvature if and only if the characteristically coefficient $J_1(p; X)$ and one of the characteristically coefficients $J_2(p; X)$ or $J_3(p; X) \neq 0$ of any Jacobi operator R_X, defined for any unit space-like or any unit timelike tangent vector $X \in M_p$, are pointwise constant functions, at a point $p \in M$.*

This result we continued by

Theorem 2. *Let (M, g) be a Petrov manifold which is one of the types T_1, T_2, T_3, and suppose one of the characteristically coefficients $J_2(p; X)$ or $J_3(p; X) \neq 0$ of any Jacobi operator R_X, are pointwise constant functions, for any unit space-like or any unit timelike tangent vector X, which belongs to one of the coordinate planes, spanned from the Thorpe basis, at any point $p \in M$. Then (M, g) is a four- dimensional Lorentzian manifold of constant sectional curvature.*

Let us we consider the special case when the characteristically coefficient $J_3(p; X)$ of any Jacobi operator R_X, defined for any unit space-like, or any unit timelike tangent vector $X \in M_p$, vanish at any point p of the Petrov manifolds, which is one of the of type T_1, T_2, T_3.

In the sequal we?ll continue our Riemannian results [8], proving

Theorem 3. *Suppose (M, g) is a Petrov manifold of T_1. Then the characteristically coefficient $J_3(p; X)$ of any Jacobi operator R_X, defined for any unit space-like, or any unit timelike tangent vector $X \in M_p$, vanish at any $p \in M$ if and only if M is one of the following types of manifolds:*

a) Minkovski space;

b) Lorentzian manifold of constant sectional curvature;

c) Reducible space with the metric which is decompose of the two quadratic metric of the forms:

$$ds = dx_1^2 + \cos(\sqrt{\lambda}x_1)dx_2^2 + dx_3^2 - \cos(\sqrt{\lambda}x_3)dx_4^2, \tag{10}$$

$$\lambda > 0 \tag{11}$$

$$ds = dx_1^2 + ch(\sqrt{-\lambda}x_1)dx_2^2 + dx_3^2 - ch(\sqrt{-\lambda}x_3)dx_4^2, \tag{12}$$

$$\lambda < 0$$

Proof. Let (M, g) is a Petrov space of type T_1, let p be a point of M, and let e_1, e_2, e_3, e_4 ($e_4 \in {}^-S_pM$) is a Thorpe basis in the tangent space M_p. If a and b are arbitrary real numbers, such that $a^2 - b^2 = \epsilon, \epsilon = \mp 1$, then the orthonormal basis

$$ae_1 + be_4, \;\; be_1 + ae_4, \;\; e_2, \;\; e_3 \tag{13}$$

is a Lorentzian basis in M_p. Then the characteristically equations of the Jacobi operators $R_{ae_1+be_4}$, $R_{ae_2+be_4}$ and $R_{ae_3+be_4}$, with respect to this basis, have the forms

$$(c+\epsilon\alpha_3)(c^2 - c(\alpha_1+\alpha_2)+\alpha_1\alpha_2 - a^2b^2((\alpha_1-\alpha_2)^2+(\beta_1-\beta_2)^2)) = 0,$$
$$(d+\epsilon\alpha_1)(d^2 - d(\alpha_2+\alpha_3)+\alpha_2\alpha_3 - a^2b^2((\alpha_2-\alpha_3)^2+(\beta_2-\beta_3)^2)) = 0,$$
$$(r+\epsilon\alpha_2)(r^2 - c(\alpha_3+\alpha_1)+\alpha_3\alpha_1 - a^2b^2((\alpha_3-\alpha_1)^2+(\beta_3-\beta_1)^2)) = 0,$$

From this equations we get

$$J_3(p, ae_1+be_4) = \epsilon\alpha_3(\alpha_1\alpha_2 - a^2b^2((\alpha_1-\alpha_2)^2+(\beta_1-\beta_2)^2)),$$
$$J_3(p, ae_2+be_4) = \epsilon\alpha_1(\alpha_2\alpha_3 - a^2b^2((\alpha_2-\alpha_3)^2+(\beta_2-\beta_3)^2)) \tag{14}$$

Using first of this representations by $a = 1, b = 0$ we obtain that

$$\alpha_1\alpha_2\alpha_3 = 0$$

First possibility of this equality is that $\alpha_1\alpha_2\alpha_3 = 0$, and then (M, g) is a Minkovski space, or we have second case when $\alpha_1 \neq 0$ and $\alpha_1\alpha_2 = 0$. Then from (14), we get

$$a^2b^2((\alpha_1-\alpha_2)^2+(\beta_1-\beta_2)^2) = 0,$$

$\alpha_1 = \alpha_2 = 0$, $\beta_1 = \beta_2$, and also from the first Bianci identity we have $\beta_3 = -2\beta_1$.

Thus if $J_3(p; X) = 0$ for any unit timelike or spice-like tangent vector $X \in M_p$, then it follows that

$$\alpha_1 = \lambda, \quad \alpha_2 = \alpha_3 = 0, \quad \beta_2 = \beta_3, \quad \beta_1 = -2\beta_2 \tag{15}$$

Suppose that $\eta_1, \eta_2, \eta_3, \eta_4, \eta_5, \eta_6$ are eigenvectors of the curvature operator $\Re$ in the second exterior product $\wedge^2(M_p)$, and suppose that $k_1, k_2, k_3, \hat{k}_1, \hat{k}_2, \hat{k}_3$ are their eigenvalues such that

$$k_j = \alpha_j + i\beta_j, \quad i = \sqrt{-1}$$

Since we assume that formulas (15) are valid, then the matrix of the curvature operator $\Re$ with respect to the orthonormal basis $\eta_1, \eta_2, \eta_3, \eta_4, \eta_5, \eta_6$ has the following form:

$$\left\{\begin{array}{cccccc} \lambda-2i\beta_2 & 0 & 0 & 0 & 0 & 0 \\ 0 & i\beta_2 & 0 & 0 & 0 & 0 \\ 0 & 0 & i\beta_3 & 0 & 0 & 0 \\ 0 & 0 & 0 & \lambda+2i\beta_2 & 0 & 0 \\ 0 & 0 & 0 & 0 & i\beta_2 & 0 \\ 0 & 0 & 0 & 0 & 0 & i\beta_3 \end{array}\right\}$$

From here it follows that there is a Lorentzian basis v_1, v_2, v_3, v_4 in the Minkovski tangent vector space M_p, at any point p of the manifold M of type T_1, such that the curvature tensor R has the following components

$$\begin{aligned} R(v_1, v_2, v_2, v_1) &= R(v_3, v_4, v_4, v_3) = \lambda - 2i\beta_2, \\ R(v_1, v_3, v_3, v_1) &= R(v_2, v_3, v_3, v_2) = -R(v_2, v_4, v_4, v_2) \\ &= -R(v_1, v_4, v_4, v_1) = i\beta_2. \end{aligned}$$

Now using the corresponding characteristic equations of the Jacobi operators R_{v_1}, R_{v_2}, R_{v_3}, R_{v_4}, with respect to the orthonormal basis v_1, v_2, v_3, v_4, in the tangent space M_p, at any point $p \in M$, we can easy obtain that Jacobi operator R_{v_1} has eigenvalues

$$\frac{\lambda - 2i\beta_2}{g(v_2, v_2)}, \quad \frac{2i\beta_2}{g(v_3, v_3)}, \quad \frac{2i\beta_2}{g(v_4, v_4)},$$

Jacobi operator R_{v_2} has eigenvalues

$$\frac{-2i\beta_2}{g(v_1, v_1)}, \quad \frac{2i\beta_2}{g(v_3, v_3)}, \quad \frac{2i\beta_2}{g(v_4, v_4)},$$

Jacobi operator R_{v_3} has eigenvalues

$$\frac{2i\beta_2}{g(v_1, v_1)}, \quad \frac{2i\beta_2}{g(v_2, v_2)}, \quad \frac{-2i\beta_2}{g(v_3, v_3)},$$

and Jacobi operator R_{v_4} has eigenvalues

$$\frac{\lambda + 2i\beta_2}{g(v_1, v_1)}, \quad \frac{-2i\beta_2}{g(v_2, v_2)}, \quad \frac{2i\beta_2}{g(v_3, v_3)}.$$

Then from our assumption

$$J_3(p; e_1) = J_3(p; e_2) = J_3(p; e_3) = J_3(p; e_4) = 0$$

it follows that

$$(\lambda - 2i\beta_2)2i\beta_2 = (\lambda + 2i\beta_2)2i\beta_2 = 0.$$

From here if $2i\beta_2 \neq 0$, then

$$\lambda - 2i\beta_2 = \lambda + 2i\beta_2,$$

which means that $\lambda = \beta_2 = 0$, hence $\alpha_1 = 0$ and since $\alpha_2 = \alpha_3 = 0$, then we have that in this case (M, g) is a Minkovski space.

If holds $\beta_2 = 0$, then

$$\alpha_1 = \lambda, \qquad \alpha_2 = \alpha_3 = \beta_2 = \beta_3 = 0,$$

and now following some results of A.Z. Petrov [6], we have that M is a Lorentzian manifold with the metric which is decompose of two quadratic forms, namely (10) and (11). Finally we remark that the converse results of this theorem is also true. □

Theorem 4. *Suppose* (M, g) *is a Petrov manifold of type* T_2 . *Then the characteristically coefficient* $J_3(p; X)$ *of any Jacobi operator* R_X, *defined for any unit space-like, or any unit timelike tangent vector* $X \in M_p$, *vanish at any* $p \in M$ *if and only if M is a maximall? speed-space with a metric*

$$ds^2 = dx_1^2 + sh^2(x_1 - x_4)dx_2^2 + sin^2(x_1 - x_4)dx_3^2 - dx_4^2. \qquad (16)$$

Proof. Let (M, g) be a Petrov manifold of type T_2 and let the characteristically coefficient $J_3(p; X)$ of the Jacobi operator R_X vanish at any point $p \in M$, for any unit spacelike and any unit timelike tangent vector $X \in M_p$. If e_1, e_2, e_3, e_4 ($e_4 \in {}^-SpM$) is a Thorpe basis in the space M_p and a, b are real numbers such that $a^2 - b^2 = \epsilon$, $\epsilon = \pm 1$, then the characteristically equations of the Jacobi operators $R_{ae_1+be_4}$ and $R_{ae_2+be_4}$ with respect to this basis have the form

$$\begin{aligned}
&(c + \epsilon K_{23})(c^2 - c(K_{12} + K_{13})\\
&\qquad + K_{12}K_{13} - a^2b^2((K_{12} - K_{13})^2 + (\beta_1 - \beta_2)^2)\\
&\qquad\qquad + a^2((a^2K_{12} - b^2K_{23}) - c)) = 0,\\
&(d + \epsilon K_{13})(d^2 - d(K_{12} + K_{23})\\
&\qquad + K_{12}K_{23} - a^2b^2((K_{12} - K_{23})^2 + (\beta_3 - \beta_1)^2)\\
&\qquad\qquad + a^2((a^2K_{12} - b^2K_{13}) - d)) = 0.
\end{aligned}$$

From here and the generally condition $J_3 = 0$ we obtain that

$$\begin{aligned}
(\alpha_2 - 1)\alpha_2^2 + (\alpha_2 + 1)\beta_2^2 = 0,\\
(\alpha_2 + 1)\alpha_2^2 + (\alpha_2 - 1)\beta_2^2 = 0
\end{aligned}$$

and then we get $\alpha_2 = \beta_2 = 0$. Following Petrov that means that (M, g) is a locally symmetric manifold and more exactly we get that a (M, g) is maximall speed- space for which is true that

$$\begin{aligned}
\alpha_2 = \beta_2 = 0,\\
R_{3114} = -R_{3224} = 1,
\end{aligned}$$

at any point of this space. From here it is trivial to prove that the converse results of this theorem is also true. □

Theorem 5. *Suppose (M, g) is a Petrov manifold of T_3 . Then it is impossible to vanish, the characteristically coefficient $J_3(p; X)$, of any Jacobi operator R_X, defined for any unit space-like, or for any unit timelike tangent vector $X \in M_p$, at a point $p \in M$.*

In the end of this paper we remark that the Einstein Lorentzian IP-manifolds was characterized by the following [10]

Theorem 6. *Let the characteristical polynomial of the Stanilov curvature operator $S_\alpha(u)$ is a constant for any non-isotropic plane α in the tangent space M_p, at any point p of the 4-dimensional Lorentzian manifold (M, g). Then it is equivalent (M, g) to be one of the following types of manifolds:*

a) Minkovski space;

b) Lorentzian manifold of constant sectional curvature;

c) maximall speed-spacewith a metric of type (16).

Aknowlegment. The authors would like specially to thank Professor Stancho Dimiev, for the help of the preparing and the publication of the present paper, either to Professors Grosjo Stanilov, Peter Gilkey, and Lieven Vanchecke for the more useful discussion on the material and methods for obtained results.

References

1. P. Gilkey, Algebraic curvature tensor which are p-Osserman, Differential Geometry and its applications (accepted to appear in August in Differential Geometry and its applications).
2. P. Gilkey, J. Leahy, K. Sadofsky, Riemannian manifolds whose skew- symmetric curvature operator has constant eigenvalues, Indiana Univ. Math J. 48 (1999), 615–634.
3. P. Gilkey, G. Stanilov, V. Videv, Pseudo-Riemannian manifolds whose generalized Jacobi operator has a constant characteristic polynomial, Journal of Geometry(Basel), 62 (1998), 144–153.
4. P. Gilkey, A. Swann, L. Vancheke, Isoparametric geodesic spheres and a conjecture of Osserman concerning the Jacobi operator, Quart. J. Math. Oxford (1995), 299–320.
5. R. Osserman, Curvature in the eighties, Journal Amer. Math. Monthly, **97** (1990), 731–756.
6. A.Z. Petrov, Einstein spaces, Moskva, GIFML(1961), in russian.
7. G. Stanilov, Differential geometry, Nauka i izkustvo. Sofia (1998), in Bulgarian.
8. V. Videv. Characterization of a three- and four-dimensional Riemannian manifolds by a pointwise conditions on the three and the determinant of Jacobi operator, Journal of Geometry (Basel), **71** (2001), 197–205.
9. ______, Metrical characterization at a basepoint of a four-dimensional Einstein Lorentzian manifolds by Jacobi operator, Mathematics and Education in Mathematics (1999), 167–192.

10. ______, Characterization of a four-dimensional Einstein Lorentzian manifolds using a skew-symmetric curvature operator, Mathematika Balkanika, **15**, (2001), fasc. 1–2.

ON INTEGRABILITY OF ALMOST QUATERNIONIC MANIFOLDS

AKIRA YAMADA*
Nagaoka National College of Technology,
888 Nishikatakai, Nagaoka,
Niigata 940-8532, JAPAN
E-mail: ayamada@nagaoka-ct.ac.jp

This article is a brief survey of our recent works ([4], [5]) on the integrability of almost quaternionic manifolds and the related topics.

1. Introduction

A smooth manifold M admitting (1,1)-tensor field J satisfying $J^2 = -id.$ is called an almost complex manifold with the almost complex structure J. The concept of almost complex manifold is a natural generalization of the one of complex manifold, which was introduced by C. Ehresmann in 1948. Almost complex structure J of an almost complex manifold $M = (M, J)$ is said to be *integrable* if M admits a complex structure and the derived almost complex structure coincides with the almost complex strucure J. An almost complex manifold (M, J) equipped with a Riemannian metric g satisfying $g(JX, JY) = g(X, Y)$ for all $X, Y \in \mathfrak{X}(M)$ ($\mathfrak{X}(M)$ denotes the Lie algebra of all smooth vector fields on M) is called an almost Hermitian manifold. Kähler manifolds ($\nabla J = 0$) are the most typical almost Hermitian manifolds and studied by many authors. A. Gray and L.M. Hervella [1] defined sixteen classes of almost Hermitian manifolds including the class of Kähler manifolds. In § 2, we shall recall several classes of almost Hermitian manifolds out of the sixteen classes.

An *almost quaternionic manifold* is a quadruple $M = (M, I, J, K)$, where I, J and K are almost complex structures on a smooth manifold M satisfying $K = IJ = -JI$. It is well-known that an almost quaternionic manifold is $4m$ ($m \geq 1$)-dimensional and orientable. An almost quaternionic manifold $M = (M, I, J, K)$ equipped with a Riemannian metric g is called a *quaternionic almost Hermitian*

*Work partially supported by Grants-in-Aid for Scientific Research, No. 14740056, the Ministry of Education, Culture, Sports, Science and Technology, Japan.

manifold (briefly, a *q-almost Hermitian manifold*) if (I, g), (J, g) and (K, g) are simultaneously almost Hermitian structures on M. Especially, if (I, g), (J, g) and (K, g) are simultaneouly Kähler structures on M, then $M = (M, I, J, K, g)$ is called a *q-Kähler manifold* (also known as a *hyper-Kähler manifold*). Concernig the integrability of q-almost Hermitian manifolds, N.J. Hitchin [3, Lemma 6.8] has obtained the following

Theorem 1.1. *Any q-almost Kähler manifold is a q-Kähler manifold.*

The main purpose of the present article is to give generalizations of the above theorem 1.1 and related notes. The results introduced in § 3 is the joint work with N. Murakoshi and K. Sekigawa [4]. The results introduced in § 4 is the joint work with T. Nihonyanagi and K. Sekigawa [5].

Throughout this article, we assume that all manifolds are connected and smooth, and further that all tensor fields are smooth, unless otherwise specified.

2. Classes of almost Hermitian manifolds

Let $M = (M, J, g)$ be an almost Hermitian manifold with almost Hermitian structure (J, g) and Ω the Kähler form of M defined by $\Omega(X, Y) = g(JX, Y)$, for $X, Y \in \mathfrak{X}(M)$. We denote by ∇ the Riemannian connection of M. The Nijenhuis tensor field N is defined by $N(X, Y) = [JX, JY] - [X, Y] - J[JX, Y] - J[X, JY]$ for $X, Y \in \mathfrak{X}(M)$. Almost Hermitian manifold M is called a *Hermitian manifold* if the almost complex structure J is integrable. Almost Hermitian manifold M is Hermitian manifold if and only if the Nijenhuis tensor field N is vanished. It is known that a Hemitian manifold M satisfies

$$(\nabla_X J)Y - (\nabla_{JX} J)JY = 0 \tag{1}$$

for all $X, Y \in \mathfrak{X}(M)$. By the definition, a Kähler manifold is special Hermitian manifold such that the almost complex structure J is parallel with respect to the Riemannian connection ∇. Almost Hermitian manifold $M = (M, J, g)$ is called an *almost Kähler manifold* if the Kähler form is closed ($d\Omega = 0$) or equivalently M satisfies

$$\underset{X,Y,Z}{\mathfrak{S}} g((\nabla_X J)Y, Z) = 0 \tag{2}$$

for X, Y, $Z \in \mathfrak{X}(M)$, where $\underset{X,Y,Z}{\mathfrak{S}}$ denotes the cyclic sum with respect to X, Y, Z. It is well-known that an almost Kähler manifold with the integrable almost complex structure is a Kähler manifold. A symplectic manifold (M, Ω) ($d\Omega = 0$) is an almost Kähler manifold with respect to an almost Hermitian structure (J, g) compatible with the symplectic form Ω. The cotangent bundle T^*M of any smooth manifold M is the most typical example of symplectic manifold. The first example of compact symplectic manifold which is not a Kähler manifold was given by W.P. Thurston [10]. Almost Hermitian manifold $M = (M, J, g)$ is called a *nearly Kähler manifold* if M satisfies the condition

$$(\nabla_X J)Y + (\nabla_Y J)X = 0 \tag{3}$$

for $X, Y \in \mathfrak{X}(M)$. Further, almost Hermitian manifold $M = (M, J, g)$ is called a *quasi-Kähler manifold* if M satisfies the condition

$$(\nabla_X J)Y + (\nabla_{JX} J)JY = 0 \tag{4}$$

for $X, Y \in \mathfrak{X}(M)$. It is known that a quasi-Kähler manifold $M = (M, J, g)$ is neccessarily a *semi-Kähler manifold* ($\delta\Omega = 0$, where δ denotes the coderivative). On one hand, almost Hermitian manifold $M = (M, J, g)$ is called a *generalized locally conformal almost Kähler manifold* if there exists a 1-form ω on M satisfying

$$d\Omega = \omega \wedge \Omega. \tag{5}$$

Further, almost Hermitian manifold $M = (M, J, g)$ is called a *para-Kähler manifold* if M satisfies the condition

$$R(X, Y) \cdot J = 0 \tag{6}$$

for $X, Y \in \mathfrak{X}(M)$, where R denotes the curvature tensor field of M. We denote by $\mathcal{H}$, $\mathcal{K}$, $\mathcal{AK}$, $\mathcal{NK}$, $\mathcal{QK}$, $\mathcal{SK}$ and $\mathcal{GLCAK}$ the classes of Hermitian, Kähler, almost Kähler, nearly Kähler, quasi-Kähler, semi-Kähler and generalized locally conformal almost Kähler manifolds, respectively. Then we have the following inclusion relations among these classes ([1]):

$$\begin{gathered}
\mathcal{K} \begin{matrix} \subset \mathcal{AK} \subset \\ \subset \mathcal{NK} \subset \end{matrix} \mathcal{QK} \subset \mathcal{SK}, \\
\mathcal{K} = \mathcal{AK} \cap \mathcal{H}, \qquad \mathcal{K} = \mathcal{QK} \cap \mathcal{H}, \\
\mathcal{K} = \mathcal{NK} \cap \mathcal{AK}, \qquad \mathcal{AK} \subset \mathcal{GLCAK}.
\end{gathered} \tag{7}$$

We note that the above inclusion relations are all strict. In particular, we have $\mathcal{K} = \mathcal{NK}$ and $\mathcal{AK} = \mathcal{QK} = \mathcal{SK}$ in dimension 4.

3. Generalizations of Hitchin's Theorem

A q-almost Hermitian manifold $M = (M, I, J, K, g)$ is called a *quaternionic quasi-Kähler manifold* (briefly, a *q-quasi-Kähler manifold*) if (I, g), (J, g) and (K, g) are simultaneously quasi-Kähler structures on M. And also, if (I, g), (J, g)

and (K, g) are simultaneouly generalized locally conformal almost Kähler structures on M, then $M = (M, I, J, K, g)$ is called a *quaternionic generalized locally conformal almost Kähler manifold* (briefly, a *q-generalized locally conformal almost Kähler manifold*).

N. Murakoshi, K. Sekigawa and A. Yamada [4] studied generalizations of N.J. Hitchin's theorem (Theorem 1.1) and obtained the following two results.

Theorem A. *Any q-quasi-Kähler manifold is a q-Kähler manifold.*

Theorem B. *Any* $4m$ $(m \geq 2)$*-dimensional q-generalized locally conformal almost Kähler manifold is a q-Kähler manifold.*

About Theorem B, we remark that we can not omit the restriction on the dimension of the manifold. So, we shall study about 4-dimensional case in the next section § 4. In fact, we may observe that any 4-dimensional q-almost Hermitian manifold is necessarily a q-generalized locally conformal almost Kähler manifold.

M. Tahara, L. Vanhecke and Y. Watanabe [7] discussed several new almost Hermitian structures on tangent bundles TM over Riemannian manifolds M and constructed new interesting examples of q-almost Hermitian structures including a family of q-Kähler structures on TM over complex space form M. On one hand, Y. Tashiro [9] studied the relationships between almost contact metric structure of almost contact metric manifold (M, ϕ, ξ, η, g) and the corresponding almost Hermitian structure on the product manifold $\tilde{M} = M \times \mathbb{R}$ ($\mathbb{R}$ denotes a real line) equipped with the Riemannian metric $\tilde{g} = e^{-2t}\begin{pmatrix} g & 0 \\ 0 & 1 \end{pmatrix}$, $t \in \mathbb{R}$. For example, we see that Sasakian manifold M (resp. contact metric manifold (or almost Sasakian manifold)) corresponding to a Kähler manifold $\tilde{M}$ (resp. almost Kähler manifold). By enlarging Tashiro's idea, we may introduce some special classes of the $(4m-1)$-dimensional manifolds with almost contact metric 3-structure $((\phi_a, \xi_a, \eta_a)_{a=1,2,3}, g)$ including the class of the manifolds with Sasakian 3-structures. As an application of the Hitchin's result, we may easily see that a manifold with contact metric 3-structure is necessarily a manifold of Sasakian 3-structure. This result was announced also by T. Kashiwada in a private communication with K. Sekigawa. As applications of Theorems A and B, we may obtain slight generalizations of this result.

Now, we shall introduce an example of a q-almost Hermitian manifold such that two of three almost Hermitian structures are almost Kähler ones and the remaining one is not almost Kähler one.

Example 3.1. Let $M = (M, J_0, g_0)$ be a Kähler manifold and TM the tangent bundle of M equipped with the Riemannian metric g and almost complex structures $I = I(f,h)$, $J = J(f,h)$, $K = K(f,h)$ defined respectively by

$$\begin{cases} g(X_u^H, Y_u^H) = f g_0(X,Y) \\ \qquad + \dfrac{h-f}{t}\{g_0(X,u)g_0(Y,u) + g_0(X,J_0u)g_0(Y,J_0u)\}, \\ g(X_u^V, Y_u^V) = \dfrac{1}{f} g_0(X,Y) \\ \qquad + \dfrac{f-h}{tfh}\{g_0(X,u)g_0(Y,u) + g_0(X,J_0u)g_0(Y,J_0u)\}, \\ g(X_u^H, Y_u^V) = 0, \end{cases} \tag{8}$$

$$\begin{cases} I(X_u^H) = fX_u^V + \dfrac{h-f}{t} g_0(X,u)u^V \\ \qquad + \dfrac{h-f}{t} g_0(X,J_0u)(J_0u)^V, \\ I(X_u^V) = -\dfrac{1}{f} X_u^H + \dfrac{h-f}{tfh} g_0(X,u)u^H \\ \qquad + \dfrac{h-f}{tfh} g_0(X,J_0u)(J_0u)^H, \end{cases} \tag{9}$$

$$\begin{cases} J(X_u^H) = f(J_0X)_u^V + \dfrac{h-f}{t} g_0(X,u)(J_0u)^V \\ \qquad - \dfrac{h-f}{t} g_0(X,J_0u)u^V, \\ J(X_u^V) = \dfrac{1}{f}(J_0X)_u^H + \dfrac{f-h}{tfh} g_0(X,u)(J_0u)^H \\ \qquad - \dfrac{f-h}{tfh} g_0(X,J_0u)u^H, \end{cases} \tag{10}$$

$$\begin{cases} K(X_u^H) = -(J_0X)_u^H, \\ K(X_u^V) = (J_0X)_u^V, \end{cases} \tag{11}$$

for $X, Y \in T_{\pi(u)}M$, $t = g(u,u)$ $(u \in TM)$ and $f = f(t)$, $h = h(t)$ are smooth positive-valued functions on $[0,\infty)$ such that $\dfrac{h-f}{t}$ is differentiable at $t = 0$, here X_u^H (resp. X_u^V) denote the horizontal lift (resp. vertical lift) of X at u.

In the above example $TM = (TM, I, J, K, g)$, we may check that (I,g) and (K,g) are almost Kähler structures and (J,g) is not almost Kähler one for generic functions f, h and generic Kähler manifold M (cf. [8]). In fact, M. Tahara, L. Vanhecke and Y. Watanabe [7] showed that, the tangent bundle $TM = (TM, I, J, K, g)$ over a Kähler manifold M of constant holomorphic sectional curvature $c > 0$ equipped with the q-almost Hermitian structure (I, J, K, g)

defined by (8) $\sim$ (11) is a q-Kähler manifold if and only if the functions f and h are given respectively by

$$f = \frac{1}{2}\left(a + \sqrt{4ct + a^2}\right), \qquad h = \sqrt{4ct + a^2}, \tag{12}$$

where a is a positive constant.

4. 4-dimensional almost quaternionic manifold

In this section, we shall investigate some fundamental structures of 4-dimensional q-almost Hermitian manifolds. Any 4-dimensional q-almost Hermitian manifold is regards as a q-generalized locally conformal almost Kähler manifold.

T. Nihonyanagi, K. Sekigawa and A. Yamada [5] studied 4-dimensional q-almost Hermitian manifold and obtained the following results.

Theorem C. *Let (M, I, J, K, g) be a 4-dimensional compact q-almost Hermitian Einstein manifold. Then (M, I, J, K, g) is a Ricci-flat and $*$-Ricci-flat q-para-Kähler manifold. Furthermore, (M, g) is either flat or its universal covering is a Ricci-flat K3 surface with a q-Kähler structure.*

Theorem D. *Let $M = (M, I, J, K, g)$ be a 4-dimensional compact q-Hermitian manifold with non-positive scalar curvature. Then M is a q-Kähler manifold.*

Remark 4.1. For any 4-dimensional q-almost Hermitian manifold $M = (M, I, J, K, g)$, we may easily observe that, if any two almost complex structures of I, J, K are integrable, then the remaining one is also integrable.

Concerning Theorems C and D, we shall introduce an example of 4-dimensional compact locally flat q-almost Hermitian manifold which is not q-Kähler manifold (cf. [11]).

Example 4.1. Let G be a connected Lie subgroup of $GL(4, \mathbb{R})$ consisting of the matrices

$$\begin{pmatrix} x & y & z & 0 \\ -y & x & w & 0 \\ 0 & 0 & 1 & 0 \\ 0 & 0 & 0 & e^t \end{pmatrix}, \qquad x^2 + y^2 = 1. \tag{13}$$

Let $\mathfrak{g}$ be the Lie algebra of G. Then $\mathfrak{g} = \mathrm{span}_{\mathbb{R}}\{e_1, e_2, e_3, e_4\}$, where

$$\begin{gathered}
e_1 = \begin{pmatrix} 0&0&1&0\\ 0&0&0&0\\ 0&0&0&0\\ 0&0&0&0 \end{pmatrix}, \quad e_2 = \begin{pmatrix} 0&1&0&0\\ -1&0&0&0\\ 0&0&0&0\\ 0&0&0&0 \end{pmatrix}, \\
e_3 = \begin{pmatrix} 0&0&0&0\\ 0&0&1&0\\ 0&0&0&0\\ 0&0&0&0 \end{pmatrix}, \quad e_4 = \begin{pmatrix} 0&0&0&0\\ 0&0&0&0\\ 0&0&0&0\\ 0&0&0&1 \end{pmatrix}.
\end{gathered} \tag{14}$$

We define a G-invariant Riemannian metric g on G by $g(e_i, e_j) = \delta_{ij}$ $(1 \leqq i, j \leqq 4)$, and almost quaternionic structure (I, J, K) by

$$\begin{array}{llll}
Ie_1 = e_2, & Ie_2 = -e_1, & Ie_3 = e_4, & Ie_4 = -e_3, \\
Je_1 = e_3, & Je_3 = -e_1, & Je_2 = -e_4, & Je_4 = -e_2, \\
Ke_1 = e_4, & Ke_4 = -e_1, & Ke_2 = e_3, & Ke_3 = -e_2.
\end{array} \tag{15}$$

In the above Example 4.1, we may observe that (G, I, J, K, g) is a 4-dimensional locally flat q-almost Hermitian manifold which is not q-Kähler manifold. We may check $N_J = 0$, $N_I \neq 0$, $N_K \neq 0$ on (G, I, J, K, g). Let Γ be the discrete subgroup of G consisting of matrices

$$\begin{pmatrix} 1&0&l&0\\ 0&1&m&0\\ 0&0&1&0\\ 0&0&0&e^n \end{pmatrix} \qquad l, m, n \in \mathbb{Z}. \tag{16}$$

Then, we may see that the Γ-orbit space $\Gamma \backslash G$ is compact.

Next, we shall state a way to construct a q-almost Hermitian manifold from a manifold equipped with almost contact metric 3-structure.

Example 4.2. Let $N = (N, \varphi, \xi, \eta, g_0)$ be a $(4m-1)$-dimensional almost contact metric manifold and $M = N \times \mathbb{R}$ (resp. $N \times S^1$) be the direct product manifold of N and a real line $\mathbb{R}$ (resp. a unit circle S^1). We define almost complex structure J on $M = N \times \mathbb{R}$ (resp. $N \times S^1$) by

$$J = \begin{pmatrix} \varphi & \xi \\ -\eta & 0 \end{pmatrix}, \tag{17}$$

that is, $JX = \varphi X$, for $X \in \mathfrak{X}(M)$, $J\xi = -\nu$ and $J\nu = \xi$, where ν is a unit tangent vector field on $\mathbb{R}$ (resp. S^1). Further, we denote by g the canonical product Rimannian metric on M. Then, we may easily see that (M, J, g) is an almost Hermitian manifold. A $(4m-1)$-dimensional manifold N equipped with three almost contact structures (ϕ_a, ξ_a, η_a) $(a = 1, 2, 3)$ satisfying the following for an even permutation (i, j, k) of $(1, 2, 3)$,

$$\begin{aligned} \phi_k = \phi_i\phi_j - \eta_j \otimes \xi_i &= -\phi_j\phi_i + \eta_i \otimes \xi_j, \\ \xi_k = \phi_i\xi_j &= -\phi_j\xi_i, \\ \eta_k = \eta_i \circ \phi_j &= -\eta_j \circ \phi_i, \end{aligned} \tag{18}$$

then $\{(\phi_a, \xi_a, \eta_a)\}$ $(a = 1, 2, 3)$ is called an almost contact 3-structure. Let N be a $(4m-1)$-dimensional manifold equipped with almost contact metric 3-structure $\{(\phi_a, \xi_a, \eta_a)\}$ $(a = 1, 2, 3)$, and g_0 be a Riemannian metric compatible with three almost contact structure (ϕ_a, ξ_a, η_a) $(a = 1, 2, 3)$. Then we see that (M, I, J, K, g) $(M = N \times \mathbb{R}$ (or $N \times S^1$)) is a $4m$-dimensional q-almost Hermitian manifold with q-almost Hermitian structure (I, J, K, g), where I, J and K are almost complex structures (ϕ_1, η_1, ξ_1), (ϕ_2, η_2, ξ_2) and (ϕ_3, η_3, ξ_3), respectively, and g is the canonical product metric on M.

Especially, it is well-known that 3-dimensional unit sphere S^3 admits a Sasakian 3-structure $(\phi_a, \xi_a, \eta_a, g_0)$ $(a = 1, 2, 3)$. Thus, we see also that the corresponding product manifold $M = S^3 \times \mathbb{R}$ or $S^3 \times S^1$ equipped with the canonical product metric g is a 4-dimensional q-Hermitian manifold (M, I, J, K, g). For example, we may easily see that a manifold (M, I, J, K, g) for $M = S^3 \times S^1$ is a compact 4-dimensional q-Hermitian manifold with constant positive scalar curvature 12. In stead of the product metric g on $S^3 \times \mathbb{R}$, considering the metric $\bar{g} = e^{-2t}g$ $(t \in \mathbb{R})$, we may also see that $(S^3 \times \mathbb{R}, I, J, K, \bar{g})$ is a locally flat q-Kähler manifold.

Now, we shall give another example concerning Theorems C and D.

Example 4.3. Let $M = \mathbb{R}^4_+ = \{(x_1, x_2, x_3, x_4) \in \mathbb{R}^4 \mid x_4 > 0\}$ and $e_i = x_4 \dfrac{\partial}{\partial x_i}$ $(1 \leqq i \leqq 4)$. We define a Riemannian metric g on M by $g(e_i, e_j) = \delta_{ij}$ $(1 \leqq i, j \leqq 4)$, then we may observe that (M, g) can be regarded as a solvable Lie group space and $\{e_i\}_{i=1,2,3,4}$ is an orthonormal basis of the corresponding Lie algebra. Further, we define the almost quaternionic structure (I, J, K) by

$$\begin{array}{llll} Ie_1 = e_2, & Ie_2 = -e_1, & Ie_3 = e_4, & Ie_4 = -e_3, \\ Je_1 = e_3, & Je_3 = -e_1, & Je_2 = -e_4, & Je_4 = -e_2, \\ Ke_1 = e_4, & Ke_4 = -e_1, & Ke_2 = e_3, & Ke_3 = -e_2. \end{array} \tag{19}$$

In the above Example 4.3, we may see that (M, I, J, K, g) is a 4-dimensional q-Hermitian, non-q-Kähler manifold of constant sectional curvature -1. However, we may also observe that (M, I, J, K, g) does not have any compact quotient as a q-Hermitian manifold.

At the end of this article, we shall give the following question which will be naturally set concerning the result of Theorem C.

Question 4.1. Investigate the more precise structures of Ricci flat and $*$-Ricci-flat q-para-Kähler manifolds by comparing the ones of q-Kähler manifolds.

References

1. A. Gray and L.M. Hervella, *The sixteen classes of almost Hermitian manifolds and their linear invariants*, Ann. Mat. Pura Appl. (4) **123** (1980), 35–58.
2. N.J. Hitchin, *Compact four-dimensional Einstein manifolds*, J. Differential Geom. **9** (1974), 435–441.
3. ———, *The self-duality equations on a Riemann surface*, Proc. London Math. Soc. (3) **55** (1987), 59–126.
4. N. Murakoshi, K. Sekigawa and A. Yamada, *Integrability of almost quaternionic manifolds*, Indian J. Math. **42** (2000), 313–329.
5. T. Nihonyanagi, K. Sekigawa and A. Yamada, *Notes on 4-dimensional almost quaternionic manifolds*, (submitted).
6. K. Sekigawa, *On some 4-dimensional compact almost Hermitian manifolds*, J. Ramanujan Math. Soc. **2** (1987), 101–116.
7. M. Tahara, L. Vanhecke and Y. Watanabe, *New structures on tangent bundles*, Abstract of the meeting on differential geometry at Nippon Institute of Technology 1998, 20–25.
8. M. Tahara and Y. Watanabe, *Natural almost Hermitian, Hermitian and Kähler metrics on the tangent bundles*, Math. J. Toyama Univ. **20** (1997), 149–160.
9. Y. Tashiro, *On contact structure of hypersurfaces in complex manifolds, I, II*, Tôhoku Math. J. **15** (1963), 62–78, 167–175.
10. W.P. Thurston, *Some simple examples of symplectic manifolds*, Proc. Amer. Math. Soc. **55** (1976), 467–468.
11. F. Tricerri and L. Vanhecke, *Flat almost Hermitian manifolds which are not Kähler manifolds*, Tensor N. S. **31** (1977), 249–254.
12. B. Watson, *Almost Hermitian submersions*, J. Differential Geometry **11** (1976), 147–165.
13. W.T. Wu, *Sur les classes caractéristiques des structures fibrées sphériques*, Actual. Sci. Ind. **1183** (1952), 1–89.